Industrial Applications of Surfactants IV

Industrial Applications of Surfactants IV

Edited by

David R. Karsa

Akcros Chemicals Ltd, Manchester, UK

The proceedings of the meeting on the Industrial Applications of Surfactants IV held at the University of Salford on 31 March – 2 April 1998.

Special Publication No. 230

ISBN 0-85404-773-5

A catalogue record for this book is available from the British Library

Published by The Royal Society of Chemistry,
Thomas Graham House, Science Park, Milton Road,
Cambridge CB4 4WF, UK

For further information see our web site at www.rsc.org

Printed by MPG Books Ltd, Bodmin, Cornwall

Preface

Well over 4.5 K tonnes of surface active agents are used worldwide in industrial applications, ranging from industrial and institutional cleaning applications to the formulation of agrochemicals and through to many applications which utilise surfactants as *process aids*. The latter include, for example, the manufacture of emulsion polymers for paints and surface coatings, auxiliaries for textile and fibre processing, dispersants for both organic and inorganic pigments, oil field chemicals and lubricant additives. No single volume can provide a comprehensive insight into this subject because of the diversity of both surfactant types and applications. This current volume of conference proceedings should be considered in combination with the three previous Royal Society of Chemistry publications on this topic ("The Industrial Applications of Surfactants": Volumes I – III).

Topics of particular note in Volume IV include up-to-date marketing and environmental information, a focus on surfactants derived from natural/renewable raw materials (including starch, sucrose and kelp), nonionic surfactants derived from secondary alcohols and more innovative surfactants including a chelating surfactant based on an alkyl ethylenediamine triacetic acid, alkyl phosphate esters with low irritancy, twin-chained ("gemini") surfactants and sulphobetaine derivatives. Nonionic surfactants are also considered in terms of optimum selection to meet current environmental demands, as well as a thought provoking paper questioning the usefulness of the HLB value concept for nonionics. Application areas considered include surfactant uses in the textile industry, silicone-based surfactants in flame retardant polyurethane foams and products to prevent gas hydrate formation in oil field applications.

Once again, it is hoped that these proceedings both illustrate the versatility of both new and established surfactants but also confirm the assertion that there are still new surfactants and applications to be discovered and commercialised.

Dr David R Karsa

Contents

Overview: A Decade of Change in the Surfactant Industry

D. R. Karsa, R. M. Bailey, B. Shelmerdine and S. A. McCann

AKROS CHEMICIALS, BENTCLIFFE WAY, PO BOX 1, ECCLES, MANCHESTER M30 0BH, UK

1 INTRODUCTION

The 1990s have witnessed continuous change in the Surfactant Industry from the raw material supplier base to the surfactant producers and through to the formulators and end-users. Both petrochemical and oleochemical raw materials continue to be readily available, speared on by the discovery of new oil fields or the further development of oil bearing crops. Petrochemical majors have made strategic moves in or out of key surfactant raw materials and intermediates, while acquisition and divestment continues apace amongst surfactant manufacturing companies as they either withdraw from commodity business in order to concentrate on high added value, niche markets or invest in low production cost, high volume markets, sometimes by "over the fence" production deals with appropriate raw material suppliers.

Likewise the surfactant producers' company base continues to evolve, particularly as majors refocus on their core businesses, attempt to reduce their product inventory and seek out "global supplier" arrangements. At the other end of the customer spectrum, the shear size and diversity of smaller surfactant end-users, particularly in the industrial and institutional (I&I) cleaning sectors and the process industries, has led to the increased use of agents and distributors to more effectively service those businesses.

A limited number of surfactant types remain dominant (e.g. alkylbenzene sulphonates and alcohol ethoxylates, sulphates and ethersulphates) as these are components of laundry detergents, household and personal care products which account for over half the use of surfactants. The more 'speciality' surfactants are generally to be found in industrial and agrochemical applications. Hence any quantum change in the use of existing surfactants or from the introduction of new ones will almost exclusively emanate from major formulation trends in the detergent sector. This may be illustrated in the 1990's by the increased use of alcohol ethoxylates at the expense of alkylbenzene sulphonates (LAS), the developing use of alkane sulphonates and the methyl esters of sulphonated fatty acids and investment in very large scale production of alkylpolyglucosides.

Environmental and legislative issues are also driving forces for change. For example, in Europe, the agreed removal of 'non-biodegradable' components and voluntary cessation of alkylphenol ethoxylate usage both in I&I products plus the reformulation of preparations to avoid the requirement for hazard labelling have all led to significant product re-formulation initiatives.

The following sections look at the past and predicted future growth of surfactants both geographically and by product type, with special reference to the split between household and industrial applications. The supply and demand for major surfactant raw materials is also considered together with the changes that have occurred in the raw material supplier base over the last decade. Finally some of the key legislative and environmental drivers are briefly discussed.

2 SURFACTANT MARKET SIZE

The following data is based on a range of Market Surveys, articles from the technical press and Akcros data on particular product and market segments. All statistics should be considered as a close guide to the volumes of materials stated, variation occurring when nameplate capacities and actual sales and captive use data are not always clearly stated in some of the literature. For example, the global surfactant market in 1995 has been quoted between 8.5 and 10.3 Million tonnes and, likewise, the Western European Market between 1.75 and 2.2 Million tonnes. In the same year Western European nonionic capacity was only 80-85% utilised.

Table 1 illustrates the estimated global market for surfactants by region for the period 1995-2005.

Table 1   *Global Surfactant Market 1995-2005 (Ktonnes)*

Area	1995	2005	% Increase p.a.
Japan	565	655	1.5
Western Europe	2,100	2,165	0.3
North America	1,800	1,960	1.0
Asia-Pacific	2,690	4,340	6.1
Latin America	1,575	1,785	2.6
Rest of World	1,645	2,765	6.8
Total (Ktonnes)	10,220	13,870	3.6

Table 1 clearly illustrates the maturity of the Western European/North American/Japanese markets and the growth potential in the rest of the world.

For example, recent predicted growth patterns for South Asia and Latin America are as follows:

Surfactant Consumption in South Asia 1995 - 2010

1995	1.80M tonnes
2000E	2.30M tonnes
2010E	2.94M tonnes

This equates to an annual growth 5% p.a. which is mainly driven by the household/laundry product sector.

Surfactant Consumption in Latin America 1995-2000

1995	1.27M tonnes
2000E	1.46M tonnes

Growth of 2.8% is predicted of which Brazil represents 50% of the market.

Examination of the Western European Market (1996-2005) shows the following usage of surfactants by ionic type with the only changes of note to be some reduction in alkylbenzene sulphonate usage and a corresponding rise in ethoxylate production. The latter contains some reduction in alkylphenol ethoxylate production and a significant increase in fatty alcohol ethoxylates even in this mature market (Table 2).

Table 2 *Sales and Captive Use of Surfactants by Ionic Type in Western Europe 1996-2005 (Ktonnes)*

Product Group	1996	2005
Alkylbenzene sulphonates	290	225
Alcohol ether sulphates	230	245
Alcohol sulphates	95	100
Alkane Sulphonates	75	85
Other anionics	90	80
Total all Anionics	760	765
All Ethoxylates	825	935
Other Nonionics	225	235
Total all Nonionics	1,015	1,140
Amphoterics	50	60
Cationics	190	200
TOTAL	**2,070**	**2,165**

The market can also be defined by application area (Table 3).

Table 3 *Surfactant usage in Western Europe (Ktonnes) by Application Area*

	1989	1990	1995	2000
Household Products	969	1,033	1,226	1,300
Cosmetics & Toiletries	92	95	106	110
Industrial & Institutional Cleaning	154	160	185	210
Textiles	133	133	137	140
Pulp & Paper	229	233	260	250
Construction	116	120	140	140
Other	159	162	185	190
Total Ktonnes	1,852	1,936	2,239	2,340
% used in "Industrial Applications"	43%	42%	41%	40%
Volume (Ktonnes)	791	808	907	930

The "industrial" sector represents just over 40% in Western Europe.

This is reinforced by other surfactant consumption data collated for Western Europe, the Asia-Pacific region and the United States shown in Table 4.

The higher use in "industrial" applications in North America is generally associated with high usage in the oil field and petroleum industries.

It is interesting to note whereas 80% of the Western European surfactants are petroleum based, 55-65% of all surfactants in the Asia-Pacific region are oleochemical based.

The latest estimate (1998) of the value of the global surfactant business is in excess of US$14 billion.

The growth of the surfactant market closely follows the world demand for detergents, as this sector commands over 55% of surfactant usage. Today, almost equal volumes of soap and synthetic surfactant are used in detergents. Since 1980, soap production has been static, whereas surfactants have shown steady growth.

Table 4 *Surfactant Consumption by Region 1995*

Region \\ Application	Asia-Pacific	Western Europe	North America
Household Products	58%	56%	40%
I&I Cleaners	2%	9%	10%
All other Industrial Uses	40%	35%	50%
% used in All Industrial Applications	42%	43%	60%
Tonnes (total)	2.8M	1.9M	2.5M
Tonnes (Industrial Use)	1.18M	0.82M	1.50M

Source: Petresa

Table 5 illustrates the relationship between detergents, soap and surfactants between 1960 and 2000.

Table 5 *World Market for Soap, Surfactants and Detergents 1960-2000 (Million Tonnes)*

	1960	1970	1980	1990	2000
Total Detergents	10	17	28	37	52
Detergents Excluding Soap	5	10	20	28	43
Soap	7	7	9	9	9
Surfactants	N/A	N/A	6	7.5	10

The actual weight of surfactants in formulated detergents has fallen steadily on a global basis, i.e.

1978	26%
1996	23%
2002(E)	21%

However, washing machines, washing temperatures and laundering procedures vary tremendously from region to region. In Western Europe, for example, with the increased sale of compact heavy duty detergents, surfactant usage is expected to increase, particularly with higher levels of nonionic surfactants. Nonionics are much more effective in removing oily soils at lower wash temperatures. Table 6 illustrates the composition changes expected by the year 2000.

Table 6 *Surfactant content (%) of European Heavy Duty Detergents*

Year	1992	2000
% Anionic	9.3	10.0
% Nonionic	4.5	12.5
% Total	13.8	22.5

In 1992, the European laundry detergent, dishwashing detergent and cleaner markets was valued at DM 21.3 Billion with a 9M tonne market of which ca 1M tonne is surfactant (two thirds anionic/one third nonionic). By contrast, the United States market was 0.9M tonnes (three quarters anionic/a quarter nonionic).

The size of the household and personal care sector and its dominance by the "soapers" (major detergent companies such as Procter & Gamble, Unilever, Henkel etc) dictates most of the quantum changes in surfactant usage. The volume of product used in this sector masks the trends to be observed in more specialist applications. Hence, even in the developed countries where surfactant usage shows limited overall growth, there are niche areas where exception growth can still be found.

The dominance of the "soapers" in the detergent and toiletry area is well illustrated by consideration of the USA market in 1997 (Table 7).

Here, Procter & Gamble and Unilever have 78% of the Household Detergents market and 55% of the Personal Care market. Growth in the USA soaps and detergent market is currently running at 4-5% p.a. and some of the main drivers in this sector are:

- In 1996, Sales of Laundry Liquids up 10-11%, powders down 0.5% (Liquids 49% of market/powders 51%).
- Growth in Liquids putting pressure on Alcohol/Alcohol ethoxylate supply (e.g. Shell plan to increase alcohol/ethoxylate capacity at Geismar/LA by ~100K tonnes by 2000).
- Move from top loading washing machines to "front loaders" which may lead to European style formulations
- LAS stagnant thro' to 2001/APG usage growing/detergent Surfactant volume growing at 2.1% p.a. to 7.1 billion lbs in 2001 (Freedonia Group/Cleveland)
- Over capacity in other product area is leading to Acquisitions/Consolidation/Plant Closure

• Alpha-Olefin Sulphonates are growing at 4-5% p.a. thro' to 2005, according to
  Chevron who are investing in a US$750 million plant in Texas. Start-up in 2000.
  Mainly driven by Far East use.

Table 7 *USA Household Detergents and Personal Care Markets*

Household Detergents		Personal Care
US$ 4.3 billion		US$ 2 billion
Procter & Gamble	58%	24%
Unilever	20%	31%
Colgate Palmolive	4%	14%
Dial	6%	15%
Church & Dwight	5%	-
USA Detergents	3%	-
Huish	1%	-
Other	3%	10%
Kao	-	4%
Johnson & Johnson	-	2%

Information Resources (Chicago)

3 SURFACTANT RAW MATERIALS

During the 1990's, world-wide surfactant demand has been predicted to grow by about
30% from 6.1 Million tonnes per year in 1990 to over 10 Million tonnes by the year
2000. Likewise, it is predicted that about 75% of these surfactants will be derived from
petrochemical-based intermediates.

Table 8 illustrates the growth in key surfactant raw materials during this period.

Table 8 *Evolution of basic surfactant raw materials 1995-2000*

	1995	2000
Linear Alkylbenzene (LAB)	53%	58%
Branched Alkylbenzene (BAB)	9%	5%
Oleochemical alcohols	20%	18%
Petrochemical alcohols	18%	19%
Tonnes	3.91M	4.59M

Petrochemical-derived surfactant raw materials should remain readily available for
the foreseeable future.

Annual crude oil consumption has remained almost unchanged since 1980 at three
billion tonnes, whereas petrochemical usage between 1970 and 2000 will have doubled

from 4 to 8% of crude oil production. In 1980, surfactant products only represented ca 1.1% by weight of all petrochemical uses and 1.5% by 1995 (which represents a change from 0.05 to 0.10% of crude oil usage over that period). Hence it is unlikely that raw material supplies will be a cause for concern in the coming decades.

Oleochemicals can be manufactured from natural fats and oils through various chemical processes, for example via methyl fatty esters (see Figure 1).

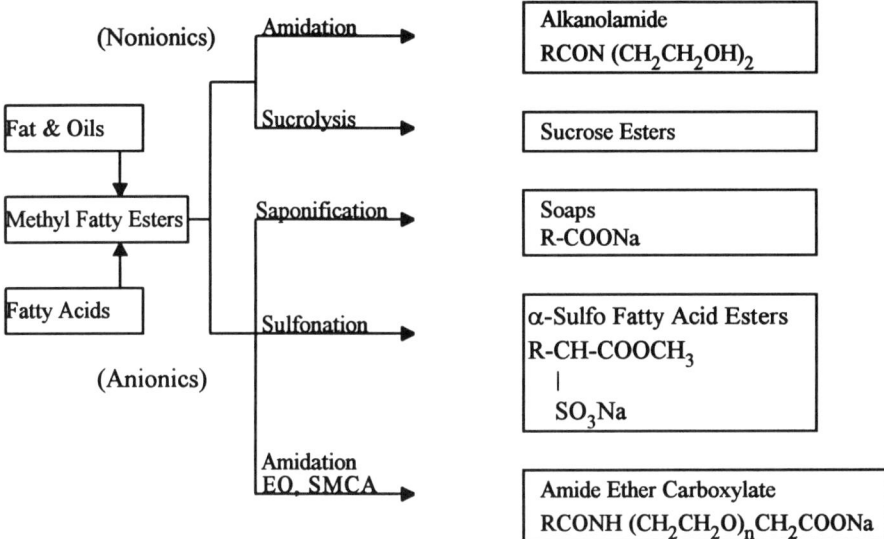

Fig. 1 *Surfactants Derived from Methyl Fatty Esters*

Traditionally such materials have been produced from tallow and coconut oils. However, significant production increases are scheduled for the manufacture of palm oil in the Far East, particularly in Malaysia and Indonesia (see Figure 2). In fact, in Malaysia alone, palm oil production increased by 10.7% in the two year 1996/7 to reach 8.7 Million tonnes.

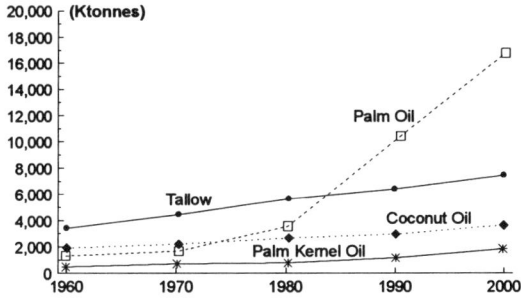

Source: INFORM

Fig. 2 *World Production of Fats and Oils*

This is expected to lead to much wider availability of oleochemicals, although adverse seasonable weather conditions can have devastating effects on the oil supply from year to year. The recent currency crisis in SE Asia has also led to suspension of palm oil exports.

The key raw materials for production of high volume surfactants and many specialities are derived from oleochemical and/or petrochemicals e.g.

Source of Hydrophobe
- Fatty alcohols   -   petrochemical derived oxo- and Ziegler alcohols
                                          -   oleochemical derived via fatty acid methylesters
                                              (from fats and oils)
- Linear alkyl benzene (LAB) -   petrochemical derivative with over 75% of
                                              production via dehydrogenation of n-paraffins and
                                              alkylation of benzene with the resultant olefin

Source of Hydrophiles
- Ethylene oxide   -   petrochemical derivative for production of
                                              hydrophilic poly (oxyethylene) units

The following sections consider the above raw materials in greater depth.

3.1 Fatty alcohols

Fatty alcohols are a basic feedstock for a whole range of both nonionic and anionic surfactants as illustrated in Figure 3.

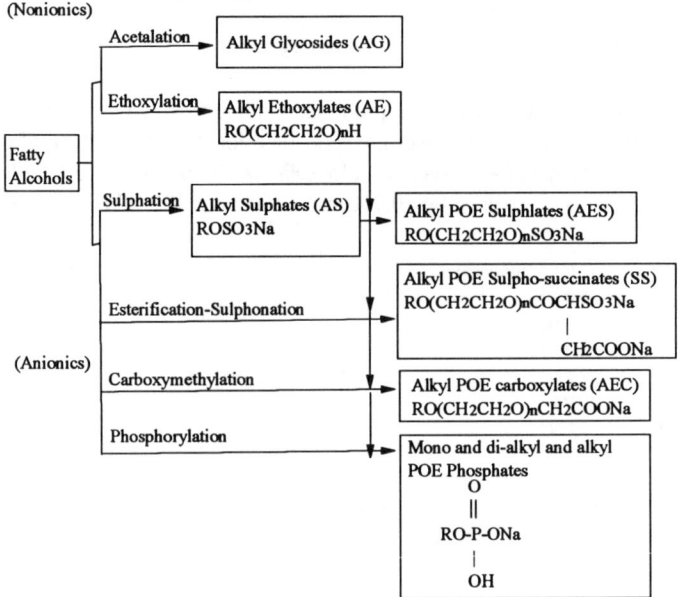

Fig. 3 *Surfactants derived from Fatty Alcohols*

There are three main classes of higher alcohol used in surfactant production -

- Alcohols derived from fatty acid methyl esters
- Alcohols derived from alpha-olefins (so called 'oxo' alcohols)
- Alcohols derived from ethylene (Ziegler alcohols).

Hence, fatty alcohols may be derived from both petrochemical and oleochemical feedstocks with over 70% of world-wide fatty alcohol production (ca 1.2 Million tonnes in 1997) being used to produce surfactants.

Demand is predicted to increase by 150-200,000 tonnes by the year 2000, but this is not a problem as capacity is either already available or will be by the year 2000 (see Table 9).

Table 9 *Fatty alcohol capacity (Ktonnes)*

	1992	1993	1996	2000	2005
Western Europe	520	520	580	N/A	N/A
USA	550	630	630	N/A	N/A
Asia	290	490	490	N/A	N/A
Total	1360	1640	1700	1,875(E)	2,170(E)

This is also illustrated in Figure 4.

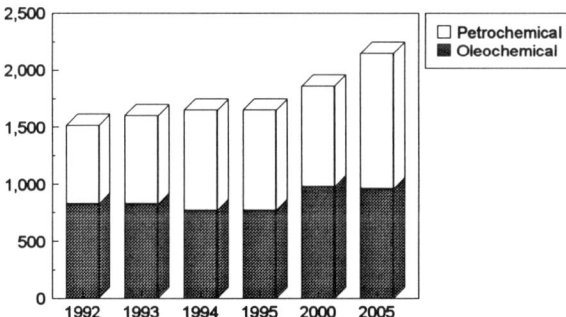

Fig. 4 *World Surfactants Alcohol Capacity*

Most additional capacity will result from investment in Asia with a subsequent increase in alcohol production based on renewable resources (Table 10).

Set against predictions of long term supply readily meeting demand, recent reports show that in the near term the market for detergent-range alcohols remains tight in advance of the posted capacity expansions. The situation has been exacerbated by plant outages, equipment problems and high natural alcohol feed prices in 1997. In fact there was said to be between 130 and 175,000 tonnes of spare oleo-alcohol capacity in 1997. A quantum jump in capacity is expected by the year 2000.

Similarly, after several years of inactivity in the petrochemical area several producers have announced increased capacity plans. Shell have recently announced a 90,000 tonnes/y increase in capacity at Geismar/USA to 273,000 tonnes/y. Condea has also signalled upto 120,000 tonne/y increase in world-wide alcohol capacity (Zeigler, oxo- and/or natural) to 470,000 tonnes/y by debottlenecking plant in Germany and Lake Charles, LA. Sasol in South Africa is said to be planning an additional 100,000 tonnes/y capacity on a $C_{12,13}$ oxo-alcohol and BASF are said to be extending oxo-alcohol capacity at Ludwigshafen to ca 170,000 tonnes/y.

Table 10 *Fatty Alcohols for Surfactant Production Natural versus Synthetic*

	K tonnes	% Synthetic	% Natural
1980	720	60	40
1990	1,300	53	47
2000	1,680	40	60

3.2 Linear Alkylbenzene (LAB)

Linear alkylbenzene, frequently referred to as 'LAB' or 'Alkylate', was introduced in the early 1960's as a replacement for the 'hard' (non-biodegradable) branched chain alkylate (DDB), derived from propylene tetramer. Upto 1980, almost all production was located in the western world, but after 1980 most new plants have been constructed elsewhere.

Western capacity, mainly in W. Europe, USA and Japan, was 1.2M tpa in 1993. World 'nameplate' capacity in the year 2000 is expected to be 3M tpa. Apart from Petresa's Canadian investment recently announced, additional capacity has been located in developing countries such as India, China, Latin America, the Middle and Far East and Russia, as illustrated in Table 11.

Table 11 *LAB expansion/New Capacity announced in 1997/Q4*

Producer	Location	Capacity Change/Total
Petresa	Canada	Increase from 80,000 to 100,000 te
Nirma	Gujarat/ India	Increase from 75,000 to 100,000 te by April 1998
Quimica Venoco	Guacara/ Venezuela	100/110,000 te (proposed second LAB plant)
Tamil Nadu Petroproducts Ltd	India	85,000 te to 100,000 tpa by April 1999
Fusshan Petrochemical Corp	China	72,000 te to 144,000 tpa
Deten	Brazil	135,000 te to 175,000 tpa
		Total ca 270,000 tpa extra capacity

Long term growth in the demand for LAB is forecast at 3.7% p.a., whereas the already rapidly declining market for the "hard" (non-biodegradable) branched chain dodecylbenzene based on propylene tetramer is falling at 8.8% p.a.

Linear alkylbenzene consumption is shown in Table 12.

Table 12 *LAB Consumption (K tonnes)*

	1980	1985	1990	2000(E)
Western Europe	310	345	380	300
Eastern Europe	105	125	180	280
North America	205	250	295	300
Asia/Pacific	280	325	485	885
Middle East/Africa	65	100	210	300
Latin America	85	105	140	200
Total	**1,050**	**1,250**	**1,690**	**2,265(E)**

By the mid 1990's over 75% of LAB was produced via dehydrogenation of n-paraffins and alkylation of benzene with the resultant olefin, the remainder being produced in similar quantities via the high purity olefin/alkylation or chlorination/alkylation routes.

LAB is converted by continuous SO_3 sulphonation to linear alkylbenzene sulphonate, the use of which in Western Europe is given in Table 13.

Table 13 *Western European Consumption of LAS (K tonnes)*

	1991	1994	1999	Average annual growth rate 1994-9 (%)
Heavy-Duty Laundry Powders	210	190	163	-3
Heavy-Duty Laundry Liquids	50	44	38	-3
Light-Duty Laundry Liquids	115	85	70	-4
Other Household Cleaners	20	19	20	1
Others (All I&I and other Commercial Uses)	72	71	75	1
Total	**467**	**409**	**366**	**-2**

Source: SRI International (1996)

Linear alkylbenzene sulphonate (LAS) is still the largest volume anionic surfactant (excluding natural soaps) used in Western Europe. Consumption peaked in 1989 at ca 485 Ktonnes.

LAB and LAS must be some of the most thoroughly investigated materials in terms of their fate in the environment and LCA (Life Cycle Assessment) studies, the favourable results of which will ensure that LAS will remain a major surfactant for many years to come.

3.3 Ethylene Oxide (EO) and Ethoxylates

In the mid 1990's world-wide ethylene capacity was in excess of 70 Million tonnes with demand estimated at just over 60 Million tonnes (ca 85% capacity). Approximately 15% of ethylene is oxidised to produce about 8.5 Million tonne ethylene oxide. Capacity in the mid 1990's was estimated at 10.5 Million tonnes which translates to 80% usage of capacity.

Table 14 illustrates the world-wide uses for ethylene oxide.

Table 14 *World-wide EO Markets (10.5 Million tonnes capacity)*

Monoethylene glycol	60%
Polyethylene glycol	10%
Surfactants	15%
Others (glycol ethers, ethanolamines, PET etc)	15%

Capacity growth in the short term is predicted at 7% per annum; long term 2-3%, with projections that demand will not exceed capacity until the early 2000's.

Monoethylene glycol (MEG), the main end-use segment for ethylene oxide has seen stocks reach low levels in late 1997, resulting in high prices with antifreeze manufacturers delaying purchase of raw material. The start-up of the 250,000 tpa Equate MEG plant in Kuwait has not brought the expected relief to tight MEG markets, due to technical difficulties and no commercial availability until Q1, 1998 at the earliest.

The European ethylene oxide capacity is given in Table 15.

Table 15 *W. European Ethylene Oxide Capacity*

Manufacturer	1992 Capacity
Enichem Anic/Sicily	75,000
Union Carbide/Wilton, UK	240,000
Inspec/Belgium	200,000
BP Chemicals/France	150,000
Hoechst (Gendorf and Kelsterbach)	205,000
BASF/Ludwigshafen	200,000
Akzo Nobel/Sweden	42,000
Shell Nederland Chemie BV	200,000
Dow Benelux, Terneuzen	130,000
Buna AG, Schkopau	104,000
Erdölchemie GmbH, Köln	150,000
Hüls AG, Germany	150,000
Industrias Quimicas Asociadas SA, Tarragona	100,000
TOTAL	**1,856,000**

Source: SRI International

Ethoxylation capacity in Western Europe has increased in spite of several smaller ethoxylators exiting from this business, i.e.

1986 :   950 Ktonnes
1991 : 1,120 Ktonnes
1996 : 1,320 Ktonnes

An estimate of world-wide ethoxylate production is given in Table 16.

Alcohol ethoxylate consumption in Western Europe, USA and Japan was 467 Ktonnes in 1987.

Table 16 *World-wide Production of Ethoxylates (Ktonnes)*

	1993
Alcohol ethoxylates	750
Alcohol ether sulphates	750
Alkylphenol ethoxylates	600
Ethoxylated nitrogen compounds	50
(excluding alkanolamides)	50
EO/PO block co-polymers	
Total (Ktonnes)	**2,200**

3.3.1   Alkylphenol Ethoxylates (APE).   The current on-going concern regarding this widely used class of nonionic with respect to biodegradability, the aquatic toxicity of metabolites from biodegradation and current studies on oestrogenic activity warrants separate attention in terms of usage trends.   Table 17 affords some insight in past and recent world demand for APE's.

Table 17 *Annual World Demand for Alkylphenol Ethoxylates 1987-95*

Country	Tonnes	
	1987	1995
North America	200,000	210,000
Asia (including Japan)	-	200,000
Japan	35,000	-
South America	-	100,000
Western Europe	150,000	75,000
Africa	-	70,000
Total	**N/A**	**655,000**

Source: Hüls Market Research

Production in Western Europe (1994) according to CESIO (Comité Européen des Agents de Surface et leurs Intermédiaires Organiques) was 109,800 tonnes of which 35,400 tonnes were exported to non-Western European countries, affording a Western European usage of 74,800 tonnes. Table 18 illustrates estimated W. European production. Unfortunately, no information is available on what percentage was exported outside the EU each year.

Table 18 *Alkylphenol Ethoxylate Production in Western Europe 1993-97 (Tonnes)*

1993	90,000
1994	110,000
1995	100,000
1996	117,000
1997	128,000

In spite of a voluntary W. European ban on the use of APE's in Industrial and Institutional Cleaners from 1 January 1998, the current production trend fails to show a decline in volume.

In the United States, alkylphenol ethoxylates continue to be widely used. Growth in I&I cleaning products is continuing at 2.5-3% per annum, whereas overall growth is ca 5% p.a.

The United States Chemical Manufacturers Association (CMA) have established a 23 company strong advisory committee, including major manufacturers such as Dow, Ciba, Rhodia, Union Carbide and Rohm & Haas, to investigate the health and environmental effects of APE's.

A recent study by the National Insitute of Environmental Health Sciences (NIEHS) reinforces results of a 1997 study by the US Chemical Manufacturers Association's APE panel, concluding that nonylphenol is a male and female reproductive toxicant at concentrations >650 ppm. Both studies showed decreased body weight in rats at 2000 ppm. Although the environmental science is mixed, APE producers such as Huntsman Corporation and Union Carbide Corporation expect "a long future of modest growth" for APE's. In the United States APE growth has been 2.5 to 3% per annum over recent years.

It will be interesting to observe whether the momentum in Europe to phase-out alkylphenol ethoxylate usage in several application areas is reflected elsewhere in the coming years.

Table 19 contrasts the growth of nonionic surfactants in Western Europe and the USA between 1990 and the year 2000.

Table 19 *Nonionic Surfactant Production Western Europe and USA (KTonnes)*

	W. Europe			USA		
Product	1990	1996	2000	1990	1996	2000(E)
Alcohol ethoxylates	330	} 845	} 875	207	230	} 560(E)
Alkylphenol Ethoxylates (APE's)	90			207	244	
Others	180	225	225	326	N/A	N/A
	600	1070	1100	740	740(E)	890(E)

In 1990 alcohol ethoxylates represented 55% of the total nonionics market and APE's 15%. By contrast both had a 28% market share in the United States.

4 SPECIALITY SURFACTANT MARKETS: PROSPECTS FOR GROWTH

It has already been amply illustrated that 'macro' changes in surfactant consumption are dominated by major changes in the detergent and personal care markets. There are only a limited number of industrial applications which could potentially utilise very large volumes of surfactants. Examples would include:

- Enhanced Oil Recovery (E.O.R.)
- Soil Washing/Contaminated Land Reclamation
- Coal-water or Coal-oil mixtures (for power plants)
- Fuel Additives (from "detergents" for petrol through to power station fuels)
- Mineral Processing//Ore Recovery

Few, if any, of these areas employ vast quantities of surfactant. With current oil prices E.O.R. is not commercially viable. Remediation of contaminated land is very much at the development stage albeit using large volumes of surfactants in trials. Coal-water mixtures (CWM) have not developed to the extent predicted in the early 1990's. Nevertheless, such opportunities will continue to develop. For example, it was reported in the UK press in late March 1998 that Elf Aquitaime, the French oil group, had developed a diesel/water emulsion named 'Aquazole' as a direct replacement for standard diesel fuel which utilised a highly specific organic surfactant as the emulsifier. The resultant "green fuel" can cut nitrogen oxide emissions from diesel buses by 30%, particulates by 50% and black smoke upto 80%. Even at ppm levels, the potential usage of surfactant is large. It will be interesting to see if this 'new' fuel gains wide acceptance.

In the industrial and agrochemical sectors, even if a selected application area exhibits low overall growth or limited growth in a particular geographic region, there can still be niche areas which exhibit tremendous growth potential. This can only be ascertained by a focused approach to a particular market. Dedicated techno-commercial resources are essential to develop collaborative partnerships with major customers and to gain in depth market knowledge of a selected area. The value of selected journals and the trade press as a source of information should not be under-estimated, backed-up by prudently selected market research reports.

This may be illustrated by briefly considering two markets; the Industrial & Institutional Cleaning Market (the so-called 'I & I' sector) and the Agrochemical Formulation Market.

4.1 Surfactants for Industrial & Institutional Cleaners

As in every surfactant application area, there are uses for both higher volume commodity surfactants, purchased on price and readily interchangeable with competitive materials and more specialised, performance-based speciality products.

Figure 5 illustrates the I & I Cleaner market in 1993 by both geographical region and by end-use. Nearly 80% of all products were sold in Western Europe and North America, once again suggesting the tremendous growth potential in the Asia-Pacific and Latin American areas.

Fig. 5 *Global Industrial & Institutional Cleaning Market 1993*

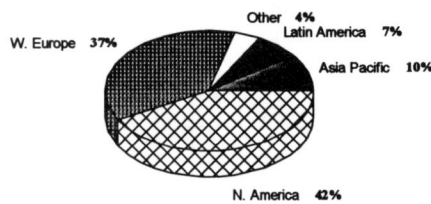

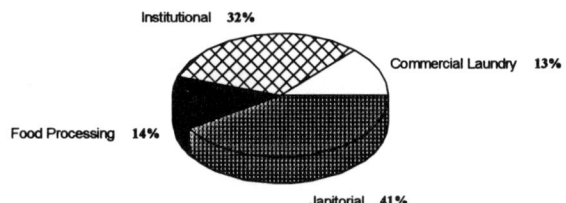

More recent data is available for the United States for 1997 usage of "speciality surfactants" in the I & I and Household Product Areas (Table 20).

Table 20 *Estimated US Consumption of Speciality Surfactants for I & I and Household Cleaning Products 1997*

Product	$ Million	% of Total
Quats [a]	190	41
Low-foaming nonionics	55	12
Speciality ethoxylates	50	11
Amine oxides and speciality alkanolamides	40	9
Phosphate esters, sorbitan esters and betaines	35	8
Alpha olefin sulphonates	20	4
Other	70	15
TOTAL	460	100

Source: Kline & Co (Fairfield NJ)

Note [a]: - includes dialkyl/dimethyl/quaternaries, diamidoamino quaternaries and benzalkonium salts.

At least $250 Million worth of these speciality surfactants will be used in I & I cleaners, which equates to 20-25% of the US Speciality Surfactants market. In 1996, the US Surfactants Market was valued at US$1,175 Million of which US$880 Million was classified as 'speciality'. It was interesting to note that over half this value came from just four classes of surfactants:

- Silicone surfactants ($243M)
- EO/PO Block Co-polymers ($151M)
- Phosphate Esters ($136M)
- Ethoxylated Amines ($126M)

Market leaders were said to be Witco (22% market share), BASF (11%) and Henkel (7%). Hence, even within a well established and mature market such as this, attractive niche areas are still to be found.

4.2 Surfactants for Agrochemical Formulations

Surfactants are used to formulate crop protection chemicals such as herbicides, insecticides and fungicides to enable ease of application or, as adjuvants, to optimise activity and reduce dosage levels.

In order to understand this market it is necessary to identify the key formulators, the main 'actives' (particularly those exhibiting high growth), the type of formulation and also the geographic area, as crops and crop protection requirements vary depending on climatic conditions.

Table 21 summarises the relative world market share of the three main classes of crop protection chemical by region.

Table 21 *World Agrochemical Market by Region 1996 (% Share)*

Region	Herbicides	Insecticides	Fungicides
Western Europe	24.0	16.2	42.6
Eastern Europe/CIS	3.4	4.2	3.1
North America	41.7	23.3	11.5
Latin America	13.5	11.5	8.8
Japan	8.4	15.0	20.7
Rest of Asia	6.4	16.6	10.7
Rest of World	2.6	13.2	2.6

In Western Europe, there has been a general perception of an overall decline in the agrochemical market, due to overproduction and environmental pressures to reduce dosage levels of crop protection products. This is not upheld by recent facts. Western Europe represents 26 per cent of the world agrochemical market with a value of ECU 25,000 million. In 1996 demand rose for the third consecutive year due to increased grower confidence and increased acreage. This represented a value increase of 8.3% (back to 1991 levels) and in real terms the market grew by 1.4% in 1994 and 5.5% in 1995 (E.C.P.A. report). Some areas witnessed even higher growth in 1997, e.g. Spain 7.3% and Portugal 8%.

Nevertheless, it must be remembered that the overall volume of surfactants in such industries is relatively low. In Western Europe this is estimated to be about 50,000 MT (expressed as '100% active' material). Table 22 illustrates the split in surfactant usage by formulation type.

Table 22 *Western European Market for Surfactants used in Agrochemicals (1995)*

Formulation Class	Volume of Surfactant (tonnes)
Emulsifiable Concentrates	12,000
Suspension Concentrates	10,000
Wettable Powders	12,500
Granules and Dusts	4,000
Others	11,500
TOTAL	**50,000**

Market share data, end-of-year results and highlighted growth areas in the trade press and agrochemical-related publications can provide a valuable 'snap-shot' of the industry, particularly by following the fortunes of some of the market leaders (see Table 23).

Table 23 *Agrochemical Formulators (1996)*

Company	Estimated Market Share in Europe (%)
Novartis	20
AgrEvo	13
Bayer	12
Rhône Poulenc	12
Zeneca	10
BASF Agrochem	7

These six companies represent nearly 75% of the agrochemical formulators business in Europe. End-of-year trading results are always reported in Q1 or Q2 of the following year and are a valuable guide to areas of product growth (both by geographical area and 'active' type). This may be illustrated by the 1997 sales of Zeneca Agrochemicals (Table 24).

Table 24 *Zeneca Agrochemical Sales (1997)*

Region	£M	% Change [1] (Value)	% Change [1] (Volume)
North America	527	0	+ 5.0
Europe	508	- 7.0	+ 7.0
Africa/Asia/ Australasia	310	- 10.0	- 7.0
Latin America	286	+ 8.0	+ 14.0
TOTAL	**1,631**	**- 3.0**	**+ 4.0**

[1] vs 1996   Fungicides 33% value increase
Herbicides   8% value increase
Insecticides 13% value <u>decrease</u>

Each summary is usually accompanied by detailed descriptions of the major products contributing to increased sales for that year and predictions of future growth sectors, thus enabling the surfactant producer to identify potential area of co-development.

5 PRODUCT DEVELOPMENTS 1990 - 2000

A decade is a comparatively short period in terms of product development, particularly for new products to achieve significant market share. Perhaps one of the more important developments has been a concentration on the production of surfactants from vegetable/ 'natural' raw materials. This has resulted in bulk production facilities for alkyl polyglucosides ('APG's') coming on-stream in both Europe and the United States and the commercialisation of 'second generation' APG's and related products, such as glucamides.

Nevertheless, worldwide production of sucrose, glucose and sorbitol-derived surfactants remains comparatively small (see Table 25).

Table 25 *Global Production of Sucrose, Glucose and Sorbitol-derived Surfactants (1997)*

Product	KTonnes	Comments
Sorbitan Esters	20 - 25,000	
Sucrose Esters	<4,000	Poor conversion (insufficient transesterification)
Alkyl Polyglucosides	80,000	
N-methyl Glucamides	30 - 50,000	Almost exclusively used by P&G

Government-industry-academia links are being strengthened and resources applied to the whole area of 'post-harvest crops' uses for chemical production, including surfactants. This is personified by the work of ATO-DLO (Instituut voor Agrotechnologisch Onderzoek) in the Netherlands and 'ACTIN' (The Alternative Crops Technology Interaction Network) in the UK.

Other product areas worthy of comment should include:

- The growth in amphoteric surfactants, particularly coco-dimethyl and coco-amidopropyl betaines. Yet seldom has a surfactant class moved from 'speciality' to 'commodity' status so rapidly.

- The demise of the di(hydrogenated tallow) dimethyl ammonium chloride (DHTDMAC), domestic fabric conditioner market through questionable environmental pressures and the introduction of fatty acid-derived "ester quats".

- The growth of alkane sulphates, the methylesters of sulphonated fatty acid and alpha-olefin sulphonates as alternatives to linear alkylbenzene sulphonates in some detergent formulations.

- Moves to replace alkylphenol ethoxylates in a variety of application areas (see Section 3.3.1).

- The growing potential for oligoneric and polymeric surfactants and

- Lack of significant commercialisation of bissurfactants which were an area of considerable focus in the early '90s.

6 ENVIRONMENTAL AND LEGISLATIVE DRIVERS (1990-2000)

This area is addressed in the next chapter. Suffice it to say that many developments in the surfactant industry have been driven by environmental and/or legislative pressures in the 1990's. Areas to note include:

- Biodegradability

European legislation still only concerns the primary biodegradability of anionic and nonionic surfactants. This has not been extended to amphoteric and cationic surfactants, although there is now commitment to develop methods for "ultimate" or "ready" biodegradability for all ionic types.

- Life Cycle Assessment

The on-going debate on the desirability of oleochemical versus petrochemical feedstock for surfactant production has received detailed study for some of the main surfactant types, but no clear environmental advantage has been shown.

- Surfactant labelling (CESIO Guidelines)

The review of surfactant labelling has led to debate regarding the subsequent labelling of "preparations" (formulated products), sometimes causing reformulation to avoid the use of hazard labels on domestic or I&I products.

• By-products/Risk Assessment

A growing awareness of minor components and by-products in surfactants has necessitated risk assessment studies and process modification to minimise levels. Areas include 1,4 - dioxane, in ether sulphates, free ethylene oxide and propylene oxide levels in nonionics and nitrosamines in alkanolamides. The 1990s have seen surfactant producers being far more proactive in this area, anticipating potential problems and recommending suitable action when appropriate.

• BSE/Restrictions on Tallow-Derived Products

The latter half of this decade has witnessed the UK B.S.E. "crisis" and the increased demand for products of vegetable origin rather than those based on animal-derived tallow. It could be another decade or more before animal tallow derivatives are rehabitated.

These are a small selection of the many regulatory and environmental issues affecting our industry today.

CONCLUSION

The surfactant market of the 1990's is a mature market with regard to developed countries but with good growth potential in the Asia-Pacific and Latin American regions. Worldwide, the use of large volume "commodity" nonionics and anionics in household, consumer and personal care products over shadows other product trends and the more specialised industrial sectors require careful study in order to identify "niche" product and application areas. Even in market sectors exhibiting limited growth, there can still be sub-sectors capable of considerable development.

Realignments both in the raw material supplier base and amongst surfactant producers will continue. With the latter, a growing scenario is that of "consolidation" leading to a limited number of large scale surfactant players, a range of small niche market companies and very few medium size players. With these medium size companies strategic alliances, either with raw material suppliers or co-producers, may be essential to survive. With multi-national companies, the move is to the identification of "global suppliers" for the bulk of their requirements, in part eliminating the need for smaller, regional producers. Many of the changes observed within the industry particularly at the high volume end of the market have been driven by the downward pressure on prices forced mainly by the "soapers" and multi-national formulators, leading to totally unacceptable low margin business. Even when "speciality surfactants" are considered, they do not always yield the margins afforded in some other performance chemical sectors. In the short term this will lead to surfactant producers undertaking less development and offering far less technical service support unless the "quality" of the business can be improved. The customer base will have to decide whether it is willing to pay a modest premium to develop a collaborative approach with its suppliers: it cannot have it both ways! This is particularly true for example where regulatory or environmental requirements necessitate extended process times to reduce or eliminate minor by-products, higher quality raw materials are required to meet FDA or BGA compliance or increased analytical data has to be generated for quality control purposes, all leading to additional costs. A partnership approach in the industrial sector must be the way forward for the future - a marriage of surfactant and customer application expertise. On a more positive note, new applications continue to develop for surfactants, many of them highly specific in their structure - performance requirements. Oligomeric surfactants and surfactants derived from "natural" raw materials are examples of new surfactants which offer tremendous scope for development. Likewise regulatory and environmental requirements can be used to commercial advantage by developing materials to meet a particular need. Hence the industrial sector of the surfactant market will continue to offer both a technical challenge to the surfactant and colloid chemist and a commercial opportunity to the more entrepreneurial marketeer.

Surfactants and the Environment – An Overview

H. Thomas

ICI SURFACTANTS, WILTON, MIDDLESBOROUGH, CLEVELAND TS90 8JE, UK

1 INTRODUCTION

Surface active agents, or surfactants, are extremely important materials in our modern industrial society. There are very many different types of surfactant and they may conveniently be classified according to the electrical charge associated with the surfactant species when dissolved in water, thus :

Anionic surfactants	- ve charge
Cationic surfactants	+ ve charge
Nonionic surfactants	no charge
Amphoteric surfactants	both + ve and - ve charges

Because of their particular chemical structures, surfactants are attracted to interfaces where they can provide a wide range of effects including wetting, cleaning, emulsification, foam control, lubricity and control of static electricity. Table 1 shows some of the market sectors where surfactants are employed and the typical effects provided.

The Western European production and use volumes for surfactants in 1995 and 1996 are shown in Table 2 (1). These figures do not include soap which probably accounts for an additional 500,000 te.

Surfactants are almost invariably used in aqueous solution and thus have the potential to enter the terrestrial and aquatic environments. It is thus particularly important to understand the environmental hazard profile of these high-tonnage products. Within the European Union, rules for the environmental hazard classification and labelling of all chemical substances are contained in the 7th Amendment (92/32/EEC) and the 12th and 18th Adaptations to Technical Progress, (91/325/EEC) and (93/21/EEC) respectively, of the so-called Dangerous Substances Directive (67/548/EEC). Criteria are best developed for the aquatic environment, where toxicity and biodegradability are the principal features, and the classification and labelling scheme is shown in Table 3.

2 ENVIRONMENTAL TOXICITY

As a consequence of their affinity for interfaces, surfactants may be relatively toxic to aquatic organisms by interaction with biological membranes. In the case of fish, for example, the surfactant may adsorb on the gill membranes and interfere with oxygen uptake leading to suffocation (2). For hazard classification purposes the EU requires short-term or acute toxicity data for the three taxonomic groups algae, Daphnia and fish, which represent key stages in the aquatic food chain. The toxicity is measured according to OECD methods 201, 202 and 203, respectively, which are contained in Annex V to Directive 97/548/EEC. For environmental risk assessment purposes, the results of long-term or chronic exposure tests are more desirable; especially the so-called No Observed Effect Concentration, abbreviated to NOEC.

TABLE 1   *Some Typical Uses of Surfactants*

USE SECTOR		EFFECTS PROVIDED
Personal Care Products	:	Cleaning, emulsification, lubricity, antistatic
Household & Industrial Laundry & Cleaning Products	:	Cleaning, wetting, foam control, emulsification, antistatic
Agrochemicals	:	Emulsification, wetting, performance enhancement
Engineering	:	Cleaning, emulsification, lubricity
Fibre Processing	:	Lubricity, antistatic
Food Additives	:	Emulsification, stabilisation
Food Processing	:	Cleaning, wetting, foam control
Oilfields	:	Emulsification/demulsification, cleaning, tertiary oil recovery
Paint & Latex	:	Emulsifiers, stabilisers, pigment dispersion
Polymers	:	Production/processing aids, antistats
Textile	:	Scouring, wetting, dye levelling

Acute toxicity data published by Schöberl et al (3) for various types of anionic surfactants are shown in Table 4. The range of results reported in the table probably reflects differences in alkyl chain distributions of the substances tested together with inter- and intra-species variability typical of tests involving biological systems and inter-laboratory variability.

Although it may be dangerous to draw general conclusions from this data, the sulphates and sulphonates appear to have similar toxicological profiles. It also appears that fish are the most sensitive organisms to these anionic surfactant types. Soap is an interesting substance since it forms essentially insoluble Ca and Mg salts in surface waters. At zero degrees of water hardness (dH) the LC_{50} to fish is reported as 6.7mg/l but the toxicity rapidly decreases as the hardness increases and the substance ceases to be bio-available.

There are many different families of nonionic surfactants, derived principally from essentially linear and highly branched fatty alcohols; from alkylphenols and from fatty acids. The most widely used nonionic surfactants on the European market are polyethoxylate derivatives of essentially linear fatty alcohols, where the alkyl chain is a mixture of linear and monobranched species. These products are versatile since it is possible to vary both the average carbon chain length in the hydrophobe and the degree of ethoxylation to achieve a range of properties and performance effects.

TABLE 2   *Surfactants In Western Europe*

| | (in thousands of m.t.) | | | |
| | PRODUCTION | | SALES AND CAPTIVE USE | |
CATEGORY	1995	1996	1995	1996
ANIONICS				
LAS	420	400	410	390
Alcohol ether sulfates	192	229	193	210
Alcohol sulfates	101	111	94	103
Alkane sulphonate	77	77	74	74
Other anionics	82	82	87	87
Total	872	899	858	864
NONIONICS				
Ethoxylates	817	844	780	800
Other nonionics	213	224	190	200
Total	1030	1068	970	1000
CATIONICS	190	170	181	145
AMPHOTERICS	38	43	39	40
TOTAL	2130	2180	2048	2049

TABLE 3   *Environmental Hazard Classification*

Acute Toxicity (mg/l)	"Ready" Biodegradability	Risk Phrases	Symbol of Danger
≤ 1	No	R50, R53	Yes
	Yes	R50	Yes
$1 \leq x \leq 10$	No	R51, R53	Yes
	Yes	R51	No
$10 \leq x \leq 100$	No	R52, R53	No
Other features which may present a danger to the aquatic environment		R52 &/or R53	No

R50   Very toxic to aquatic organisms
R51   Toxic to aquatic organisms
R52   Harmful to aquatic organisms
R53   May cause long-term adverse effects in the aquatic environment

TABLE 4   *Acute Toxicity of Some Anionic Surfactants*

	Fish LC$_{50}$ (mg/l)	Daphnia EC$_{50}$ (mg/l)	Algae IC$_{50}$ (mg/l)
C$_{11.6}$ LAS	3-10	9-14	10-300
C$_{13}$-C$_{18}$ sec alkane sulphonate	3-24	8.7-13.5	-
C$_{12}$-C$_{18}$ alcohol sulphate	3-20	5-70	60
C$_{12}$-C$_{14}$ alcohol +2EO sulphonate	1.4-2.0	1-50	65
C$_{14}$-C$_{18}$ α-olefin sulphonate	39	33	-
di-C$_8$-sulphosuccinate	39	33	-
Soap 0 dH	6.7	-	-
3-23dH	20-150	-	-

FIGURE 1   *Relationship Between Toxicity To Fish And Ethoxylate Chain Length For C$_{12}$-Alcohol Ethoxylates*

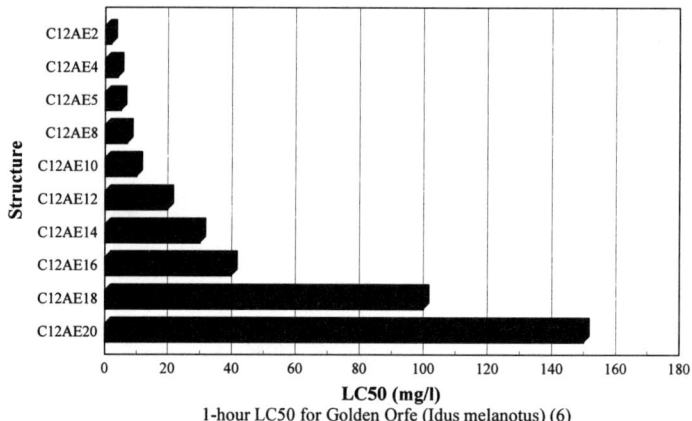

LC50 (mg/l)
1-hour LC50 for Golden Orfe (Idus melanotus) (6)

Acute toxicity data for some typical products (4,5) are shown in Table 5 and some structure-related trends are immediately apparent. The toxicity decreases with an increase in the degree of ethoxylation and with a reduction in the alkyl chain length of the hydrophobe.   The variation in aquatic toxicity with the degree of ethoxylation was demonstrated many years ago by Gloxhuber et al (6) for a pure C$_{12}$ alcohol (Figure 1). More recently the availability of computer software has allowed development of quantitative structure-activity relationships for surfactants which predict physical properties (7), performance (8) and aquatic toxicity (9, 10).   Figure 2 shows the relationship between observed and calculated EC$_{50}$ values of nonionic surfactants to Daphnia developed by Roberts & Marshall (10).

From an understanding of the relationships between surfactant structure, performance and toxicity it is possible to develop products with minimum environmental impact.

TABLE 5   *Acute Toxicity of Some Nonionic Surfactants (Essentially Linear Alkyl Chain Hydrophobes)*

		Fish LC_{50} (mg/l)	Daphnia EC_{50} (mg/l)	Algae IC_{50} (mg/l)
C_9-C_{11} alcohol	2.5EO	4-7	2.5-4	1.4
	5EO	8-12	5-7	7
	8EO	12-24	9-17	47
C_{12}-C_{15} alcohol	3EO	0.6-1.8	0.14	0.4-0.75
	7EO	0.5-2.4	0.4-0.95	0.9-2.9
	11EO	1.2-2.8	-	-
C_{16}-C_{18} alcohol	14EO	0.4-3.4	-	-
	18EO	2.8	-	-

FIGURE 2   *Observed And Calculated Ec_{50} Values Of Nonionic Surfactants To Daphnia (After Roberts & Marshall, 1995)*

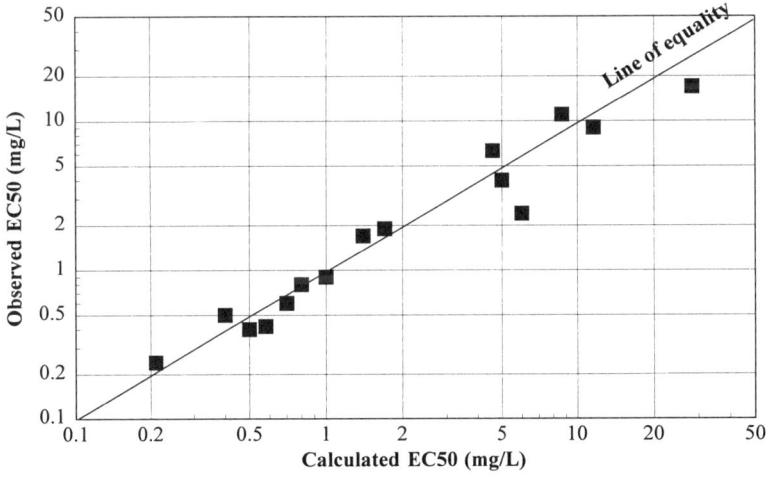

3 BIODEGRADABILITY

The fate of a substance in an environmental compartment is related to its susceptibility to degradation. This degradation may occur by abiotic physical chemical processes such as hydrolysis or photolysis or by the action of living organisms, in the process known as biodegradation. In practice, this is by far the most important process by which the majority of substances are degraded in the environment. Biodegradation may take place in the presence of oxygen (aerobic) or in the absence of oxygen (anaerobic). The aquatic and terrestrial environments are generally aerobic although anaerobic conditions may exist in aquatic sediments and sub-soil environments.

The degree of biodegradation will lie on a continuum ranging from a slight modification of the parent substance to complete mineralisation (Figure 3). The term "Primary Biodegradability" is given to the initial process in which the intact surfactant is broken down. The process may be monitored by observing the loss of surface active properties such as surface tension or by specific analytical techniques. Primary biodegradability is the criterion by which the suitability of surfactants for use in detergent products is currently assessed in the European Union under the so-called Detergents Directives 82/242/EEC and 82/243/EEC. At the other end of the continuum, ultimate biodegradability reflects the complete breakdown or mineralisation of the substance into simple molecules such as carbon dioxide, methane, ammonia, nitrate, nitrite or sulphate ions. In between, there are degrees of partial degradation giving rise to metabolites which may be relatively stable. Their impact on the environment may be assessed by an appropriate bio-assay to show whether or not the level of degradation is environmentally acceptable.

Three other concepts of biodegradability need to be considered in the context of EU Law :

3.1.1 "Ready" Biodegradability.   A substance may be said to be "readily" biodegradable if it passes a test in the OECD 301- series. Characteristic features of these tests are the relatively low bacterial density and the test substance is the sole source of organic carbon. No allowance is made for acclimation of the bacterial inoculum. The tests are considered to approximate the situation in surface waters and are used in Directive 67/548/EEC for environmental hazard classification.

3.1.2 "Inherent" Biodegradability.   A substance may be said to be "inherently" biodegradable if it passes a test in the OECD 302- series. Here the test substance need not be the only source of organic carbon and co-metabolism processes may thus take place. The bacterial density is higher than in "ready" tests and the chance of inhibition by the test substances is reduced. The methods also allow for bacterial acclimation to occur and thus demonstrate whether, given a fair chance, a substance is capable of being degraded.

3.1.3 "Simulation" Tests.   These are tests in the OECD 303- series which are designed to simulate the conditions in sewage treatment plant and allow assessment of the "treatability" of a substance in the effluent treatment process.

As mentioned earlier, "ready" biodegradability information is required for hazard classification purposes and Table 6 shows the key features of the six OECD 301- series methods. The pass level for oxygen uptake and carbon dioxide evaluation is set at 60% of the theoretical value in recognition of the fact that much of the organic carbon in the test substance may be assimilated by the biomass as it grows. The pass level for methods which monitor removal from solution of dissolved organic carbon is set at 70% to allow for the fact that, in addition to mineralisation, some of the test substance may be absorbed on the biomass. The test substance concentrations also vary across the test methods and there is the possibility, with surface active agents, of inhibiting bacterial activity at high test substance concentrations. The test guidelines also require that the test must reach the pass level (60% or 70%) within 10 days of exceeding 10% degradation; the so-called "10-day window". Thus, care may be necessary in interpretation of the results of "ready" test if they are used to rank products in a "beauty contest" fashion.

FIGURE 3   *Biodegradation of Surfactants*

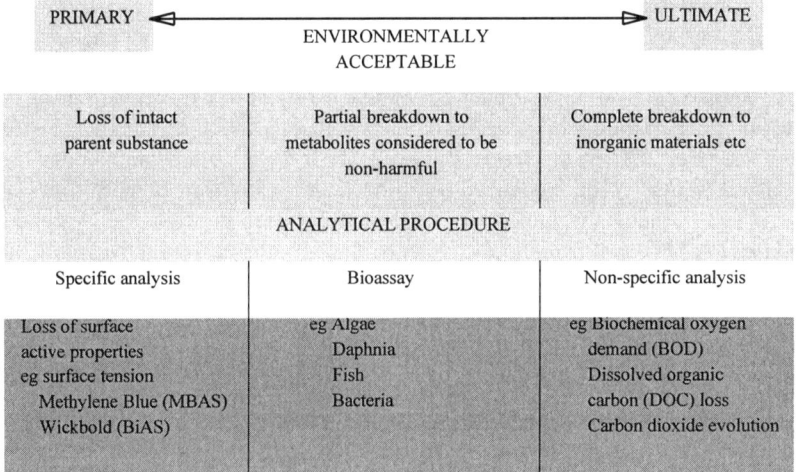

TABLE 6   *"Ready" Biodegradability Methods*

TEST METHOD	INOCULUM	BACTERIAL DENSITY (per ml)	TEST SUBSTANCE CONCENTRATION	ANALYTICAL PARAMETER	PASS LEVEL
DOC Die-Away (modified AFNOR) 301A	15-30 mg/l SS	10^7-10^8	10-40 mg DOC/l	DOC	70% removal
C02 Evolution (Modified Sturm) 301B	15-30 mg/l SS	10^7-10^8	10-20 mg DOC/l	Respirometry : C02 evolution	60% C02/ThC02
Modified MITI 301C	30 mg/l SS	10^7-10^8	100 mg/l	Respirometry : 02 consumption	60% BOD/COD
Closed Bottle 301D	Up to 5 ml secondary effluent per litre	10^4-10^6	2-10 mg/l or 5-10 mgThOD/l	Respirometry : dissolved oxygen	60% BOD/COD
Modified OECD Screening 301E	0.5 ml secondary effluent/litre	10^5	10-40 mg DOC/l	DOC	70% removal
Manometric Respirometry 301F	15-30 mg/l SS	10^7-10^8	100 mg/l 50-100 mgThOD/l	Oxygen consumption	60% BOD/COD

Table 7 summarises the biodegradability data collated by industry for a variety of anionic surfactants in the 301- series tests, opposite the three measurement parameters of carbon dioxide production, oxygen uptake and removal of dissolved organic carbon. In all cases, a high level of biodegradability is observed.   Table 8 summarises data for essentially linear alcohol ethoxylates, which again shows a high level of biodegradation. It is interesting to note that the low BOD/COD results were obtained in protocol 301C at 100 mg/l test substance concentration whereas the high values for C_{16}-C_{18} AE were obtained in 301D at not more than 10 mg/l test substance.

The effect of hydrophobe type and extent of branching on the biodegradability of nonionic surfactants has been reported by Kravetz and co-workers (11, 12). Figure 4 shows the carbon dioxide production curves for a linear primary alcohol ethoxylate (LPAE), a highly branched primary alcohol ethoxylate mole from propylene tetramer (BAE), a linear secondary alcohol ethoxylate (LSAE) and a nonyl phenol ethoxylate (NPE) in a modified Sturm test (301B).   The two linear alcohol ethoxylates were converted to greater than 75% of their theoretical yields of carbon dioxide in 30 days. In comparison, the two highly branched nonionics were very much less degraded with NPE9 reaching less than 30% of the theoretical yield at 30 days. Biodegradation of the linear products was shown to be associated with the rapid production of polyethylene glycol indicating that the molecules decompose by central fission between the hydrophobe and hydrophile (13). The absence of polyethylene glycol production from the highly branched ethoxylates suggests they degrade by terminal oxidation (14, 15, 16).   Figure 5 summarises the two processes.

TABLE 7   *"Ready" Biodegradability Of Some Anionic Surfactants*

	CO_2/ ThCO_2	BOD/ COD	DOC Removal
C_{10}-C_{13} LAS	69	54-65	96
sec C_{13}-C_{17} alkane sulphonate	-	73	95
C_{12}-C_{15} alkyl sulphate	86-96	99	-
C_{12}-C_{18} alkyl 2EO sulphate	65-80	67-69	-
C_6-C_8 dialkyl sulphosuccinate	68	-	100
C_{11}-C_{16} alpha olefin sulphonate	81	78	-
Coco soap	84	-	-

In summary, the principal families of surfactants used in Europe are very rapidly biodegraded in laboratory screening tests.

4 ENVIRONMENTAL MONITORING OF LAS

In practice, most surfactants are discharged to drain after use and are ultimately subjected to some form of effluent treatment. The volumes placed on the market may be very large and, although the materials are rapidly biodegraded, the actual removal efficiency in effluent treatment is extremely important in defining environmental levels and any associated risk to aquatic organisms in receiving waters.

TABLE 8   *"Ready" Biodegradability Of Some Nonionic Surfactants (Essentially Linear Chain Hydrophobes)*

		$CO_2/$ $ThCO_2$	BOD/ COD	DOC Removal
C_9-C_{11} alcohol	5EO	73	-	95
	6EO	76	-	-
	8EO	79	56	80-90
C_{12}-C_{15} alcohol	3EO	86	-	-
	7EO	70	57	97
	11EO	68	-	-
C_{16}-C_{18} alcohol	10EO	-	77	94
	14EO	-	86	94
	18EO	88	-	-

FIGURE 4   *Ultimate Biodegradability Of Nonionic Surfactants By CO_2 Evolution*

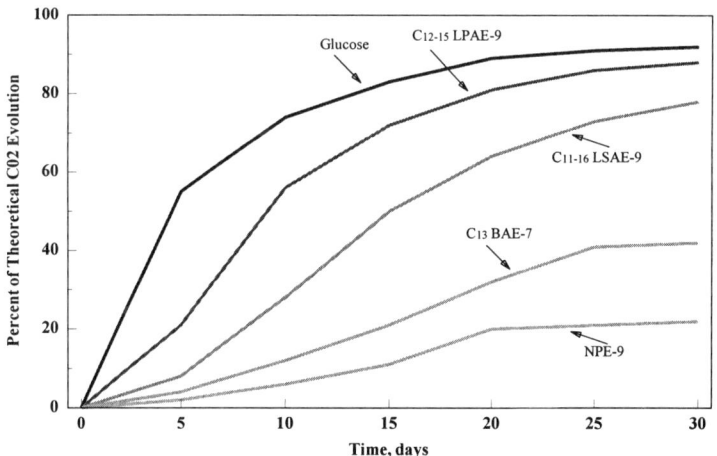

In 1990 the Dutch Ministry of Housing, Spatial Planning and the Environment (VROM) and the Dutch Soap Manufacturers' Association (NVZ) agreed a voluntary programme to assess the environmental impact of the surfactants used in detergents using a step-wise risk assessment process.

Recognising the wider relevance of such a study for the European industry the Association Internationale de la Savonnerie et de la Détergence (AIS) and the Comité Européen des Agents de Surface et leurs Intermédiaires Organiques (CESIO) set up a joint industry Task Force to develop and apply specific analytical methodology for the

Industrial Applications of Surfactants IV

environmental monitoring of the major detergent surfactants. The first phase of this AIS/CESIO surfactant monitoring programme was designed :

-        to measure LAS concentrations in relevant compartments of sewage treatment plant and receiving surface waters, and
-        to provide the necessary data to verify predictive mathematical models.

FIGURE 5   *Mechanism Of Biodegradation Of Ethoxylates*

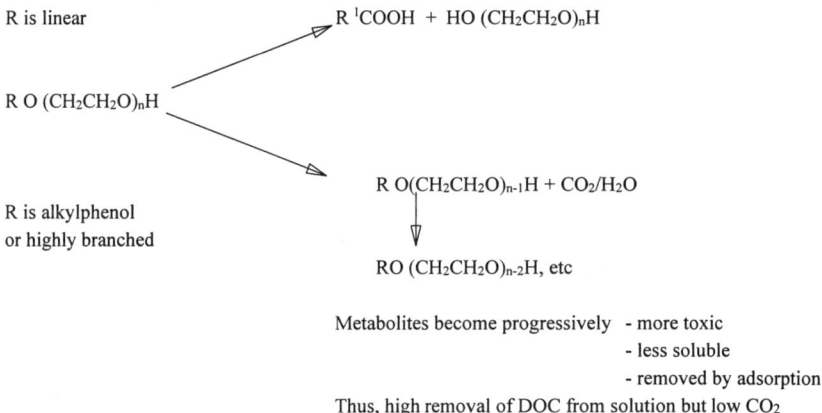

R is linear

$R^1COOH + HO(CH_2CH_2O)_nH$

$R O(CH_2CH_2O)_nH$

R is alkylphenol
or highly branched

$R O(CH_2CH_2O)_{n-1}H + CO_2/H_2O$

$RO(CH_2CH_2O)_{n-2}H$, etc

Metabolites become progressively - more toxic
                                                           - less soluble
                                                           - removed by adsorption
Thus, high removal of DOC from solution but low CO_2
production or oxygen uptake

The results of five national pilot studies on LAS, carried out in Germany, Italy, the Netherlands, Spain and the United Kingdom, have been reported by Waters and Feijtel (17). A very high average LAS removal of 99.2% in sewage treatment was found under normal operating conditions and the national data is shown in Table 9. The differences observed in the main operating characteristics of the sewage treatment plant, ie treatment type, plant size, sludge retention time, hydraulic retention time and temperature, were not found greatly to influence the removal of LAS. At 98.5-99.9% removal, the elimination of LAS was always greater than that of the average burden of degradable organics as measured by BOD, which lay in the range 91-98% (Table 10).

The measured LAS concentrations in the composite raw sewage from the UK study were found to agree closely with those predicted and thus implied little or no degradation of the surfactants in the sewers before reaching the treatment plant. (Holt et al, (18)). The Spanish and Dutch findings, however, suggested that between 35-60% of the LAS load had been removed/biodegraded in sewers and in-line holding tanks (19, 20).

The LAS monitoring data for the five pilot studies were used by the AIS/CESIO "Monitoring and Model Validation" Task Force to check the capability of the SIMPLETREAT (21) and WWTREAT (22) models to predict the fate of LAS in the respective sewage treatment plants. It was observed that the SIMPLETREAT model consistently underestimates the LAS removals whereas WWTREAT predicts them more closely once calibrated with one of the data sets from the pilot study (23). The average

LAS removal figures are compared with the predicted SIMPLETREAT and calibrated WWTREAT removal figures in Figure 6.

TABLE 9   *AIS/CESIO European Surfactants Monitoring Study*

Concentrations of LAS found in sewage treatment liquors and river waters

Pilot Study	Raw Sewage[1] (mg/l)	Effluent[1] (mg/l)	Removal (%)	River Above[1]	River Below[1]
				(µg/l)	
Germany	8.0	0.067	99.2	9	11
UK	15.1	0.010	99.9	30-130	9-47
Netherlands[2]	4.0	0.009	99.8	<2.1-2.9[3]	<2.1-7.1[3]
Spain	9.6	0.140	98.5	27[3]	30[3]
Italy	4.6	0.068	98.5	8.6[4]	9.7[4]

[1] Averages/range for 24h composite samples unless otherwise indicated
[2] Under normal operating conditions
[3] Grab samples
[4] Average for composite and grab sample results

FIGURE 6   *AIS/CESIO European Surfactants Monitoring Study*

Comparison of average measured LAS removal (%) with those predicted by WWTREAT (calibrated) and SIMPLETREAT

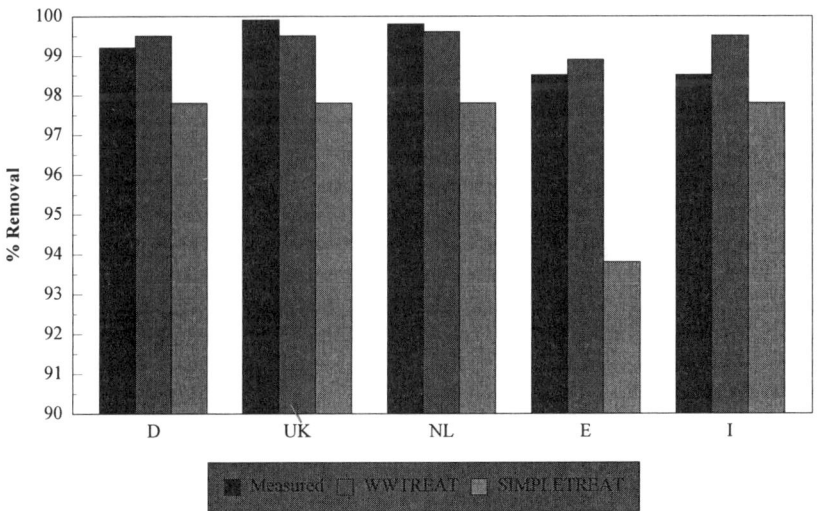

Key: D, Germany; UK, United Kingdom; NL, the Netherlands; E, Spain; I, Italy.

The high removal levels in sewage treatment processes observed for the high volume surfactant LAS lead to final treated effluent concentrations in the range 9-140 µg/l. Allowing for dilution of the treated effluent in the receiving waters these correspond to environmental concentrations in the range 2-47µg/L, which are significantly below the mesocosm no observed effect concentrations (NOEC) of 100-350µg/L for aquatic ecosystems (24). This industry pilot study supports the view that LAS is intrinsically environmentally safe.

TABLE 10   *AISE/CESIO European Surfactants Monitoring Study*

Main features of the pilot study sites

Country	Germany	UK	Netherlands	Spain	Italy
STP site	Munich II	Owlwood, Yorkshire	De Meern	Manresa	Roma Nord
Study period (1993)	16/2 - 24/2	30/3 - 7/4	1/7 - 8/7	1/6-3/6, 17/7	2/6 - 8/6
Treatment Type	Two stage AS, 1° and 2° settlement, P removal	Conventional AS, 1° and 2° settlement, TF for polishing effluent	Carrousel AS, 1° and 2° settlement	Conventional AS, 1° and 2° settlement	Conventional AS, 1° and 2° settlement
Population Served	1,000,000	30,525	32,000	80,000	700,000
Average daily flow (m³ d⁻¹)	216.000^s	7,400 8,200^s	8,000 6,500^{ds}	30,500	276,500
Contribution of industrial source (%)	30	<5	<10	30	<5
Current STP loading (%)	80-100	c 80	c 80	-	c 80
HRT (h)	1st stage 1.3 2nd stage 2.3	6	12	4	7
SRT (days)	1st stage 0.9 2nd stage 19.8	8.8	15	5	10-12
Average BOD removal (%)^s	98	98	98	93	91
River	Isar	Sheffield Beck	Leidsche Rijn	Cardener	Tiber
Sampling sites	above outfall 1.4 km below	0.15 km above 0.1 km below	0.1 km above 0.7 km below 1.2 km below 2.0 km below	above outfall 1.0 km below	0.1 km above 1.0 km below
Average river flow above STP (m³ d⁻¹)	1,468,000	2,300-18,100^s	43,200^e	358,000	8,208,000^d
Effluent dilution factor	9	0.2-3	3-4	13	33

Key : s - measured during study   P - phosphate removal   ds - measured during dry period
        d - dry weather flow   e - estimate   of study
                                                                                                AS - activated sludge
                                                                                                TF - trickling filter

5 SURFACTANTS RISK ASSESSMENT - THE NETHERLANDS

Building on the experience gained in the five-nation LAS environmental monitoring study AIS/CESIO, acting through NVZ, co-operated with VROM, with the Dutch National Institute of Public Health and Environmental Protection (RIVM) the National Institute of Inland Water Management and Waste Water Treatment (RIZA), with universities and with BKH Consulting Engineers in a major project to develop risk assessments for LAS, alcohol ethoxylates (AE), alcohol ethoxysulphates (AES) and soap. The overall project was split into a number of discrete stages and the results and conclusions were presented at the AISE/CESIO Limelette III Workshop in November 1995 (25).

Predicted environmental concentrations of the four surfactants were calculated using the sewage treatment process models SIMPLETREAT and WWTREAT to predict removal and a national dilution database (26) to obtain the predicted total surfactant concentration at 1000m below the sewage outfall. An extensive monitoring study was carried out in parallel to measure removal data at seven sewage treatment works, selected by the authorities to represent typical treatment situations in the Netherlands. The monitoring results are summarised in Table 11, and once again, showed the models to be conservative. Using the dilution model, predicted 90th percentile river water concentrations at 1000m below the outfall were obtained assuming three different in-stream removal rates. These results are shown in Table 12.

TABLE 11 *Summary Of The Key Monitoring Data For Surfactants In The Netherlands*

Surfactant	Influent Range (mg/L)	Influent Average (mg/L)	Effluent Range (µg/L)	Effluent Average (µg/L)	Removal Range (%)	Removal Average (%)
BOD	134-285	221	2000-4300	3200	96.4-99.2	98.1
	(µg/L)	(µg/L)	(µg/L)	(µg/L)		
LAS	3400-8900	5200	19-71	39	98.0-99.6	99.2
AE (C_{12-15})	1600-4700	3000	2.2-13	6.2	99.6-99.9	99.8
AES (C_{12-15})	1200-1600	3200	3.0-12	6.5	99.3-99.9	99.6
AS (C_{12-15})	100-1300	600	1.2-12	5.7	99.0-99.6	99.2
Soap	1400-4500	2800	91-365	174	97.7-99.6	99.1

TABLE 12 *Predicted 90th Percentile River Water Concentration (µg/L) In The Netherlands*

In-stream removal (day^{-1})	LAS (µg/L)	AE (µg/L)	AES (µg/L)	SOAP (µg/L)
k = 0.00	9.2	1.3	2.9	50
k = 0.14	6.4	0.9	2.1	35
k = 0.70	3.7	0.5	1.2	20

The analytical data generated in the monitoring study allowed an assessment of the removal of the surfactants within the sewer system, before treatment. The in-sewer removal varied significantly from one plant to another and it is believed that the differences reflect the length of the sewer (and hence the residence time) and the inherent biological activity. The data recorded are shown in Table 13.

With regard to aquatic toxicity, a large body of data exists in industry and the literature, generated over many years and covering a wide range of structures within the classes of surfactants in the monitoring programme. Whilst this abundance of information should aid the risk assessment, a difficulty arises because of the variety of structures that have been tested. Thus, a "normalisation" process was adopted, based on quantitative structure-activity relationships, to predict the toxicities of the actual surfactant structures monitored in the environment. This process and the derivation of predicted no-effect concentrations (PNEC) are described in reference 25. The PNEC values for the surfactant species observed in the environment are shown in Table 14.

TABLE 13   *In-Sewer Removal Of Surfactants In The Netherlands*

Surfactant	In-sewer removal (%) (Range)
LAS	50 (10-68)
AE (C_{12-15})	42 (28-58)
AES (C_{12-15})	11 (0-40)
AS (C_{12-15})	55 (18-85)
Soap	-

TABLE 14   *Predicted No-Effect Concentrations For Surfactants In The Aquatic Environment Of The Netherlands*

Surfactant	Structure	PNEC (μg/L)
LAS	$C_{11.6}$ LAS	250
AE	$C_{13.3}$ EO$_{8.2}$	110
AES	$C_{12.5}$ EO$_{3.4}$S	400
Soap	unspecified	27

The risk characterisation is achieved by comparing the predicted environmental concentrations with the predicted no-effect concentrations, using the data presented in Tables 12 and 14. The ratios of PEC/PNEC, assuming three different rates of in-stream removal (k) are shown in Table 15. These show clearly that, in the Netherlands and with correctly functioning waste water treatment plants, the risks for the aquatic environment posed by the major surfactant families LAS, AE and AES are negligible.

TABLE 15   *PEC/PNEC For Four Major Surfactants In The Netherlands*

	k=0	k=0.14	k=0.7
LAS	0.04	0.03	0.02
AE	0.01	0.01	<0.01
AES	<0.01	<0.01	<0.01
Soap	1.90	1.30	0.74

Recognising the cost and complexity of environmental monitoring programmes to generate data for risk assessment, the Environmental Risk Assessment Steering Committee (ERASM) of AISE and CESIO proposed the development of a Geography - referenced Regional Exposure Tool for European Rivers (GREAT-ER). This will be a computer software system with which to calculate a realistic distribution of environmental concentrations of down-the-drain chemicals in European surface waters. GREAT-ER will allow an accurate prediction of aquatic chemical exposure for use within the EU environmental risk assessment schemes (27).

The GREAT-ER project began in 1996, in co-operation with the UK Environment Agency and Yorkshire Water. It is managed by a task force of ECETOC (European Centre for Ecotoxicology and Toxicology of Chemicals), in partnership with :

- the University of Osnabrück in Germany
- the University of Gent in Belgium
- the University of Milan, Italy
- the Institute of Hydrology in the UK
- the European Chemicals Bureau in Italy
- the University of Venice, Italy
- CNR-IRSA, Italy
- the Consortium Alto Lambro, Italy

The model is expected to be presented in March 1999.

6 CONCLUSION

Surfactants are very widely used in large volumes and in applications which often lead to discharge to drain. Although many families of surfactants are relatively toxic to aquatic organisms, most are readily biodegradable and are thus capable of being removed in high degree in effluent treatment processes. Real world environmental monitoring data in the Netherlands has confirmed that the risk posed to the environment by the correct use and discharge of the major families of surfactants is negligible.

7 REFERENCES

1.   Chem Week, Jan 28 1998, p37.

2.   Shell Chemical Co. 1983. The Aquatic Safety of Neodol® Products, Brochure SC : 612-83.
3.   P. Schöberl, K.J. Back and L. Huber, Tenside Surf. Det., 1988, 25(2), 86-98.
4.   S.S. Talmage, "Environmental & Human Safety of Major Surfactants : Alcohol Ethoxylates & Alkylphenol Ethoxylates". Lewis Publishers, for US Soap & Detergent Association, 1994.
5.   BKH Consulting Engineers (1993) "Environmental Hazard Assessment of Alcohol Ethoxylates - Data List". NVZ in co-operation with European surfactants industry, Delft, The Netherlands.
6.   C. Gloxhuber and W.K. Fischer. Food Cosmet. Toxicol., 1968, 6, 469-477.
7.   P.D.T. Huibers, V.S. Lobanov, A.R. Katritzky, D.O. Shah and M. Karelson, J. Colloid Interface Sci, 1997, 187, 113-120.
8.   Å. Lindgren, M.Sjöström and S.Wold, JAOCS, 1996, 73 863-875.
9.   D.C.L. Wong, P.B.Dorn & E.Y. Chai, Environ. Tox. Chem, 1997, 16, 1970-1976.
10.   D.W. Roberts and S.J. Marshall, SAR & QSAR in Environmental Research, 1995, 4 167-176.
11.   L. Kravetz, J.P.Salanitro, P.B. Dorn and K.F. Guin, JAOCS, 1991, 68, 610-618.
12.   L. Kravetz in "Agricultural & Synthetic Polymers - Biodegradation & Utilisation" ed Glass and Swift, American Chemical Society 1990.
13.   S.J. Patterson, C.C. Scott and K.B.E. Tucker, JAOCS, 1970, 47, 37.
14.   S.J. Patterson, C.C. Scott and K.B.E. Tucker, JAOCS, 1967, 44, 407.
15.   L. Kravetz, H. Chung, K.F. Guin, W.T. Shebs and L.S. Smith. Household Pers. Prod. Ind. 1982, 19, 46 and 62.
16.   L.Rudling and P. Solyom, Water Research, 1974, 8, 115.
17.   J. Waters and T.C.J. Feijtel, Chemosphere, 1995, 30, 1939-1956.
18.   M.S. Holt, J. Waters, M.H.I. Comber, R. Armitage, G. Morris and C. Newbery, Wat. Res., 1995, 29, 2063-2070.
19.   J. Sánchez Leal, M.T. Garcia, R Tomás, J. Ferrer and C. Bengoechea, Tenside Surf. Deterg., 1994, 31, 253-256.
20.   T.C.J. Feijtel, E. Matthijs, A. Rottiers, G.B.J. Rijs, A. Kiewet and A. de Nijs, Chemosphere, 1995, 30, 1053-1066.
21.   J. Struijs, J Stoltenkamp and D van de Meent, Wat, Res., 1991, 25, 891-900.
22.   C.E. Cowan, R.J. Larson, T.C.J. Feijtel, R.A. Rapaport, Wat. Res., 1993, 27, 561-573.
23.   T.C.J. Feijtel, H. Vits, R. Murray-Smith, R. van Wijk, V. Koch, R. Schröder, R. Birch and W Ten Berge, Chemosphere, 1996, 32, 1413-1426.
24.   BKH Consulting Engineers. Report prepared for ECOSOL, a Sector Group of CEFIC. "The Use of Existing Toxicity Data for Estimation of Maximum Tolerable Environmental Concentration of Linear Alkylbenzene Sulphonate", Delft 1993.
25.   "Environmental Risk Assessment of Detergent Chemicals", Proceedings of the AISE/CESIO Limelétte III Workshop on 28-29 November 1995. Available from AISE and CESIO.
26.   J. de Greef and T. de Nijs, "Risk Assessment of New Chemical Substances : Dilution of Effluents in the Netherlands", RIVM Report No 670208001, Bilthoven, 1990.

27. T.C.J. Feijtel, G. Boeije, M. Matthies, A. Young, G. Morris, C. Gandolfi,
     B. Hansen, K. Fox, M. Holt, V Koch, R. Schröder, G. Cassani, D. Schowanek,
     J. Rosenblom and H. Niessen, Chemosphere, 1997, <u>34</u>, 2351-2374.

The Surface Active Agents Market: Where Is It Going?

G. Bognolo

ICI SURFACTANTS, BELGIUM

Introduction

One of the underlying principles of the theory of relativity is that at the cosmological scale the universe we observe at any given time is not what it is then, but what it was when the electromagnetic radiation left the objects of our observations. Thus the moon we see at night is the moon as it was a second earlier, the sun we see at noon is the sun as it was 8 minutes earlier, the Proxima Centauri we see now is the Proxima Centauri of 4.2 light years ago. That is the period of time it has taken light to travel from these celestial bodies. Obvious on reflection and beautiful in its astonishing simplicity, yet gone almost unnoticed until Einstein brought it to general attention at the beginning of this century.

If one thinks about it, the same principle could apply to non-cosmological phenomena : there is always an interval between an event and its consequences, such that what happens now is the reflection of individual events or combination of events that might have happened nanoseconds or centuries ago. This is not to say that there is an automatic correlation of cause and effects : a given event may have different consequences, depending on the circumstances prevailing at that time or later on, the same way as a beam of light travelling through space may end-up in different places depending upon the gravitational field and the distribution of matter it crosses. Although it is obvious that any event bears consequences, many of these are of limited impact or nullify each other, and only a comparatively small number remain to produce long lasting effects. Thus if it would be possible to :

- identify and isolate the "consequence-bearing" events from the spurious ones
- assess the length of time it might take to produce an effect
- map-out the possible consequences and the likelihood for them to occur

we would have a powerful tool for better decision making and for planning the activities ahead. Devising a process capable of dealing with this challenge and getting it to work is not easy but not impossible. The quality of the results it might produce is in direct proportion to the rigorous application of concepts common to each scientifically sound problem-solving approach :

- careful study and analysis of the information available
- continuous challenge of hypotheses, even if these are based on well established, generally accepted principles. Few things (if any) have been more destructive for the social and scientific development than the infamous "ipse dixit" attitude, that promoted Aristotelian nonsense to the rank of universal truths.
- testing and confirmation of conclusions.

Now then, let see how all this might work when applied to the realm of surface active agents.

If I am coherent with the above reasoning, the first question to ask is : when and why did it all start, as this is an essential analysis to identify factual evidence relevant for future developments. Before answering this question however we should agree on where we are now, i.e. what are the key features of today's surface active agents market.

The surfactants market

The global market for surface active agents in 1997 was of the order of 8.6 millions metric tons, for a value of about GBP 7.1 billions, split almost equally among Western Europe, North America, Asia-Pacific, whilst Latin America represented about one third of the consumption of the other territories (Fig. 1). The consumption of anionic surfactants was about twice that of nonionics, who in turn exceeded the total combined cationics and amphoterics share by a factor of about three (Fig.2).

15 major raw materials of natural and petrochemical origin are used in the manufacture of the top 50 surfactants classes, and the number of applications is endless : to quote the Nobel Price winner Pierre Gille de Gennes "without surfactants we would be completely powerless opposite 90% of the industrial problems". Fig. 3 gives a summary of the major surfactants markets and an estimate of their size.

Historical evolution of surfactants

Back now to the original question, the starting point might be traced back 4000 or 5000 years ago, in Egypt according to certain sources, in Mesopotamia according to others, when somebody noticed that the grease dripping from the meat being roasted on an open fire turned into a white-off paste on contact with the hot ashes. The substance solidified on cooling, and it was further observed that it was particularly effective in removing dirt and fat from the hands and faces of the attendants to these paleo-barbecues. The white-off solid was to be christened SOAP in later times.

Interestingly, what is now a multi billion pound world-wide business did not originate in response to a nicely articulated demand. It was rather the result of observing a phenomenon producing an effect which was in turn perceived as a benefit. The route of the human progress is paved with such accidental discoveries; just think of the steam engine or of the thermosetting and thermoplastic resins. There is perhaps something to learn here for the scores of pseudo-marketeers that offer their services to the industry nowadays.

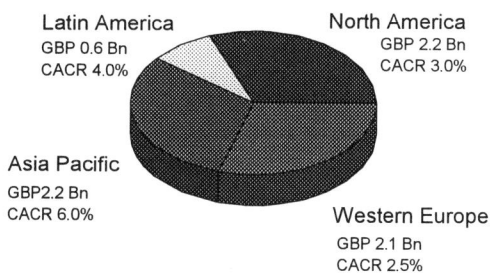

Latin America
GBP 0.6 Bn
CACR 4.0%

North America
GBP 2.2 Bn
CACR 3.0%

Asia Pacific
GBP2.2 Bn
CACR 6.0%

Western Europe
GBP 2.1 Bn
CACR 2.5%

Total GBP 7.1 billions
Average CAGR 3.8%

Figure 1 *The global surfactants market by geographical region*

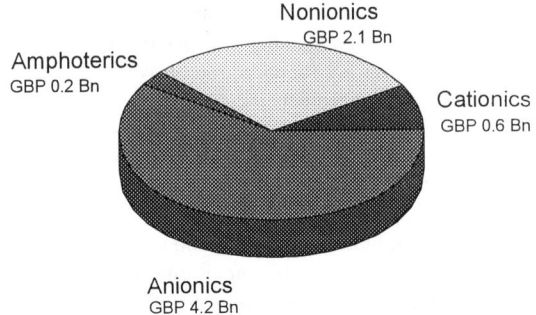

Figure 2 *The global surfactants market by product group*

	VALUE GBP millions	VOLUME metric tons '000
Household detergents	2,800	4,000
Industrial and institutional cleaning	420	530
Personal care	940	860
Crop protection	290	200
Oilfield	390	440
Paints and coatings	140	160
Textile spin finish	200	160
Textile auxiliaries	450	500
Construction	190	470
Emulsion polymerisation	240	290
Food	190	200
Leather	30	60
ORE/mineral	60	150
Plastic additives	60	40
Pulp and paper	100	120
Explosives	10	10
Other	630	380
Total	7,140	8,570

Figure 3 *The major surfactants markets*

Equally interesting is that although the original application of soap was cleaning, its diffusion was driven by cosmetics motivations, in particular hair care : Roman ladies used it to get shiny, lighter colour hairs, Gauls and Germans to fortify the scalp. But we should push the analysis somewhat further, and make sure that important evidence is not overlooked. If we do so we soon realise that, although the cosmetic use might have been the most powerful influence at that particular point in time, there were in reality four factors constantly at play :

-   availability of raw materials: the Mediterranean countries had an abundant supply of vegetable fat (olive oil), whilst central Europe could provide plenty of animal fat (pig and mutton tallow)
-   availability of bodies or linen to clean or of hair to care for
-   desire to have bodies/linen clean or strong/light colour hairs
-   possibility to pay for the above benefits

Although it is obvious that all these factors were necessary for expansion, and that the desire for a specific effect was the driver, the underlying force that made it possible was the availability of raw materials. It is important here to clarify the meaning of "availability" : fat could be obtained everywhere, but it is where it was more abundant, and where the new application could presumably extract more value than other existing ones that the soap industry grew. The concept of availability is at the root of virtually any development in the surface active market as we shall see later on .

From Roman times to the early years of this century many spectacular developments took place: steam and internal combustion engines replaced water or animals as source of energy, the geocentric theory was swept away by the Copernican revolution, Galileo set the basis of modern science, Newton anticipated the light deflection by gravitational fields, Leibniz developed the infinitesimal analysis and so forth and so on. Yet if we look at the description of a soap factory in Diderot and D'Alembert Encyclopaedia, it was not too dissimilar from what it was going to look like about one and a half centuries later, nor for that matter from what it looked like one and a half millenniums earlier. More importantly no other surface active species were exploited, apart from some root extracts (saponins) for cloth cleaning, or naturally occurring emulsifiers, e.g. lecithins, accidentally intervening in the preparation of food.

Why ? Because none of the driving forces described above could generate enough momentum for innovation and change. First, there was no surplus of fats as the productivity of a still primitive agriculture was barely sufficient to feed a growing population. Second, despite the demographic expansion (and hence the potentially accessible market), not many people were interested in personal or garment cleaning, or were prepared to pay for it. Last, there was no real demand for effects other than the ones provided by soap.

Growth of the surfactants market

All this started to change with the first world war. The generally improved standards of living and emerging industrial processes demanded significant tonnages of fats to be transformed into soap at a time when many countries, and in particular the Central Empires, badly needed to turn them into food. At the same time there had been enough progresses in synthetic chemistry and chemical engineering to enable processes like the sulphonation of aromatic substrates to be performed at industrial scale. Thus the first synthetic detergents were born, originally in the form of anionic wetting agents. Anionic surfactants in the full sense of the word were to follow shortly thereafter. Interesting to note that what had been a limiting factor to the development of the surface active agents technology, i.e. the non-availability of a key raw material turned into a driving force when the market conditions changed, i.e. when the demand for effects outstripped the capability of satisfying them with conventional products.

The issue of raw materials availability was also at the root of the development of the ethoxylated nonionics as a class : it was indeed by searching to replace fatty acids in the emulsion polymerisation of styrene-butadiene rubber that Schoeller, Wittwer, Steindorff, Balle, Horst and Michel synthesised the first ethylene oxide derivatives.

If military and strategic considerations prompted the development of these new classes of surfactants, their technological advantages over soaps ensured commercial success, in particular the consistency of performance, the reduced sensitivity to water hardness and the better surface and interfacial properties. The discovery that combinations of anionic surfactants and inorganic phosphates could provide the same or better soil peptisation, hence cleaning, than soaps started a displacement process that was eventually going to relegate soaps to an ancillary position in domestic and industrial detergency.

Raw materials availability, in the sense of utilisation of waste by-products or of attempting to extract higher added value from large volume, low price commodities was at the origin of developments like the sorbitan esters and the polysorbates (sorbitol, ethylene oxide), paraffin and olefin sulphonates (n-paraffines and alpha olefins respectively), alkyl polyglucosides (glucose, molasses, starch), sugar esters and sucroglycerides (sucrose, fats, molasses), glycerol esters (glycerol, fats) , ethylene oxide/propylene oxide copolymers. The industrial expansion and the emergence of new industrial processes, as well as the recognition that surface active agents could greatly improve yields or product quality or manufacturing economics ensured the continuing growth of the surface active agents market at rates higher than the GDP expansion. One can not avoid the impression that for a certain period of time, especially in the late fifties/early sixties everything with amphiphatic properties was turned into a commercial proposition. Eventually natural selection exacted its toll, and only the fittest molecules survived the scrutiny of cost/effectiveness. Alike falling stars, sulfoxides, phosphine oxides, benzyl, allyl and chloride capped ethoxylates, sulphonamides, carbamates, guanidines, phosphonamidic acids, ethoxylated mercaptans, sulphobetaines, borate esters, guanyl and acylguanylureas, alkoxylated dehydroabietylamines, polyoxyethylene t-alyphatic alkylamines, just to quote a few, vanished as rapidly as they had appeared.

All the conditions set forth above for a rapid market expansion were there at the same time :

- the search for alternative raw materials, originally driven by military strategy considerations, met with the availability of suitable substrates and with the technology enabling them to turn into functional products whilst creating the opportunities for the industry to generate profit in the process

- a rapidly expanding industry with new manufacturing processes demanding enhanced surface effects

- a combination of socio-economic factors leading both to higher hygiene standards and the possibility to pay for them.

The growth of the synthetic surfactants market was spectacular by any standard, even considering that there was a proportion of replacement for natural soaps. Everything was growing, sky was the limit and then....

Times are getting rough
Well, in the western hemisphere the surfactants industry did not have its black Friday, but had rather its black decade(s). When did it all start ? Difficult to single-out precisely, but the mid seventies are probably a realistic date, and the consequences begin to be heavily felt now, at least if we believe the press reports that often resemble a bulletin of war in the most dramatic days of the Passchendaele offensive. Why did it all start ? Saturation of the market is the first reason : no matter how clever and creative the surfactants marketeers are, there is a limit to the number of laundry or personal washes that can be performed in a given time, whilst the industrial consumption, although believed to grow faster than GDP do not benefit any more from the quantum leaps it had experienced earlier in the century. The second reason is the structure of the industry, that in Europe and the USA is progressively moving towards services at the expenses of the heavy industrial production. There is clearly a difference in the volume and type of surfactants that are required in a steel mill from the ones required by car paints or by the electronic industry. One might rightly argue that the basic demand for surfactants is simply shifted away from one territory to another, following the migration pattern of the industry. This is correct, but the emerging markets, from a surfactancy point of view, are less sophisticated and demanding and are much more prepared to accept cost/ effectiveness compromises.

So then, where is all this leading to, and with what time horizon.

The short term

This covers, say, the next 5 years. At present the industry is claiming overcapacity, and indeed there have been plant closures whilst not many new units have come on stream. I believe however that the real problem of the surfactants industry at the moment is not capacity, it is prices and margins. The increased consumer's awareness of supply opportunities, products interchangeability, surfactants effects and formulations (just to mention a few) have created a competitive climate unheard of only a decade ago.

If this assumption is correct, the next few years will see the industry actively busy to recover margins through re-engineering and rationalisation of products and services. Apart from what is already in the pipeline, there will be little room for major innovation : even in the presence of focused and well articulated market demands it will take some time for the industry to react, distracted as it will be by other priorities.

It is quite possible that some fusions or alliances will take place, for example the collapse of Huels's Contensio with Condea has been recently rumoured and has some credibility. Rhodia in the USA have widely publicised their search for a partner to reinvigorate their ailing surfactants business.

Surfactants manufacturers will compete essentially on commercial terms, e.g. prices and services, whilst the technical component of sales will focus on quality, consistency, SHE.

These are all features of a mature market, even for products that not long ago were positioned as "effect" providers. At the same time they spell good news for the end users who should enjoy better prices and services, with two potential issues however :

- the product rationalisation will reduce the consumers capability to achieve differentiation in their products through taylor-made surfactants or formulations

- with shrinking profits it is questionable whether industry will devote much resource to new, innovative developments.

The medium term

The SHE and regulatory initiatives that originated in the 80 ies will enter full swing. If it is true that not many surfactants users have shown readiness to change their existing formulations simply on the ground of SHE considerations, it is equally true that none will develop new formulations incorporating surfactants deemed (rightly or wrongly) to pose safety or environmental threats. The most endangered species are the ones easily detectable analytically in infinitesimal amounts, which incidentally are also the ones built around an objectionable aromatic ring. Alkylphenol alkoxylates are already facing the firing squad (at least in Europe), watch what will happen to LAB next.

SHE clearly creates opportunity for substitution products, but creativity and innovation will be the challenge for profit-minded producers, as only original chemistry providing not just substitution, but also additional effects/benefits will enable the extraction of value from a profit-degraded market.

On a more positive note, regulations on VOC reduction will cause the replacement of oil based systems with water borne formulations raising the demand on effect emulsifiers and dispersants, a trend further enhanced by the substitution of hydrocarbons with CO_2 neutral renewable oils.

There will also be a raw materials drive, motivated at one end by the exploitation of emotional ecological messages (green, renewable, natural etc.) and at the other end by the desire to move to milder hydrophile precursors. Carbohydrates will begin to replace ethylene oxide especially in the personal care industry, at a rate that will largely depend upon the commercial availability of suitable surface active molecules and the regulatory pressure on transport, handling and storage of ethylene oxide. Here the expected tightening on transport regulations has failed to materialise, mainly because of the proactive steps taken by ethylene oxide producers and users, however, just in case, a number of European medium size ethoxylators have firmed-up long-term supply deals on the back of an ethylene oxide pipe-line link.

In the USA the progression of front loading machines will in all probability continue and even accelerate, as consumers should appreciate the laundering performance, the economy of operation and the reduced demand for space. This should not cause an increase in detergent consumption, but will certainly affect the detergent formulations with a growing demand on low foaming surfactants, foam controlling agents and other formulation ingredients, e.g. optical brighteners, anti-redeposition agents, enzymes.

The consumption of nonionic surfactants will be driven by the progression of liquid laundry detergents.

The new political and economic order in the former Soviet block will favour closer integration with the western economies and the EU, resulting in the need for upgrading manufacturing technologies and higher standards in the quality of finished products. Inevitably this will lead to an increasing use of speciality and effect surfactants like differentiated fatty alcohol alkoxylates, ethylene oxide/propylene oxide block copolymers, esters, amphoteric surfactants at the expenses of fatty acids and soaps, alkylphenol ethoxylates and sulphonated aromatics.

Apart from the alkylpolyglucosides, glucamides, sugar esters and phospholipids it is unlikely that other species will have reached a sufficient level of manufacturing and market development to launch widespread commercialisation initiatives, although it is not totally unrealistic to expect that "effect" surfactants like silicone, acetylenic, polymeric and fluorinated derivatives will displace less performing conventional surfactants in high added value market segments.

The long term
What will happen around the second decade of the new millennium ? The major change is likely to occur in developing countries, where the improved standards of living will drive a progressive replacement of soap in favour of more differentiated and task-specific cleaning formulations. This should be good news for the sulphonate/sulphate technology, whilst it is still doubtful if the nonionics will follow the alkylphenol or the fatty alcohol route. It is probable however that alkylphenols will have a significant role to play, because of their versatility in application, the low cost and the increasing evidence negating the earlier suspicions of oestrogenic activity.

The industrialised countries will see a further evolution of the patterns that have emerged in the previous years : ethoxylated alkylphenols will become marginalised in some countries and markets and be replaced by ethoxylated fatty alcohols and alkylpolyglucosides. Sulphonated anionics will still represent a major proportion of the surfactant consumption volume-wise, with LAB probably loosing ground to olefin and paraffin sulphonates, fatty alcohol sulphates and ethoxylates.

It is quite possible that some original surfactancy concepts will have firmed-up in the market sectors where a demand exists and the economics can be afforded. In the front line is the low-level emulsification, a technology of general interest but particularly appealing to the personal care market.

More and more effect applications will turn to higher molecular weight surfactants, either oligomeric or polymeric because of mildness and performance, for example low-level emulsification and colloidal stabilisation of emulsions and dispersions. In this context it will be interesting to follow the technological and market progresses of :

- oligomeric surfactants : these have shown interesting physico-chemical properties, it remains to be seen if these can translate in commercially exploitable effects

- biosurfactants : still in their infancy these products offer a number of remarkable features, but are crippled by psychological barriers, uncertain costs and availability, lack of application know-how. It is worth noting that biosurfactants can offer a route to polymeric species, with unconventional hydrophilic and hydrophobic properties and therefore peculiar and differentiated effects.

Conclusions

The major transformation in the surfactants market over the next generation is likely to be of evolutionary nature in the industrialised countries; be driven in sequence by economic considerations, regulatory and evolving market demands and centred around mildness, responsiveness to legislation, public opinion movements and specificity of effects. Detergency in a broad sense (domestic, industrial, hygiene) will continue to account for the majority of the volume. Specific, effect driven non-detergent applications will provide margins, although commoditisation of former speciality surfactants will occur as competition increases and the understanding of products and application technologies grows.

Revolutionary changes will occur in developing countries, driven by improved standards of living. A progressive replacement of soap as all-purpose cleaners with more specialised formulations combine with an increasing overall demand for detergents and emulsifiers.
South East Asia, China, India, Latin America, Africa, the Middle East will see a fast growth in the consumption of synthetic surfactants, the remaining question is if these will be based on petrochemical or renewable resources. Surfactants derived from petrochemical feedstocks have the advantage of well established manufacturing processes using proven assets, and the formulation technology is well established.   However they are harsh, environmentally objectionable and economically exposed if a surplus of naturally derived hydrophobes will become available following a contraction in the consumption of soap..

Clearly these forecasts are speculative and it might well be that not all will materialise, however what is sure is that surfactants will continue to be used everywhere and as such will provide plenty of business opportunities. The issues of commoditisation, overcapacity or declining profitability can be overcome by entrepreneurial spirit and by putting in place the organisations, the marketing plans and the investments matching the needs of the sectors served.

Equally sure is that technically and commercially the surfactants market will continue to be varied, challenging and rewarding for the producers who understand it and are ready to tackle it with the appropriate strategies and resources.

Hydrotropes and Their Applications

E. H. Fairchild, J. A. Komor, A. J. Petro and G. T. Baird

MONA INDUSTRIES, INC., PATERSON, NJ 07544, USA

1 INTRODUCTION

Hydrotropes are organic compounds capable of increasing the solubility of other organic species in water or aqueous salt solutions. They have been used for decades in the form-ulation of surfactant systems and have been the object of numerous mechanistic studies. Despite their widespread use, there is little information in the literature to guide the form-ulator in selection of an appropriate hydrotrope system. This paper will provide the form-ulator with a systematic approach to hydrotrope selection, when the goal is to produce a nonionic surfactant cleaner system containing a high concentration of inorganic alkaline detergent builders. Calculation tools (sliderule and spreadsheet) were developed to aid in this process and their use will be described.

1.1 Background

Hydrotropes are widely used by formulators of industrial cleaners. These cleaners employ a wide variety of alkaline builders with various nonionic surfactants. While there have been a number of papers written on hydrotrope mechanism and numerous papers on specific hydrotrope structures, there are few papers outlining a general approach to form-ulating a cleaner system. It is the intent of this paper is to show the development work that was done on a commercial line of hydrotropes to allow the formulator to readily optimize cleaners designed with mixed builder systems. It should be noted that tradenames will be used in this paper, since many of the hydrotropes shown in the accompanying figures are proprietary compositions. It is the *approach* that is under discussion, not the specifics of any particular hydrotrope.

1.1.1. Definition. Hydrotropes are generally defined as organic compounds having hydrophobe-hydrophile properties and being capable of increasing the solubility of other organic substances or salts in water or aqueous salt solutions.[1]

1.1.2. Mechanism. The mechanism of action of hydrotropes has been studied since the first article in the field by Neuberg[2] in the early 1900's. Others to discuss mechanistic aspects were Rath[3], who in 1965 discussed individual features of the hydrotrope structural organization , Ward[4], who used deuterium NMR to investigate hydrotrope mechanism in 1988 and Friberg[5], in 1989. While this is not a mechanistic or chemical study, it is useful for the reader to have a basic feel for some of the essential elements of hydrotrope function.

They can be summarized in a very general way in the following four statements:
- Hydrotropes often have bulky short, cyclic or branched hydrophobic groups
- Hydrotropes are generally poor surfactants
- They form mixed micelles with nonionic surfactants
- They delay or inhibit the formation of liquid-crystalline phases in the medium at high surfactant concentration.

The surfactant-like behavior of a typical hydrotrope, an aryl ethoxylate phosphate, is demonstrated in Figure 1.

It should be noted that although the reduction of surface tension is characteristic of a surfactant, the critical micelle concentration (CMC) is high, in this case approximately 2%. In this particular case, the ethoxylate chain is quite short, only 6 units. The material is quite similar in structure to the non-ionic surfactant and therefore readily forms mixed micelles, but interferes with formation of the lamellar (or liquid crystalline form) characteristic of high surfactant concentrations.

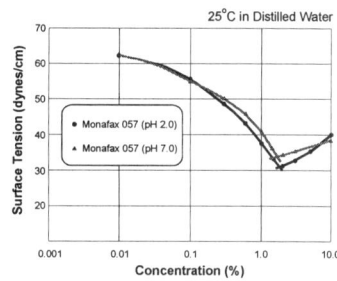

Figure 1. *Equilibrium Surface Tension vs. Concentration*

The well known tendency of surfactants to exhibit lowered cloud points and increased Krafft points in high electrolyte solutions is illustrated in Figure 2.

Throughout this paper we will be using cloud point determinations to define the use concentrations of the various hydrotropes. Nonionics are most effective at concentrations at or near their cloud point. For the industrial cleaning applications of greatest interest in this paper, we have chosen a temperature of 70°C. This is somewhat arbitrary, but useful for making comparisons.

1.1.3. Structure. Before proceeding our discussion of the formulations, it will be useful to list a number of typical materials used as hydrotropes. It is quite a varied list and mixtures are often used as well.

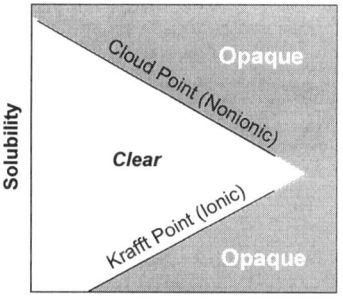

Figure 2. *Effects of Electrolytes on Surfactant Solubility*

- Xylene Sulfonate Salts {Na and K}
- Alkyl Ether Sulfonate Salts
- Alkyl Substituted Succinic Acid Salts
- Dodecyl Diphenyl Ether Disulfonate Salts
- Phosphate Esters {R-$(EO)_x$-O}$_a$-$PO_3^-Na^+$ (Alkyl and Aryl)
- Tall Oil Dimer Acid Salts
- Unbalanced Amphoteric Phosphate Salts

1.2 Formulation Considerations

1.2.1. Load Factors. When compounding liquid detergent concentrates, the formulator must balance solubilizing factors against the factors that tend to kick materials out of solution. The nonionic surfactant, builders (alkaline salts) and neutral ionic species tend to destabilize the solution and must be balanced out by selection of an appropriate hydrotrope. Fortunately, there are many to pick from. The problem is to maintain the cleaning effectiveness of the system and to produce a product that is stable. The diagram shown in Figure 3 illustrates the formulator's challenge.

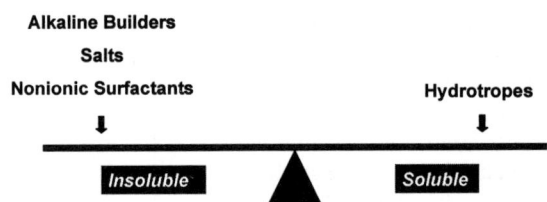

Figure 3. *Load Factors Involved in Compounding Liquid Detergent Concentrates*

1.2.2. Effect of Hydrotrope on Nonionic Surfactant Surface Tension. To illustrate the dominant nature of the nonionic surfactant on the surface active properties of a hydrotrope-surfactant mixture, the surface tension curves for a series of mixtures were determined. The resultant plots are shown in Figure 4.

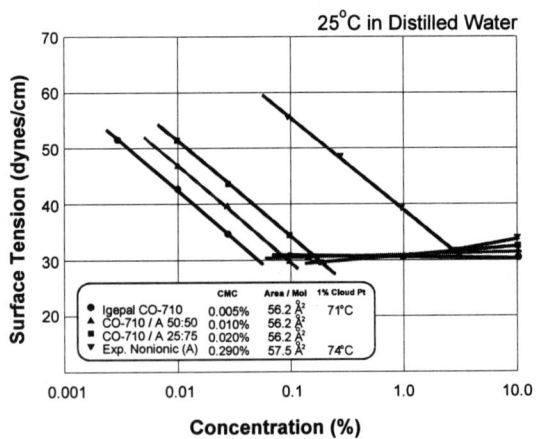

Figure 4. *Equilibrium Surface Tension vs. Concentration*

The surface activity of these mixtures are clearly dominated by the Igepal® CO-710*.

---

* Nonylphenol ethoxylate (10.5 mole) - a product of Rhone-Poulenc (formerly GAF)

1.2.3. Typical Industrial Detergent Formulations. In the formulation of industrial cleaning systems a variety of builders are used. Table 1 summarizes the typical levels seen in industrial detergent systems.

Table 1 *Typical Concentrations (%) of Alkaline Builders Found in Industrial Detergents*

Ingredient	Heavy Duty	Medium Duty	Light Duty
NaOH	30 - 80	0 - 20	–
Na_2SiO_3	0 - 40	40 - 65	–
Na_2CO_3	10 - 25	10 - 30	0 - 10
Na_3PO_4	0 - 30	0 - 30	0 - 10
$K_4P_2O_{10}$	–	–	0 - 20
$Na_5P_2O_{10}$	–	–	0 - 20
Surfactant	5 - 10	2 - 10	2 - 10

2 EXPERIMENTAL

2.1 Effects of Different Electrolytes

As previously seen in Figure 2, the addition of electrolytes lowers the cloud point of nonionic surfactants in aqueous solution. This linear decrease of cloud point has been reported to be dependent upon the hydrated radii of both the cations and anions of the electrolyte.[6,7] We have examined various electrolytes and their effects on cloud point.

2.1.1. Effect on Cloud Points of Nonionic Surfactants. Figure 5 illustrates the varying effects of a number of electrolytes on cloud point. As can be seen from the graph, sodium hydroxide produces the greatest depression of cloud point.

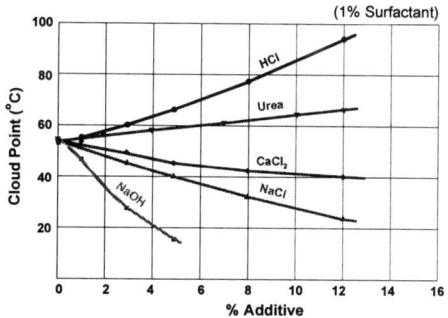

Figure 5. *Effect of Electrolytes on the Cloud Point of Igepal® CO-630*

2.1.2. Effect on Cloud Points of Hydrotropes. In a similar manner, one can look at the cloud point behavior of a typical alkyl ethoxy phosphate hydrotrope in various alkalis. The linear nature of the effect is clearly demonstrated in this series of plots (Figure 6)

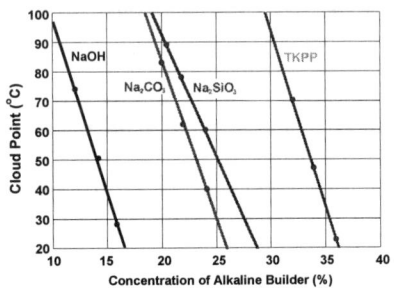

Figure 6. *Cloud Point of Monafax® 1293† in Various Alkaline Builders*

Interestingly, when hydrotropes are present with nonionics the hydrotrope raises the cloud point of the mixture. Again, it is important to remember (as in Figure 3) that the nonionic surfactant properties are dominant. The mixed micelles formed in such a mixture are apparently more resistant to ordering in the lamellar phase.

2.2 Cloud Point vs. Hydrotrope Concentration

This hydrotrope effect of increasing the cloud point of a nonionic surfactant is a very useful property in assessing hydrotrope effectiveness. The steep increases make it a sensitive assay, and it is quite easy to assess in the laboratory.

2.2.1. Distilled Water. While this is not a very useful technique, as almost all cleaners require builders for effective detergent action, it is useful in screening materials for potential hydrotrope character. The two materials shown in the next figure (Figure 7) are obviously quite different in effectiveness.

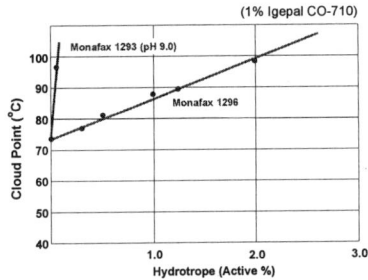

Figure 7. *Cloud Point vs. Hydrotrope Concentration in Distilled Water*

2.2.2. NaOH Solutions. Far more useful, and to the point, are plots of hydrotrope concentration vs. cloud point for various concentrations of sodium hydroxide. The resulting plots are linear in nature, although the slopes do flatten out at increased alkali concentration.

---

† A $C_{8,10}$ alcohol ethoxy phosphate produced by Mona Industries, Inc.

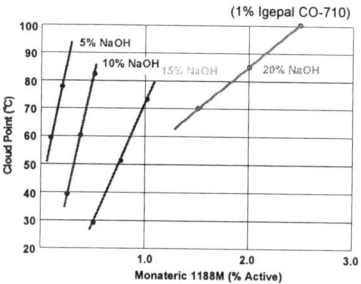

Figure 8. *Cloud Point vs. Hydrotrope Concentration in NaOH Solutions*

2.2.3. Hydrotrope to Give a 70°C Cloud Point. Another way of looking at hydrotrope action is to determine the amount of hydrotrope necessary to give a 70°C cloud point. This gives a uniform way of looking at hydrotrope effectiveness and the resulting data is of immediate use to the formulator. This is a quite useful temperature for industrial hot water cleaning systems. We have determined these values for literally hundreds of candidate hydrotropes and can provide the formulator with an array of useful hydrotropes, each with its own unique properties. In addition, we have accumulated data on a wide array of builders, a few of which are seen in the next figure.

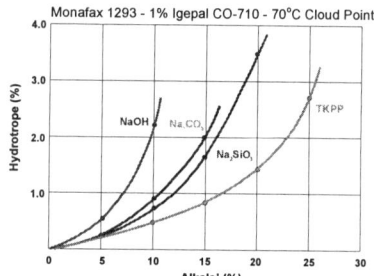

Figure 9. *Hydrotrope Requirements in Alkaline Builder Systems*

The curved nature of the lines indicates a logarithmic behavior that can be better seen by plotting on a semilog scale, as in Figure 10.

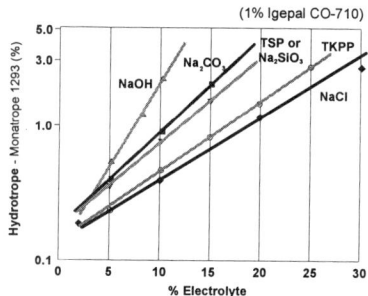

Figure 10. *Hydrotrope Requirements in Alkaline Builder Systems (Semilog Plot)*

2.3 Alkaline Builder Effects

One can see already that this systematic approach to looking at hydrotrope effects has given us good insights into the practical aspects of hydrotrope use. The semilog nature of the previous plot allows us to readily predict how much hydrotrope will be needed to produce a given cloud point in simple one builder systems. Are builder effects additive?

2.3.1. Effect of Various Builders on Hydrotrope Requirement (70°C). In order to look at the additivity of builder systems, we looked at some mixed builder systems.

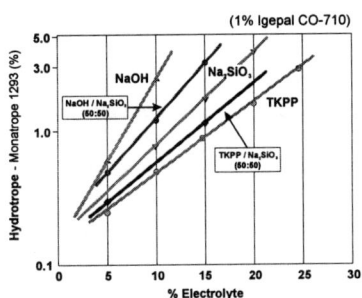

Figure 11. *Effect of Mixed Builder Systems on Hydrotrope Requirement for 70°C Cloud Point*

Note that the 50:50 blends bisect the lines for the components. The systems are additive!

2.3.2. Relative Effects on Cloud Point of a Nonionic. The slopes of the previous plots appear to be related to base strength, so we went back to our cloud point vs. electrolyte plots for the nonionic and replotted the abcissa as normality.

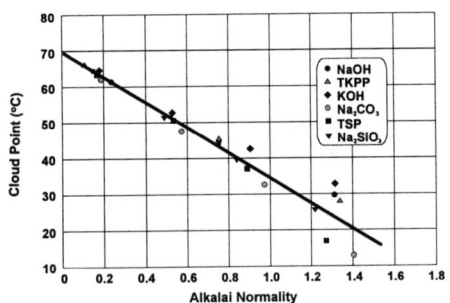

Figure 12. *Relative Effects of Alkaline Builder Concentrations on the Cloud Point of Igepal® CO-710*

It is apparent from this graph that at lower normalities the effects are clearly a function of the base normality. There does seem to be scatter at higher nomality, however. We then repeated our hydrotrope-nonionic experiments in a similar fashion.

The line in the lower left corner of Figure 13 represents the plot of the nonionic alone – no hydrotrope. It matches the similar line in Figure 12. The other lines represent the various builders shown in the legend. For any given builder, the plot is linear. At these

concentrations, however, they are not all conincident.   NaOH and KOH are colinear.
Sodium metasilicate and sodium carbonate are colinear.   The monovalent, divalent and
trivalent cations are all parallel and separated by the same increment. Tetrapotassium
pyrophosphate is not parallel.   We do not have a full explanation for this behavior, but it
makes the systems quite predictable.

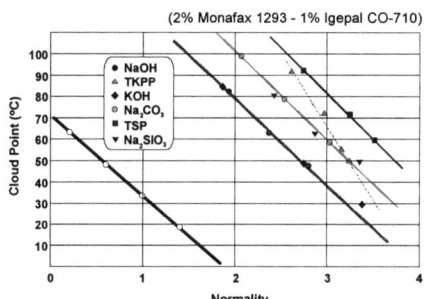

Figure 13. *Effect of Various Alkaline Builders on Solution Cloud Point*

2.4 Nonionic Content Effects

Having gained a quite good insight into the behavior of hydrotrope-surfactant
mixtures in the presence of a number of basic builders, we returned to the study of
hydrotrope- surfactant mixtures that would produce 70°C cloud points. By plotting these
as a function of builder concentration, we obtained a series of straight lines. The slope of
these lines is the ratio of hydrotrope to surfactant necessary to produce the 70°C cloud
point. This is a very useful piece of information for the formulator. It means that for a
given hydrotrope-surfactant combination, one can readily optimize the system for any
base strength. Couple this with the additivity of electrolytes and the ability to calculate an
equivalent normality of a mixed builder system and you have all the parameters necessary
to eliminate a lot of trial and error.

2.4.1. Effect of Nonionic Content on Hydrotrope Requirement (70°C). Figure 14
shows the linear behavior of the ratios of hydrotrope to nonionic surfactant for electrolyte
concentrations ranging up to 20%.

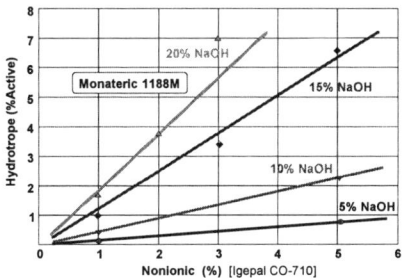

Figure 14. *Effect of Nonionic Content on Hydrotrope
Requirement for 70°C Cloud Point*

2.4.2. Formulation Cloud Points. In order to proceed on to the development of simplified calculation tools we tried many formulations out in our labs. During this period we developed charts of normalization factors. In our parlance, this means the equivalent weight of sodium hydroxide divided by the equivalent weight of the given electrolyte. Tables of these are on our slide rule and in our spreadsheet.

Table 2. *Formulation Cloud Points for Various Alkali Equivalents*

| | | | 1% Igepal | CO-710 Plus |
Alkalai	% Conc.	Normalization Factor	0.88% Monafax 1214	3.28% Monatrope 1296
NaOH	10.0	1.0	70°C	70°C
KOH	14.0	0.714	86	65
Na_2CO_3	13.25	0.755	75	71
Na_3PO_4	13.66	0.732	78	67
Na_2SiO_3	15.26	0.655	73	75
$K_4P_2O_7$	20.65	0.484	73	66

It is always useful to look at actual formulation results. The predicted cloud point were calculated using the slide rule that will be presented in the next section.

Table 3. *Formulation Cloud Points (Predicted vs Observed)*

| | % Composition | |
Ingredients	Formula A	Formula B
TKPP	9.25	9.25
KOH	1.50	1.50
IGEPAL CO-710	3.00	3.00
Monatrope 1296 (50%)	**6.00**	—
Monateric CyNa-50 (50%)	—	**3.20**
Cloud Point Predicted	70°C	70°C
Cloud Point Observed	71°C	61°C

(1) Formulations A and B, normalized, contain the equivalent of 5% NaOH.

3   SLIDE RULE AND SPREADSHEET

3.1 Side 1 – Normalized Builder

All of the work described in the earlier sections of this paper was directed at producing an easy to use system for the formulator. Often the approach is to produce an elaborate brochure, filled with the type of charts and graphs presented here. We certainly have made available data of this type, but we also wanted to produce an easy to use slide rule for the person in the development lab. We realize that formulations produced are only a starting point in the process of product development. Many of the parameters of a cleaning system lie far outside the scope of this paper. Nevertheless, the formulator going to the literature will find little information of direct applicability to the design of an appropriate hydrotrope system. This tool is to fill tht gap.

3.1.1. Slide Rule. The first side of the slide rule helps the formulator in determining the % Total Normalized Builder.

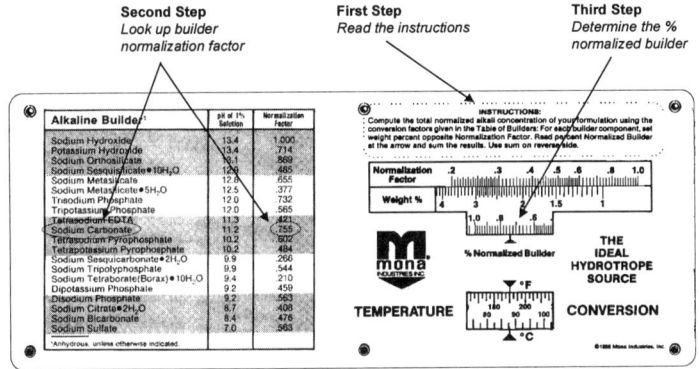

Side #1- Normalized Builder Calculation

Figure 15. *Side 1 of the Hydrotrope Slide Rule*

3.1.2. Spreadsheet. More recently we have begun development of a spreadsheet approach that will carry out formulator calculations and provide data on our hydrotropes. The same basic information is now tabulated as in Figure 16.

Alkaline Builder (Anhydrous, unless otherwise indicated)	pH of a 1% Solution	Neutralization Factor	Formula	MW	Charge
Sodium Hydroxide	13.4	1.000	NaOH	40.00	1
Potassium Hydroxide	13.4	0.714	KOH	56.11	1
Sodium Orthosilicate	13.1	0.869	Na_2SiO_4	184.04	4
Sodium Sesquisilicate * 10 H_2O	12.9	0.485	$Na_2SiO_2 \cdot 10H_2O$	162.47	2
Sodium Metasilicate	12.8	0.655	Na_2SiO_3	122.06	2
Sodium Metasilicate * 5 H_2O	12.5	0.377	$Na_2SiO_3 \cdot 5H_2O$	212.14	2
Trisodium Phosphate	12.0	0.732	Na_3PO_4	163.94	3
Tripotassium Phosphate	12.0	0.565	K_3PO_4	212.28	3
Tetrasodium EDTA	11.3	0.421	$Na_4C_{10}H_{12}N_2O_8$	380.17	4
Sodium Carbonate	11.2	0.755	Na_2CO_3	105.99	2
Tetrasodium Pyrophosphate	10.2	0.602	$Na_4P_2O_7$	265.90	4
Tetrapotassium Pyrophosphate	10.2	0.484	$K_4P_2O_7$	330.35	4
Sodium Sesquicarbonate * 2 H_2O	9.9	0.266	$Na_2CO_3 \cdot NaHCO3$	150.38	1
Sodium Tripolyphosphate	9.9	0.544	$Na_5P_3O_{10}$	367.86	5
Sodium Tetraborate (Borax) * 10 H_2O	9.4	0.210	$Na_2B_4O_7 \cdot 10H_2O$	378.57	2
Dipotassium Phosphate	9.2	0.459	K_2HPO_4	174.18	2
Disodium Phosphate	9.2	0.563	Na_2HPO_4	141.96	2
Sodium Citrate * 2 H_2O	8.7	0.408	$Na_3C_6H_5O_7 \cdot 2H_2O$	289.06	3
Sodium Bicarbonate	8.4	0.476	$NaHCO_3$	84.01	1
Sodium Sulfate	7.0	0.563	Na_2SO_4	142.04	2

Figure 16. *Table of Builders from Excel Spreadsheet*

3.2 Side 2 – Hydrotrope Selection

This side of the slide rule is a lookup table to allow the formulator to select from an array of hydrotropes designed to meet a wide variety of needs. It allows % Normalized Builder concentrations as high as 25%. We are currently revising it to include a wider variety of improved hydrotropes and builder concentrations as high as 30%.

3.2.1. Slide Rule. Figure 17 shows the other side of the slide rule

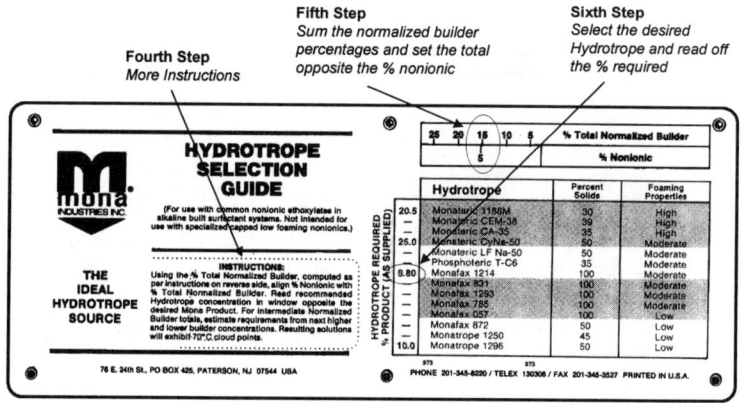

Side #2 - Hydrotrope Selection

Figure 17. *Side 2 of the Hydrotrope Slide Rule*

3.2.2. Spreadsheet. Once again, the spreadsheet mimics the slide rule. Only one of the tables is reproduced here for comparison, one for 1% Normalized Builder.

% Hydrotrope @ 1% Nonionic			1 % Nonionic				
			% Normalized Builder				
Hydrotrope	% Solids	Foaming Properties	5	10	15	20	25
Monateric 1188M	30	High	0.53	1.43	4.00	6.17	
Monateric CEM-38	39	High	0.82	2.26			
Monateric CA-35	35	High	1.09	3.43	9.71		
Monateric CYNa-50	50	Moderate	2.14	2.70	5.00		
Monateric LF Na-50	50	Moderate	4.44	6.46	11.80		
Phosphoteric T-C6	35	Moderate	1.14	3.49	9.14		
Monafax 1214	100	Moderate	0.36	0.88	2.00		
Monafax 831	100	Moderate	0.53	3.00			
Monafax 1293	100	Moderate	0.55	3.00			
Monafax 785	100	Moderate	1.24				
Monafax 057	100	Low	1.60	2.96			
Monafax 872	50	Low				9.00	6.60
Monatrope 1250	45	Low	1.56	3.22			
Monatrope 1296	50	Low	4.62	3.28	2.48	2.40	5.60

Figure 18. *Table of Hydrotropes from Excel Spreadsheet*

4   SAMPLE FORMULAE

A number of sample formulae are included:

"All Purpose Cleaner"

Ingredients	% by Weight
Water	78.0
Sodium Metasilicate Pentahydrate	8.0
Trisodium Phosphate	3.0
MONATROPE 1296 (as supplied)	**7.0**
Igepal CO-710	4.0
	100.0%

Cloud Point = 78°C

Institutional Hard Surface Cleaner System

Ingredients	% by Weight
Sodium m-Silicate (anhydrous)	2.60
Soda Ash	1.70
Tetrasodium Pyrophosphate	1.30
Igepal CO-710	1.20
Hydrotrope	0.25 Active
Water	Q. S. to 100%

	PHOSPHOTERIC T-C6	SXS	TRITON H-66	DOWFAX 2A1
Cloud Point °C	64	38	42	53
Foam ht (ml of) a 1:20 Dilution in 250 ppm H_2O	30	25	30	30

Medium Duty Alkaline Cleaner
F-468

Ingredients	% by Weight
Water	86.5
TKPP (POWDER)	5.0
Sodium Metasilicate (Anhydrous)	5.0
MONATERIC 1188M	**1.5**
Nonyl Phenol (10.5 moles EO)	2.0
	100.0%

TYPICAL PROPERTIES:
Appearance — Clear Liquid
Cloud Point — 83°C   Krafft Point — 0°C

Recommended Use Dilution — 1:10 with water

5   CONCLUSION

The previous work has been successful in providing key concepts for formulating hydrotrope-surfactant systems.

- Cloud points of nonionic surfactants vary linearly with electrolyte concentration
- Hydrotropes increase the cloud point of nonionic surfactant in a linear fashion for a given concentration of electrolyte
- The hydrotrope – electrolyte relationship is semilogarithmic
- Mixed builder systems are additive
- The cloud point – builder relationship is linear and proportionate to normality of the given builder
- The hydrotrope – surfactant ratio is constant for a give electrolyte concentration
- One can sum the normalized builder concentrations and use data generated on NaOH as a model for formulating

These facts allow the systematic development of hydrotrope – builder systems and minimize the trial and error necessary to develop effective cleaning systems.

References

[1] Chakrabarti, P.M., Wayne, N.J., and Grifo, R.A., *U.S. Pat. 4,137,190*

[2] Neuberg, C., *Biochem. Z.,* **76,** 107 (1916).

[3] Rath, H., *Tenside,* **2,** 1 (1965)

[4] Ward, Anthony J.I., *J. Dispersion Sci. Technol.,* **9(2),** 149-69 (1988).

[5] Friberg, Stig E. and Chiu, Ming, *J. Dispersion Sci. and Technol.,* **9(5&6),** 443-457 (1988/9).

[6] Nakagawa, T. and Shinoda, K. in 'Colloidal Surfactants', Academic Press, New York, 1963, p. 129.

[7] Popescu, F. and I. Ana, *Petrol. Geze* **24,** 88 (1973).

The Use of Amphoteric Surfactants to Utilise Imidazolines in Industrial Detergents

D. Horner and P. McCormack

LAKELAND LABORATORIES LTD., PEEL LANE, ASTLEY GREEN, TYLDESLEY, MANCHESTER M29 7FE, UK

1 INTRODUCTION

This paper describes the work carried out to utilise imidazolines in industrial detergents for alkaline aqueous degreasing, with rust prevention, and water-shedding properties. The work was carried out in three phases:

Phase 1 - The hydrotroping properties of salt-free alkyl iminodipropionates when used with nonionics.

Phase 2 - The alkaline stability, rust prevention, water-shedding and foam properties of imidazoline, alkyl iminodipropionate solutions.

Phase 3 - Using the information determined in phases 1 and 2 to produce stable alkaline degreasers with rust prevention properties, based on alkyl iminodipropionates, nonionics and imidazolines.

2 THE HYDROTROPING PROPERTIES OF ALKYL IMINO-DIPROPIONATES ON NONIONICS

Alkyl iminodipropionates are firstly detergents that have excellent hydrotroping properties. In this exercise we are concerned only with the hydrotroping properties. The C_{12-14} dipropionate was selected to look at the hydrotroping effect on three nonionic surfactants of the same alkyl chain length but of differing degrees of ethoxylation, ie of differing HLB values (see Figure 1).

As the Lakeland dipropionates are of different active levels, and of differing RMM, it was decided to compare them at molecular concentrations (see Table 1):

Table 1 *Calculation of Molecular Concentrations*

RMM x 10^{-3} x 100/active matter dipropionate
Lakeland AMA (C_{12-14} Alkyl Chain) RMM 366 active 30%
1.0 = 366 x 10^{-3} x 100/30 = 1.22 gms Lakeland AMA
0.2 increments = 0.244 gms Lakeland AMA

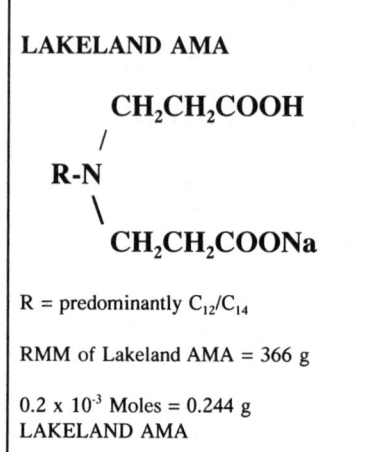

Nonionic	Alkyl Chain	Ratio of C_{13}/C_{15}	Moles EO	HLB Value
A	$C_{13/15}$	67/33	4	9.1
B	$C_{13/15}$	67/33	9	13.1
C	$C_{13/15}$	67/33	14	14.9

Figure 1 *Relationship between HLB Value and Hydrotrope Requirement*

2 gms of the nonionics plus 4 gms of NaOH were made up to 100 gms with deionised water. The NaOH was added to depress the cloud point as would builders in an alkaline detergent. The cloud points were measured 0.2 Moles x 10^{-3} increments of the C_{12-14} dipropionate were added to 2 gms of the nonionic and 4 gms of NaOH. The solutions were made up to constant mass of 100 gms and the cloud points measured. The test were continued until the cloud points approached 100°C (see Figure 2).

The results show that the greater the degree of ethoxylation the less hydrotrope required.

If we calculate from the graph the amount of C_{12-14} dipropionate required to give a cloud point of 70°C we obtain the following (see Table 2):

2% nonionic in 4% NaOH

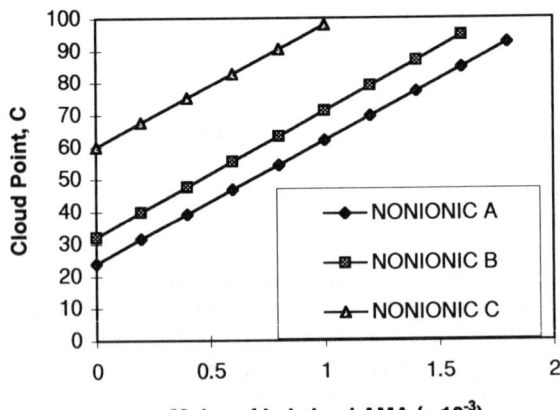

Figure 2 *Comparison of Hydrotroping with differing HLB Values*

Table 2 *Cloud Point 70°*

> 2% w/w nonionic A 4% NaOH requires 1.45 gms Lakeland AMA
> 2% w/w nonionic B 4% NaOH requires 1.10 gms Lakeland AMA
> 2% w/w nonionic C 4% NaOH requires 0.35 gms Lakeland AMA

This clearly demonstrates that the choice of nonionic used is important in terms of hydrotrope cost if high cloud points are required. The hydrotroping costs of nonionic A are four times the cost of hydrotroping nonionic C if a cloud point of 70°C is required.

The hydrotroping properties of all the Lakeland dipropionates were then compared (see Figure 3) on a selected nonionic at differing levels of NaOH. The nonionic selected was nonionic B. The dipropionates were compared at (see Table 3):

Amine Feedstock		Dipropionate Derivatives	
Amine	RMM	Dipropionate	RMM
Coconut $C_{12/14}$	200	AMA	366.0
2-ethylhexyl C_8	129	AMA LF 40	295.0
Octyl C_8	129	LF 60	289.4
Oleyl C_{18}	265	ODA	431.0

Structure

$$CH_2CH_2COOH$$
$$/$$
$$R\text{-}N$$
$$\backslash$$
$$CH_2CH_2COONa$$

Figure 3 *Structures of Alkyl Iminodipropionates and Parent Amines*

Table 3 *Calculation of Molar Quantities*

> RMM x 10^{-3} x 100/% active molecule
> Lakeland AMA (C_{12-14} Alkyl Chain)   30% Active
> 1.0 = 10^{-3} x 366 x 100/30 = 1.22 gm   0.2 Increments = 0.244 gms
> Lakeland AMA LF 40 (2 Ethyl Hexyl)   40% Active
> 10^{-3} x 295 x 100/40 = 0.737 gms   0.2 Increments = 0.147 gms
> Lakeland LF 60 (C_8 Alkyl)   60% Active
> 10^{-3} x 289.5 x 100/60 = 0.482 gms   0.2 Increments = 0.096 gms
> Lakeland ODA (Oleyl C_{18})   30% Active
> 10^{-3} x 431 x 100/30 = 1.43 gms   0.2 Increments = 0.286 gms

The percentage of NaOH in three separate series of tests was 2%, 4% and 8%. Nonionic B was constant at 2% in all the three series. The cloud points were measured and then 0.2 increments were made until 2 x RMM x 10^{-3} x 100/% active material had been added and the cloud points measured at a constant mass of 100 gms (see Figures 4, 5, 6).

The results clearly indicate that at the 2% w/w level NaOH, C_{12-14} is the most effective hydrotrope, followed by C_{18}. The $C_{8's}$ are relatively less effective. The straight chain C_8 clearly better than the branched chain. At the 4% w/w NaOH conditions, all the dipropionates were less effective hydrotropes, the C_{18} performing less well than at the 2% NaOH level. At the 8% NaOH level performance again diminished. C_8 straight chain dipropionate performs best followed by C_{12-14}. The C_{18} does not perform at all. The most interesting feature of these results is the performance of the C_8 dipropionate as the performance of this product is least effected by the higher conditions of NaOH (see Table 4).

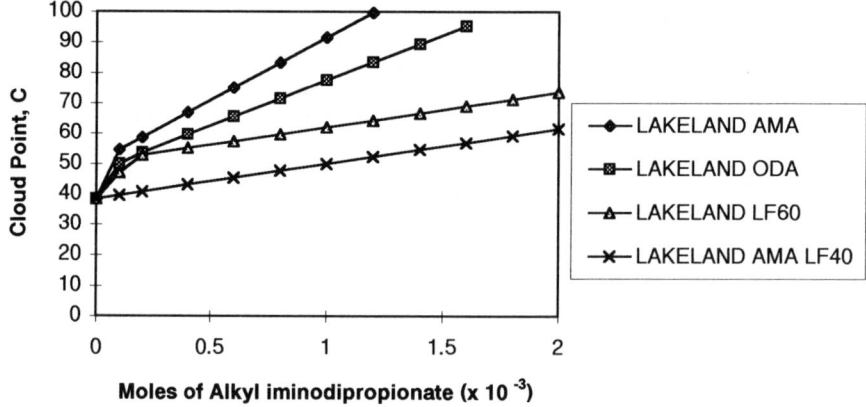

Figure 4 *Cloud Points of Alkyl Iminodipropionates - 2% Nonionic B + 2% NaOH*

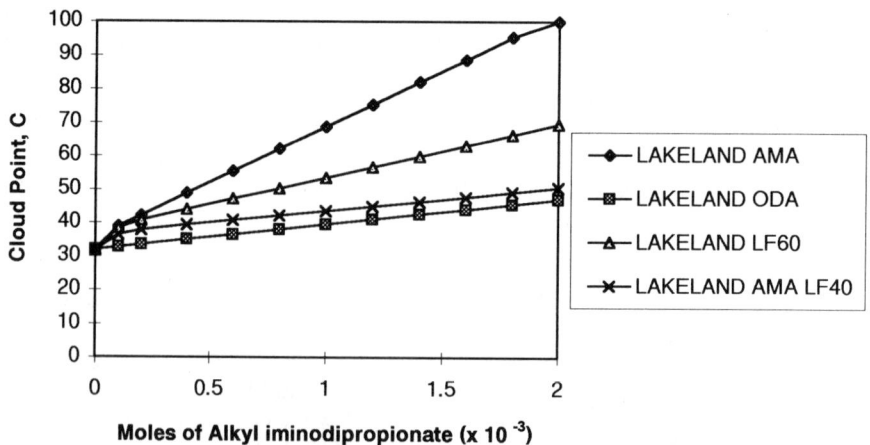

Figure 5 *Cloud Points of Alkyl Iminodipropionates - 2% Nonionic B + 4% NaOH*

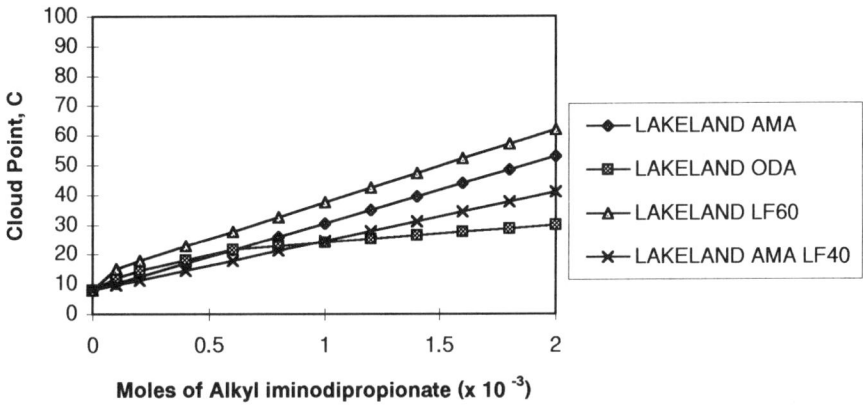

Figure 6 *Cloud Points of Alkyl Iminodipropionates - 2% Nonionic B + 8% NaOH*

Table 4 *Cloud Points of Additions of 10^{-3} x 2 Moles Lakeland LF 60*

2% Nonionic B + 2% NaOH Cloud Point 71°C = 0.96gms Lakeland LF 60
2% Nonionic B + 4% NaOH Cloud Point 63°C = 0.96gms Lakeland LF 60
2% Nonionic B + 8% NaOH Cloud Point 61°C = 0.96gms Lakeland LF 60

The ability of the C_8 dipropionate to hydrotrope efficiently at high alkali levels has been used to produce high active products in combination with the C_{12-14} dipropionate.

3 SOLUBILISING IMIDAZOLINES

Imidazolines are produced by reacting a fatty acid with short chain synthetic amines such as Amino Ethyl Ethanolamine (AEEA) and Diethylene Triamine (DETA). To produce 1 Hydroxy Ethyl 2 Alkyl Imidazoline and 1 Amino Ethyl 2 Alkyl Imidazoline. Imidazolines are perhaps best known as quick break emulsifiers for bitumen emulsions and as rust preventers in oil-based products. They are also used as precursors for further chemical reaction and rheology modifiers. When solubilised in water by acetic or mineral acids they are used as acid cleaners with rust prevention properties. Their use as detergents has been limited in that they have had to be used in acid media.

It had been noticed that the C_{12-14} dipropionate solubilised C_{18} imidazoline. This system has some useful properties that have been exploited commercially. The work described in this paper details all the combinations of Lakeland imidazolines and dipropionates to determine the optimum combinations. We were particularly interested to see if we could produce low foam systems with rust prevention properties, and water-shedding, for spray cleaning.

The following imidazolines were used to test their solubility in dipropionates. The imidazolines were minimum 70% ring closed as detailed in the sales specification (see Figure 7).

The dipropionates were the same as those used in the hydrotroping tests - see Figure 3. To 10 gms of each of the imidazolines an addition of the equivalent of 1.1 Moles of the dipropionates were made. This ratio was used as the imidazoline is

Imidazoline	Parent Acid	Amine	RMM	Imidazoline Structure
8 OH	Octanoic	AEEA	210	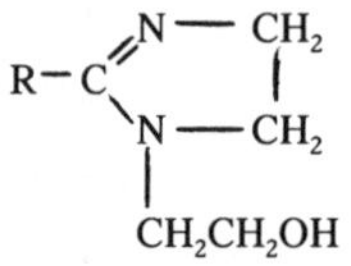
12 OH	Coconut	AEEA	250	
18 OH	TOFA(Oleic acid)	AEEA	350	
S OH	Stearic	AEEA	352	
18 NH	TOFA(Oleic acid)	DETA	310	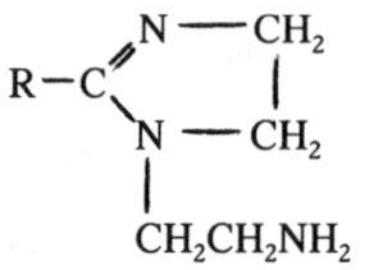

$$R-C \Big\langle {}^{N-CH_2}_{N-CH_2} \quad CH_2CH_2OH$$

$$R-C \Big\langle {}^{N-CH_2}_{N-CH_2} \quad CH_2CH_2NH_2$$

Figure 7 *Imidazolines*

cationic, the dipropionates are the mono sodium salt and have one free carboxylic group. The mixtures were made up to 100 gms with deionised water. The mixtures were stirred for approximately one hour. The following results were obtained - a tick indicates a stable solution, a cross indicates an undissolved imidazoline (see Figure 8).

The undissolved imidazolines were discarded and not considered for further evaluation. The solubilised samples were retained and used for subsequent tests.

4 ALKALI STABILITY

One of the interesting features of solubilising imidazolines with dipropionates is that they were found to have some stability in alkaline systems. The solutions of C_{18} imidazoline were selected for stability trails. The solubilising effect of the dipropionates were compared with the effect of HCl, H_2NO_3 and acetic acid. 10 gms of C_{18} imidazoline was solubilised with the equivalent of 1.1 Moles of the dipropionates and the acids, then additions of 1 gm, 2 gms and 3 gms NaOH were made. The mixtures made up to 100 gms constant mass with deionised water, stirred for ten minutes and then observed for over a week (see Figure 9).

Imidazoline	ODA	AMA	AMA LF 40	LF 60
18 OH	✗	✔	✔	✔
S OH	✗	✗	✗	✗
12 OH	✔	✔	✔	✔
8 OH	✔	✔	✔	✔
18 NH	✗	✗	✗	✗

Figure 8 *Compatibility of Imidazolines and Dipropionates*

10 g Imidazoline 18 OH	A) 1% sodium hydroxide	B) 2% sodium hydroxide	C) 3% sodium hydroxide
1 HCl	separates instantly	separates instantly	separates instantly
2 HNO$_3$	separates instantly	separates instantly	separates instantly
3 CH$_3$COOH	separates instantly	separates instantly	separates instantly
4 ODA	N/A	N/A	N/A
5 AMA	remains clear	remains clear	remains clear
6 LF 60	remains clear	remains clear	remains clear
7 AMA LF 40	separates on standing	separates on standing	separates on standing

Figure 9 *Stability in Alkaline Conditions*

The result clearly shows that the C$_{12-14}$ dipropionate and the straight chain C$_8$ dipropionate enable the imidazoline to remain in solution at moderate NaOH levels. This effect allows imidazolines to be used in alkali cleaners where their corrosion inhibition and water-shedding properties can be exploited.

5 CORROSION INHIBITION

Imidazolines are used in both the oil and water phase as rust preventers. To determine if the imidazoline dipropionate solutions have rust prevention properties, tests were carried out using the Institute of Petroleum 287/83 cast iron chip on filter paper test. This test was selected as it is accepted in the lubricant industry and is simple and pragmatic in that very little equipment is required.

Firstly the C$_{18}$ imidazoline C$_{12-14}$ dipropionate solution was selected and tested against deionised water, and the C$_{12-14}$ dipropionate of the same activity level. The C$_{18}$ imidazoline C$_{12-14}$ dipropionate gave a break point between 10 and 25 x dilution. The C$_{12-14}$ dipropionate had no measurable break point rusting greater than 50%. The deionised water blank showed over 50% rusting in one hour. Further tests were carried out on C$_{18}$ imidazoline C$_{12-14}$ dipropionate. The C$_{12-14}$ imidazoline/C$_{12-14}$ dipropionate, C$_8$ imidazoline C$_8$ branched chain dipropionate, and straight chain C$_8$ dipropionate. Solutions were tested as rust preventers using the IP 287/83 cast iron chips on filter paper test. The break points were between 10 and 20 times dilution of the original solutions (see Figure 10). The C$_{18}$ imidazoline C$_{12-14}$ dipropionate gave the best results followed by the C$_{12-14}$ imidazoline. C$_{12}$ dipropionate and then C$_8$ imidazoline C$_8$ diproprionate. It was noticed that at the lower dilution the C$_8$ solutions were cloudy. Ross Miles Foam Heights were carried out as low foam properties are desirable for spray cleaning. The C$_{18's}$ and the C$_{12-14's}$ gave considerable foam heights. The C$_{8's}$ straight and branched were cloudy and gave no foam at all. This dilution effect was not understood and further work is planned. This was disappointing as it was hoped that C$_8$ imidazoline with the C$_8$ dipropionate would produce low foam systems with rust prevention properties and water-shedding.

dilution	10x	25x	100x
10g Imidazoline 18OH + AMA(1.1 moles equiv)	0%	2%	9%
10g Imidazoline 12OH + AMA(1.1 moles equiv)	2%	2.5%	>50%
10g Imidazoline 8OH + AMA LF40(1.1 moles equiv)	0%	5.3%	>50%
10g Imidazoline 8OH + LF60(1.1 moles equiv)	4%	5.2%	>50%
Water		>50%	>50%	>50%
AMA		>50%	>50%	>50%

Figure 10 *Summary of Corrosion Inhibition Testing IP 287/83*

6 WATER-SHEDDING

All the above solutions showed water-shedding properties.

7 FORMULATING IMIDAZOLINES, DIPROPIONATES AND NONIONICS

Before we look at formulating imidazolines, dipropionates and imidazolines we looked at the effect on cloud point that an addition of imidazoline would make to an alkaline, nonionic system hydrotroped with a dipropionate. Firstly we examined the C_{12-14} dipropionate. From Figure 4 we calculated the amount of the C_{12-14} dipropionate required to give a cloud point of 70°C when added to 2% nonionic B + 2% NaOH at constant mass of 100 gms. This required 0.54 gms. We then calculated the amount of the solutions of C_{12-14} dipropionate with C_{18}, C_{12-14} and C_8 imidazolines that had been retained after the solubility tests (Figures 7 and 8) that contained 0.54 gms of the C_{12-14} dipropionate. This amount was added to the 2% nonionic B + 2% NaOH of constant mass of 100 gms. The cloud points were then measured. The results are detailed below (see Figure 11).

The results show that the addition of C_8 imidazoline gives no reduction in cloud point, the C_{12-14} imidazoline very slight. The C_{18} imidazoline is much more pronounced.

The tests were repeated on the C_8 dipropionate. From Figure 4 it was calculated 0.82 gms of C_8 dipropionate was required to give a cloud point of 70°C (see Figure 12).

The results show that the C_{18} imidazoline depresses the cloud point to 50°C, the C_{12-14} unchanged and the C_8 a cloud point increase.

The above tests indicate that mixtures can be produced that will operate at reasonable temperatures with relatively small additions of hydrotrope.

To determine if stable products can be produced at reasonable active levels the C_{18} imidazoline with C_{12-14} and C_8 dipropionate were scaled up to approximately 30% with and without 2% NaOH (see Table 5).

Solutions analysed based on 1.1 Moles Amphoteric + 1 Mole Imidazoline	
System: 2% C$_{13/15}$ alcohol 9EO + 2% NaOH	
Product	**Cloud Point**
1. AMA 0.54%*	70°C
2. AMA 0.54%* + Imidazoline 18 OH	61°C
3. AMA 0.54%* + Imidazoline 12 OH	67.5°C
4. AMA 0.54%* + Imidazoline 8 OH	70°C
(0.54% = 0.44 x 10^{-3} Moles)	

Figure 11 *Changes in Cloud Point of Nonionic/Dipropionate Solutions with Imidazoline Additions*

Solutions analysed based on 1.1 Moles Amphoteric + 1 Mole Imidazoline	
System: 2% C$_{13/15}$ alcohol 9EO + 2% NaOH	
Product	**Cloud Point**
1. LF 60 0.82%*	70°C
2. LF 60 0.82%* + Imidazoline 18 OH	50°C
3. LF 60 0.82%* + Imidazoline 12 OH	67.5°C
4. LF 60 0.82%* + Imidazoline 8 OH	76°C
(0.82% = 1.70 x 10^{-3} Moles)	

Figure 12 *Changes in Cloud Point of Nonionic/Dipropionate Solutions with Imidazoline Additions*

Table 5 *Imidazoline/Iminodipropionate Formulations*

A		B
22.3%	Nonionic B	22.3%
6.0%	Lakeland AMA (C$_{12-14}$)	6.0%
1.7%	Lakeland Imidazoline 18 OH	1.7%
0.0%	NaOH	2.0%
To 100% Water		To 100% Water

Stable, clear, free-flowing liquids after one month.

A		B
16.5%	Nonionic B	16.5%
6.8%	Lakeland LF 60	6.8%
4.4%	Lakeland Imidazoline 18 OH	4.4%
0.0%	NaOH	2.0%
To 100% Water		To 100% Water

Stable, clear, free-flowing liquids after one month.

8 CONCLUSIONS

The work carried out has only scratched the surface. Further work is intended to fill some of the gaps that have been left. However, the work shows that aqueous alkaline degreasing products can be produced from nonionics, imidazolines and dipropionates. The dipropionates recommended are C_{12-14} and straight chain C_8. The imidazoline recommended is the C_{18}. The choice of nonionic is left to the formulator.

High active nonionic-based detergents 40% plus can be produced by using a mixed hydrotrope system of C_{12-14} and straight chain C_8.

C_{18} imidazoline/C_{12-14} dipropionate systems have found commercial uses in water-shedding in traffic film removers, aqueous degreasing with rust prevention properties, spray cleaners with anti-static properties for office equipment, and industrial wet wipes.

Although this paper has been concerned with alkaline systems, it must be remembered that acid systems can be produced for aluminium and stainless steel cleaning using C_8 dipropionates and C_{18} imidazolines. The acid recommended would be methyl phosphate ester, Lakeland PA 100.

9 ACKNOWLEDGEMENTS

Many thanks to Lucy Speight, student Sheffield Hallam University, as her hard work made this paper possible.

Application of Sucrose Fatty Acid Esters as Food Emulsifiers

S. Nakamura

RESEARCH DEPARTMENT – FOOD ADDITIVES, LIVING & ENVIRONMENTAL MATERIALS DIVISION, DAI-ICHI KOGYO SEIYAKU CO., LTD., JAPAN

Sucrose fatty acid esters (hereinafter referred to as "SE") being ester type non-ionic surface active agents consist of sucrose residue as a hydrophilic group and fatty acid residue as a lipophilic group. With their remarkable safety justified by the detailed studies on the metabolism and the digestion and absorption[1,2,3], SE has been approved as a food additive in many countries in the world.

SE was approved in 1959 as a food additive in Japan first before any other countries in the world and a majority of consumption were limited to Japan and East Asian countries those days. After the approval in EC Directive in 1978 (E 473) and the subsequent approval by the FDA in the US in 1983 (21 CFR § 172.859), the demands for SE as food additives expanded to Europe and the US. At present, the total consumption in the world is estimated to exceed 5,000 metric tons per year with a rapid growth in EU and the US and a steady growth in Japan. Two SE manufacturers in Japan are now taking steps to increase the production capacity of SE to catch up with the strong demand.

Sucrose has eight hydroxyl groups as the hydrophilic group which allow to manufacture a variety of SE ranging widely from low HLB (hydrophilic-lipophilic balance) value to high HLB value by controlling the degree of esterification.

The main characteristics of SE are: O/W and W/O emulsifying properties, solubilizing property, foaming property, inhibition of crystal growth in fat, retrogradation in starch, releasing property, and reduction of flat-sour spoilage. These characteristics are quite useful in the production of foods, such as dairy products and baked goods.

Author overviews the development of SE on their chemistry and manufacturing technology in Japan. Several practical applications are also explained somewhat in detail and the future trend is finally touched.

1. SURFACE ACTIVITY OF SE

Structural dependency of surface activity has been studied on SE with various degrees

Table 1 Surface activities of SE with various degree of substitution and fatty acid constituent[1)]

SE sample	Composition (%)		Fatty acid residue (Purity of main composition. %)		Surface tension[*3] (mN/m)		Wetting power[*4] (sec.)		Foaming power[*5] (mm)			
	SME[*1]	SDE+SPE[*2]			0.1%	0.3%	0.1%	0.3%	0.1%		0.3%	
									0 min.	5min.	0 min.	5 min.
Series 1[*6]												
SE-7	74.3	25.7	Hydrogenated Tallow		35.0	34.2	> 1000	865	31	29	35	33
SE-6	61.6	38.4			34.5	34.4	> 1000	> 1000	24	23	25	23
SE-5	50.2	49.8			36.7	35.0	> 1000	> 1000	12	9	14	13
SE-4	40.8	59.2			40.6	38.5	> 1000	> 1000	7	4	8	6
SE-3	32.4	67.6			46.8	44.4	> 1000	> 1000	4	2	6	5
Series 2[*7]												
C$_8$SE	64.3	35.7	Caprylate	96.9	27.0	26.5	115	89	31	29	55	54
C$_{12}$SE	68.7	31.3	Laurate	92.1	28.5	29.9	201	157	127	124	136	134
C$_{14}$SE	65.8	34.2	Myristate	91.8	33.0	31.2	357	285	62	59	92	89
C$_{16}$SE	66.0	34.0	Palmitate	90.2	34.0	32.0	405	387	29	26	36	36
C$_{18}$SE	63.8	36.2	Stearate	92.1	34.8	33.9	> 1000	965	29	28	34	32
C$_{22}$SE	70.0	30.0	Behenate	86.0	38.5	37.4	> 1000	> 1000	22	21	25	23
C$_{18:1}$SE	68.4	31.6	Oleate	90.4	35.0	33.6	540	437	24	21	25	22
C$_{18:2}$SE	67.2	32.8	Linoleate	89.8	35.0	37.5	230	163	20	25	31	27
Blank[*8]					72.8		> 1000		0	0	0	0

*1: sucrose monoester.   *2: sucrose di- and polyester.   *3: measured by the Du'Nouy method.   *4: measured by the canvas disk method.

*5: measured by the Ross-Miles method.   *6: SE sample differed in degree of substitution.   *7: SE sample differed in fatty acid residue.

*8: Distilled water was determined as blank.

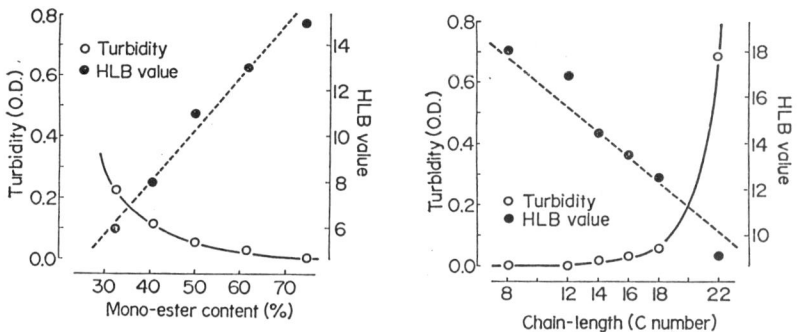

Fig.1 *Effect of degree of substitution and fatty acid constituent on turbidity of 0.3% solution and HLB value of SE[4]*

of substitution and fatty acid constituent[4].

As shown in Fig. 1, hydrophilic capacity of SE is dependent on mono-ester content and fatty acid constituent. In case fatty acid constituent is identical, high mono-ester content SE shows higher HLB than SE with low mono-ester content. SE with a longer-chain or a more saturated type of fatty acid shows lower hydrophilic capacity.

Table 1 shows surface tension, wetting power, and foaming power of SE with various degree of substitution and fatty acid constituent. SE with a shorter-chain fatty acid or with the lower degree of substitution exhibits lower surface tension and higher foaming power. A longer-chain or more saturated type of fatty acid gives less wetting power.

2. BIOCHEMISTRY OF SE

Digestibility of SE has been studied systematically using rats[3]. Table 2 shows SE digestibility by rats. It is observed that the digestibility of SE monoester is high. The ester linkage of SE is known to be hydrolyzed by enzyme preparations such as liver homogenate and pancreatic juice[5].

Daniel et al. made metabolic studies of SE using [14C]SE in rats[2]. The disposition of radioactivity after the administration of [14C]SE suggested that SE was hydrolyzed before absorption, and no evidence was obtained for the accumulation of SE in the adipose tissue of rats by repeated daily oral administration of [14C]SE. They considered that the use of SE as food additives in human foods would not appear to present a significant toxicological hazard, because SE is hydrolyzed easily under physiological conditions into sucrose and fatty acid.

As described earlier, results of multiple studies on the safety of SE have been reported. These studies have been evaluated by an international organization, the joint FAO/WHO Expert Committee on Food Additives[1]. No toxicity at relevant SE levels has been recognized, and it has been confirmed that SE poses no problems as a food additives.

Table 2 *Digestibility of sucrose stearate in rat* [3)]

SE*[1] Content In feed (%)	Digestibility (%)		
	MS*[2]	DS+PS*[3]	Total
1 %	83.7 ±1.4	33.0 ±2.8	61.3 ±2.0
2.5%	87.2 ±1.7 -N.S.*[4]	29.0 ±2.4 - N.S.	58.6 ±2.0 - N.S.
5%	87.9 ±2.8	32.1 ±2.2	60.8 ±1.6
Average	87.5	31.4	60.2

*1: Sucrose stearate whose composition is 51% of monoester, 38% of diester and 11% of polyester
*2: Digestibility of sucrose monostearate
*3: Digestibility of sucrose di- and polystearate
*4: No significant difference

3. MANUFACTURING PROCESS

Manufacturing methods of SE are roughly classified into the direct esterification method, with fatty acid chloride or fatty acid anhydride, the interesterification method, using a lower alcohol ester of fatty acid as raw material, and the enzymatic method, in which enzyme such as lipase is used as the catalyst.

The direct esterification method, which was mainly attempted in laboratory scale testing during early stages of development, is considered uneconomical and is not taken into consideration today at all.

The enzyme method is attracting keen attention as a new SE manufacturing method with steady studies in progress with commercialization expected for sometime in the future, but has not been realized in a practical use yet.

The method now used in the industry is the interesterification method, in which sucrose and methyl esters of fatty acids are used as raw materials. SE products sold in market today are manufactured using methyl esters of long-chain fatty acids originating from plants who fatty acid composition mainly consists of laurate, palmitate, stearate, oleate, erucate, and behenate.

In view of the fact that main fields of application of SE are in foods and cosmetics, purification is of primary importance in the SE manufacturing process.

As in conventional industrial purification methods, organic solvents such as ethyl acetate, methyl ethyl ketone and isobutyl alcohol are used as extraction solvents to improve SE purity; but from a safety point of view, removal of the organic solvents is an important subject. Residual levels of such organic solvents in products is strictly restricted by a law in the US, Europe, Japan, and so on.

Dai-Ichi Kogyo Seiyaku Co., Ltd. Japan had developed a new system, and started commercialization of the product with a unique purification process as shown in Flow sheet diagram.

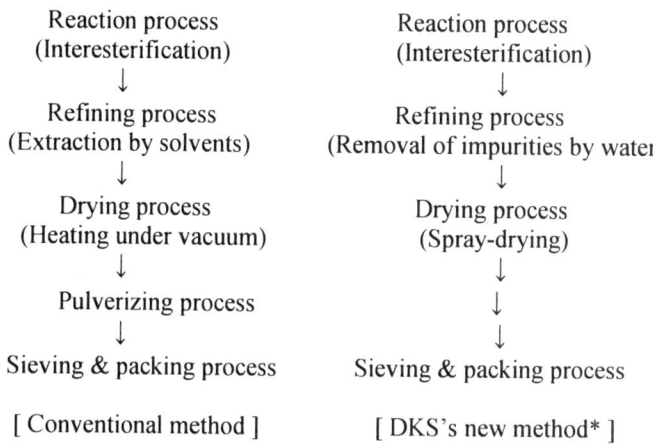

Flow Sheet Diagrams of Industrial Manufacturing Methods of SE
(*: DKS means Dai-Ichi Kogyo Seiyaku Co., Ltd.)

A new plant in Shiga prefecture was completed in 1995 with an annual production capacity of 2,000 MT. A steady operation started in 1996 after conducting trial runs.

The purification process developed by the company is featured first in that not a single organic solvent but water is used as a purification solvent.

Because the new process does not use any organic solvents, such as ethyl acetate, methyl ethyl ketone and isobutyl alcohol, which can be highly explosive and represent a fire hazard, the plant can be operated quite safely. In addition, less expensive non-explosion-proof specifications are adequate. Thus, plant construction costs can be relatively low. Another advantage is that construction work and inspection or modification of the facilities can be done without stopping plant operation.

It should be noted that use of a large amount of water in the purification process can lead to contamination with microorganisms, such as bacteria and molds. The new plant is equipped with many sterilizing filters installed one after another to treat process water with an ultraviolet sterilizer. Steam sterilization is used to prevent contamination of water in piping.

The second feature of the new process is adaptation of a spray drier to dry SE. It is difficult, in general, to dry hydrous substances of comparatively low melting points, such as SE, without using a builder. However, our company has been manufacturing detergents since the foundation and the technologies acquired for many years allowed to develop a spray-drying process for manufacture of SE.

Advantages of applying spray-drying technology to the SE manufacturing process are:

(1) Thermal deterioration of SE products, including decomposition, is inhibited, thus preventing development of off-colors or odors because of short drying time (of several seconds).
(2) SE powder produced by spray drying consists of spherical particles, with high fluidity and little dust formation, which facilitates easy handling.
(3) The hydrous slurry can be powdered directly and no grinding process is required. This eliminates the risk of dust explosion.

The disadvantages are not so serious as to nullify the above advantages. The spray drier adapted for the new process requires a comparatively high equipment cost and high operating costs, owing to high heat consumption as a result of the high water content in the raw spray slurry.

Dai-Ichi Kogyo Seiyaku Co., Ltd. filed patent applications for technologies obtained through development of the new process in the United States (U.S. Patents 5,011,922/5,017,697/5,008,387/4,995,911/4,996,3099) as well as in Japan. Details of the technologies are described in these patents.

4. PERFORMANCE AND APPLICATIONS OF SE

Sucrose has eight hydroxyl groups constituting the hydrophilic moiety, so it is possible to manufacture SE ranging from low HLB to high HLB values by controlling the degree of substitution. Dai-Ichi Kogyo Seiyaku Co., Ltd. manufactures SE commercially under the trade name DK ESTERs with a wide range of HLB values (see table 3), which are widely used for both O/W and W/O emulsified foods by selecting an appropriate SE. In addition, it is known that SE interacts with the main food components to affect physical properties of foods. This attribute of SE is utilized in a number of processed foods.

Table 4 is a list of SE applications including its performances.

Table 3 *Examples of typical commercial sucrose fatty acid ester products*
(Dai-Ichi Kogyo Seiyaku Co., Ltd.)

DK ESTER	HLB value	Ester composition (%)		Appearance
		Mono-ester	Di-/tri- poly-ester	
SS	Approx.19	Approx. 100	Approx. 0	Powder
F-160	15	70	30	Powder
F-140	13	60	40	Powder
F-110	11	50	50	Powder
F-90	9. 5	45	55	Powder
F-70	8	40	60	Powder
F-50	6	30	70	Powder
F-A50*	-	0	100	Solid
F-20W	2	10	90	Powder
F-10	1	0	100	Powder
F-A10E*	-	0	100	Flake

＊ : With free hydroxyl group acetylated

Table 4 *Performance and Food Applications of DK ESTER*

Performance	Application	DK ESTER
O/W emulsifier	Coffee whiteners	F-110～160
	Whipped toppings	F-110, 20W, 10
	Frozen dairy desserts	F-110～160
	Milk drinks	F-110～160
	Candies	F-20W～70
W/O emulsifier	Margarines	F-10, 20W
	Fat spreads	F-10, 20W
Solubilizer	Beverages	SS[b)]
Foamer	Cakes	SP-Gold[a)]
	Fish pastes	SP-Gold[a)]
Crystal Growth inhibitor	Margarines	F-10, 20W, A10E[c)]
	Shortenings	F-10, 20W, A10E[c)]
	Chocolates	F-10～50, A-10E[c)]
Starch retrogradation inhibitor	Baked goods	F-110～160
	Flour pastes	F-50～90
	Fish pastes	F-70～110
Lubricant (prevent sticking)	Tablet candies	F-20W, 50
	Candies, Chewing gums	F-20W, 50
	Biscuits, Cookies	F-20W～70
Prevention of flat-sour spoilage	Canned coffee drinks	F-140, 160
	Canned tea drinks	F-140, 160

a);Compound product containing SE
b);Pure sucrose monostearate
c);Acetylated sucrose fatty acid esters

4.1 O/W emulsifying performance of SE

With regard to O/W type emulsions, it is important to note that oils and fats are taken into micelles of surfactant to form a dispersed phase in the continuous phase (water), and that emulsifiers form stable micelles in water. The effect of fatty acid constituent and degree of substitution in SE on stability of aqueous emulsion has been studied[6], and it was found that the emulsifying performance of SE with high mono-ester content and with saturated-type long-chain fatty acids used as the lipophilic group

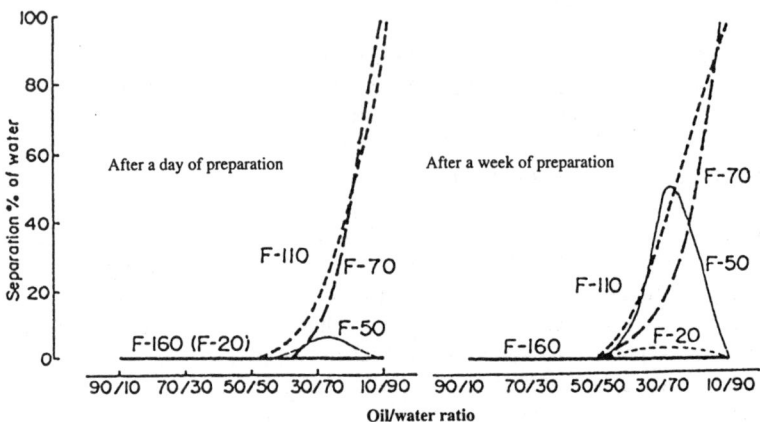

Fig.2 *Effect of degree of substitution in SE on stability of aqueous emulsion*[6]
SE sample : DK ESTER (see Table 2)
SE Conc.   : 5 %
Fats & oils used : Soybean oil

is high. Fig.2 shows the water separation rates one day and one week after preparation of O/W ratio varying from 10/90 to 90/10. This indicates that DK ESTER F-160 of high mono-ester content, i.e., a high HLB value, shows stable emulsifying performance at all O/W ratios.

Typical applications for SE with such a high HLB value would be in dairy products, such as coffee whiteners and whipped topping creams.

4.2 W/O emulsifying performance of SE

SE with high poly-ester content such as DK ESER F-10 did not show any phase inversion at more than 50/50 of the W/O ratio, and stable W/O type emulsified products were obtained.

Currently, EC and the FDA don't permit use of low-HLB value SE for typical W/O type emulsified food such as margarines; but in Japan, low-HLB-value SE such as DK ESTER F-10 and F-20 are used widely for fat and oil foods.

4.3 Inhibition of starch retrogradation

Surface active agents possessing fatty acid residues such as SE are known to form a complex with starch, which is similar to the formation of an iodine and starch complex[7], and are effective for inhibiting the deterioration of bread and other foodstuffs which have starch as a main constituent. The technology incorporated into bread application is described in detail in Subsection 5.2.

4.4 Inhibition of crystal growth in fat

Oil-soluble SE is known to contribute to inhibition in crystal growth of fats[8], and are used to improve and preserve the stability of fat-based foods. Further descriptions are given in detail in Subsection 5.3 .

4.5 Preventing stickiness (Lubricants)

Low-HLB value SE also gives excellent performance as a lubricant in the manufacture of tablet candies. Raw materials, primarily powdered sugar, are put in a mold of approximately 1 cm in diameter at a constant rate and compressed. Tablets thus prepared are then taken out of the mold. Therefore, it is necessary that the raw materials be free-flowing and have excellent release properties.

· Low-HLB value SE are widely used as a lubricant for such purposes. Even a very small quantity of SE produces tablets with good luster.
· The SE lessens the influence of tableting speed and long-term tabletings.
· The SE produces tablets almost free from deterioration with time.
· The SE produces tablets with good disintegration properties.

4.6 Preventing flat-sour spoilage

In Japan, a large quantity of canned coffee is sold from vending machines. At first, coffee sold from vending machines was primarily cold coffee, and canned coffee sterilized in an autoclave posed no quality problems. Recently, however, canned hot coffee drinks are sold from hot vending machines. This leads to the possible growth of thermophilic bacteria that are not destroyed during sterilization after a long-term storage of coffee in hot vending machines. This spoilage is usually called flat-sour spoilage.

Nakayama et al. studied the antimicrobial effects of SE on the causative bacteria of flat-sour spoilage[9], and found that SE monoesters were effective. Further, they discovered the following facts, as a result of experiments using six strains of flat-sour deteriorating bacteria separated from canned coffee and others (Table 5).

· Growth of all strains is inhibited by 100 ppm SE composed of either stearic or palmitic acid as the fatty acid residue.
· Germination and out-growth are inhibited by sucrose stearate at 10 ppm and by sucrose palmitate, depending on the applicable strain, at 10 ppm or 100 ppm.

This performance of SE can be used to prevent deterioration caused by thermophilic bacteria in foods exposed to high temperatures for a long time, such as canned coffee sold in hot vending machines.

5. NEW APPLICATION TECHNOLOGIES OF SE

As described above, SE are used as food emulsifiers in the manufacture of various foods. Further, their application is expanding gradually. New application technologies of SE are described below.

Table 5 *Effects of SE on Germination and/or Out-growth of the Strains (causative bacterias of flat sour spoilage) in the TSiF Medium[9]*

Sucrose ester	Concn.of sucrose ester	Strain No.					
		24-1	13-1	26-11	27-8	28-3	28-4
$C_{16}SE^{*2}$	0 ppm	+,+	+,+	+,+	+,+	+,+	+,+
	0.01	+,+	+,+	+,+	+,+	+,+	+,+
	0.1	+,+	+,+	+,+	+,+	+,+	+,+
	1.0	+,+	+,+	+,+	+,+	+,+	+,+
	10	+,+	-,-	-,-	+,+	+,+	+,+
	100	-,-	-,-	-,-	-,-	-,-	-,-
$C_{18}SE^{*3}$	0	+,+	+,+	+,+	+,+	+,+	+,+
	0.01	+,+	+,+	+,+	+,+	+,+	+,+
	0.1	+,+	+,+	+,+	+,+	+,+	+,+
	1.0	+,+	+,+	+,+	+,+	+,+	+,+
	10	-,-	-,-	-,-	-,-	-,-	-,-
	100	-,-	-,-	-,-	-,-	-,-	-,-

Modified thioglycolate media with L-Cystine removed and Na_2SO_3 and ferric citrate added were used.

*1 +: Growth seen, -: Growth inhibited

*2 SE composed mainly of palmitic acid as fatty acid residue

*3 SE composed mainly of stearic acid as fatty acid residue

5.1 Solubilizing and dispersing performance

Fig.3 is a diagram of adsorption of SE of individual ester types on surfaces of calcium carbonate particles[10]. Sucrose monoester is excellent, but sucrose polyester is poor in configuration because of its chemical structure. This affects dispersing and solubilizing performance of SE. Sucrose monoester works as a dispersant or a solubilizer at a concentration above a certain level, but polyester acts as a flocculant.

Dai-Ichi Kogyo Seiyaku Co., Ltd. recently developed pure sucrose monoester (Product name: DK ESTER SS) that can be used as dispersant and solubilizing agent. For example, its use facilitates preparation of transparent aqueous solutions of oil-soluble vitamins such as vitamin E. Thus, the ester is utilized in the manufacture of beverages requiring transparency, such as sports drinks. Further, pure sucrose monoester also is attracting wide attention today as a dispersant or solubilizing agent in medical formulations.

In addition, an example of a characteristic application of SE with high monoester content that takes advantage of its dispersing performance would be calcium-enriched beverages. Not a few people suffer from osteoporosis in Japan and East Asian countries, because, unlike Europe and USA, people there live on rice, a good resource of phosphorus and this results in a chronic shortage of calcium intake. Under these circumstances, enrichment of calcium has been posed as an effective way to maintain health and calcium enriched food has been developed and introduced in market. To

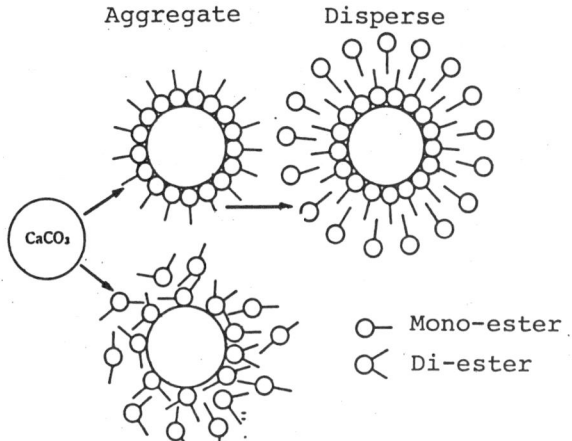

Fig.3 D*iagram showing SE adsorption on surfaces of calcium carbonate particles*[10]

manufacture beverages enriched with calcium, Houjou et al. developed a technology to prepare a slurry of calcium carbonate that would remain stable for a long period by using SE with high monoester content[11].

5.2 Application in baked goods

The quality of dough depends on the gas-generating force (yeast activity) and the gas-holding force (dough structure). When frozen, however, dough is damaged with a reduction of its baking performance, resulting in bread with poorer-quality of internal layer and texture. To cope with the problem, attempts have been made for improvement of dough structure, utilizing oxidants (ascorbic acid), emulsifiers, enzymes (amylase), etc. as well as development of yeasts specifically for frozen dough[12]. SE is known to inhibit retrogradation of starch, the main component of bread[7], and also increases bread volume[13]. Further, it is reported that SE is effective in inhibiting retrogradation of frozen dough[14].

Our company has developed an emulsifier for baked goods that is being sold under the brand name "DK Foamer FD-30 (FD-30)" (see Table 6), in which SE are used in combination with glycerin fatty acid esters. This product is a type of micro-encapsulated compound (emulsifier form micelles in water and are then coated with saccharides) that disperses and dissolves easily in tap water. Conventionally, emulsifiers with high HLB values have had to be heated at 60-70°C to dissolve them in water. For the baking process, in which the dough is manufactured at low temperature, such emulsifiers are difficult to use, although their effectiveness is known. FD-30 is designed to overcome this problem. Photo 2 shows the bread fermented and baked, the dough of which is frozen for 4 weeks after addition of DK Foamer FD-30. DK Foamer FD-30 gives a large loaf volume and maintains a good loaf volume even after frozen storing. An excellent soft mouth feel is obtained as a result.

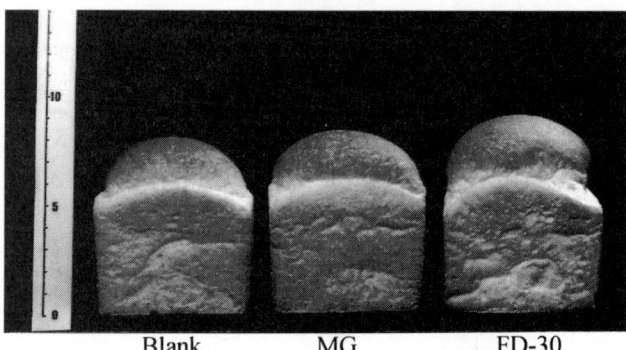

Blank   MG   FD-30

Photo 1 Bread fermented and baked after freezing dough for 4 weeks
      FD-30   : Dough with 1% FD-30 addition for flour
      MG   : Dough with 1% Glycerine mono-stearate addition for flour
      Blank   : Dough with no addition

Table 6 *DK Foamer FD-30*

Component and weight ratio	
Sucrose fatty acid esters	25%
Glycerin fatty acid esters	25%
Natural ingredients (Saccharides)	50%

5.3 Application to processed fat and oil foods

Effects of SE on crystal growth in fats were studied by Niiya et al[8,15]. It is now known that SE of high polyester content and acetylated SE (DK ESTER F-A10E) inhibit crystal growth in fats and oils (Photo. 3.1 and 3.2).

The effects of emulsifiers, including SE, on crystal growth in cacao butter were examined by DSC(differential scanning calorimeter) analysis, and acetylated SE with high polyester content inhibited transformation of cacao butter crystals from Type II to Type III with a high melting point (Fig. 3)[16]. Thus, SE with high polyester content and acetylated SE can be utilized to prevent fat bloom on chocolate and to maintain the quality of processed fat and oil foods.

In addition, Niiya et al. examined the effect of emulsifiers on deterioration of hardened coconut oil and reported that the addition of SE was effective in preventing deterioration. Further, Oishi et al. confirmed that SE enhanced the antioxidative and synergistic activities of antioxidants[17]. In recent years, health benefits of fats and oils consisting of poly-unsaturated fatty acids such as DHA (docosahexanoic acid) have been explored, but preventing of autoxidation is an important factor in the use of such materials in foods. The effect of SE on antioxidative activity is useful in coping with autoxidation.

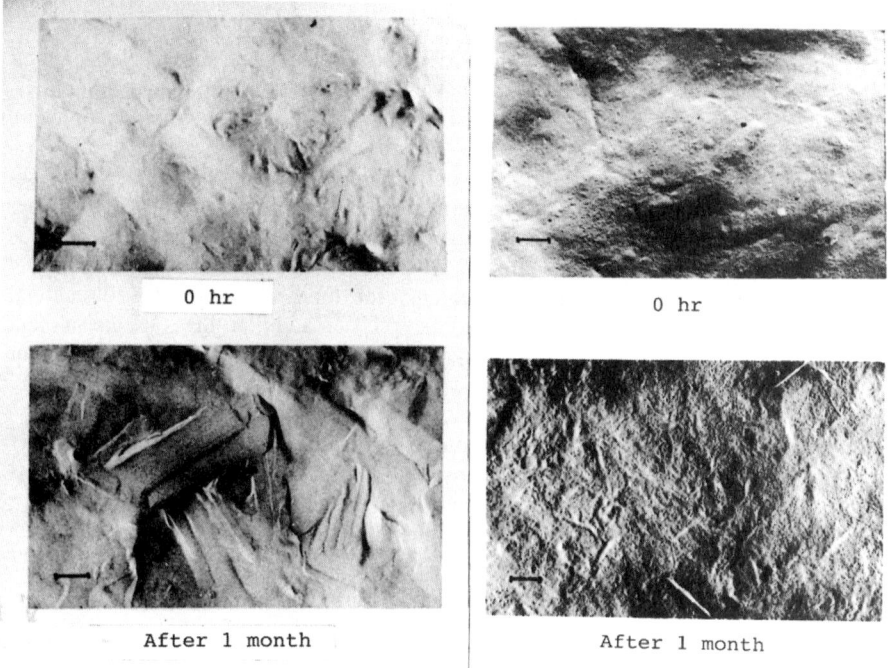

Photo. 2.1 *Surface of hardened palm kernel oil by electron microscope photograph at 15℃[8)]*

Photo. 2.2 *Hardened palm kernel oil containing SE at 15℃[8)]*

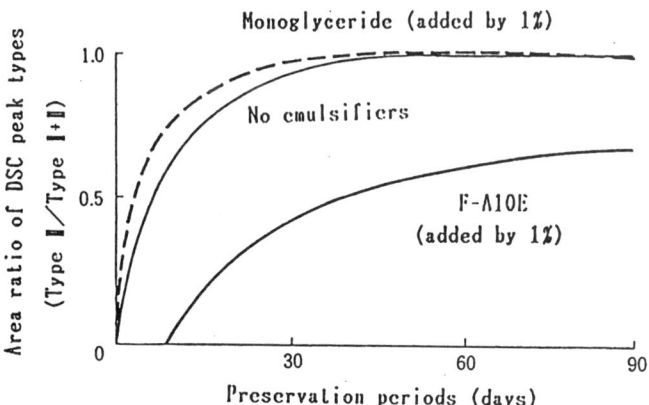

Fig.4 *Effect of emulsifiers on migration of endothermic peaks in DSC analysis of cacao butter[16)]*

(material stored at 25℃)

5.4 Application to nonfood products

As described above, SE are applied mainly to foods at present. In view of their safety and low irritation to the skin, however, applications in medical supplies, cosmetics and toiletries also should be considered. Examples of such applications are described below.

5.4.1 Detergents.   With the popularization of packaged foodstuffs and cut vegetables, the importance of their sanitation is increasing, leading to higher expectations of detergents of higher safety. SE, sucrose laurate in particular, which has the highest HLB value among the surfactants approved as food additives, is attracting attention as a main component of detergents for food washing, and commercial products already are on the market ("Sunny-Safe S"). In addition, SE is used as a main component of shampoos and rinsing agents for hair, as well as of detergents for dishwashing machines.

5.4.2 Cosmetics.   SE with low skin irritation and wide emulsification performance are used to manufacture various O/W- or W/O-type creams or lotions.

5.4.3 Oral care products.   SE is used in the manufacturing oral-care products such as toothpaste, due to the following features.

- Low irritation to the oral membrane and the skin
- Little effect on enzyme activity
- Appropriate foamability
- Little effect on the sense of taste after washing

5.4.4 Medical supplies.   As described above, development of SE with high HLB values is promoted for use in the manufacture of solubilizing agents for medical supplies. On the other hand, SE with low HLB values are used as lubricants for tablets. In addition, studies are in progress for use of such SE as the base for ointments.

5.4.5 Others.   SE is being used in anti-fogging agents for plastic films, polymerization emulsifiers, and feed additives.

Epilogue

The usage in the food constitutes the large majority of the demand for SE. SE products on the market are mostly saturated long-chain fatty acids, such as sucrose palmitate or sucrose stearate, which are tasteless. These SE have excellent properties as emulsifiers; however, their applications are mainly in the food industry and not in other industries as surfactants.

Introducing unsaturated fatty acids or short-chain fatty acids yields SE with a bitter taste that would be unsuitable for use as food emulsifiers. However, their properties as surfactants, such as wetting power, probably will be improved. If SE based on various types of fatty acids, in addition to the currently available palmitate and stearate derivatives are developed in the future, increased demand can be anticipated in the other fields, especially in cosmetics and toiletries.

SE are based on naturally derived sucrose with edible fatty acids and not from petrochemicals; being a type of esters, they are biodegradable. Thus, SE are considered to be inoffensive to the human body and also to environment, conforming to the "green" trend which is being experienced worldwide.

Because two Japanese companies are the major SE manufacturers in the world, there has not been much promotion of application studies in countries other than Japan.

On the other hand, eating habits vary among nationalities, and a food has to have a recipe designed to provide the taste and texture its potential consumers expect. The application technologies of SE as food emulsifiers developed in Japan may not be accepted in other countries; however, demand will expand if the technology is adapted to fit the recipes of other countries.

Dai-Ichi Kogyo Seiyaku Co., Ltd. established Sisterna CV under a joint-venture with Cooperatie Cosun U.A., a Dutch sugar manufacturer. Sisterna is aiming at SE production in the Netherlands in the near future after exploring good market demands in Europe through application developments to fit European requirements. The brand name being manufactured and sold by Sisterna CV is "SISTERNA". Markets for SE can be expected to expand widely into many countries and fields in the future.

References

1) The Joint FAO/WHO Expert Committee on Food Additives: F.A.O. Nutrition Meeting Report No. 599, p.100 to 114 (1976), No. 653, p.140 to 145 (1980) & No. 859, P.12 to 14 (1995), in Rome.

2) J. M. Daniel, C. J. Marshall, H. F. Jones & D. J. Snodin, *Food Cosmet. Tox. (G.B.R.)*, 1979, **17**, 19 to 21.

3) T. Ishizuka, S. Nakamura, *J. Japanese Society of Nutrition and Food Science*, 1974, **27**, 65 to 70, 71 to 75 & 289 to 203.

4) T. Ishizuka, T. Watanabe, I. Sasaki, S. Nakamura, *J. Japanese Society of Nutrition and Food Science*, 1974, **27**, 449 to 453 .

5) J. F. Berry & D. A. Turner, *J. Am. Oil Chem. Soc.*, 1960, **37**, 302 to 305.

6) T. Ishizuka, S. Matsushita, T. Takahashi, S. Nakamura, *J. Japanese Society of Nutrition and Food Science*, 1974, **27**, 455 to 459 .

7) T. Ishizuka, S. Nakamura, *J. Japanese Society of Nutrition and Food Science*, 1974, **27**, 221 to 224 (1974).

8) R. Niiya, H. Kanematsu, M. Imai, M. Okada, T. Matsumoto, *J. Japan Oil Chemists' Soc.*, 1972, **21**, 888 to 892.

9) A. Nakayama, J. Sonobe & R. Shinya, *J. Food Hyg. Soc., Japan*, 1982, **23**, 25 to 32.

10) T. Baba, K. Namiki, S. Maeda, *J. Chem. Soc. Japan, Ind. Chem. Sec.*, 1964, **76**, 2081 to 2085.

11) Y. Houjou, k. Hashimoto, M. Takahashi, T. Motoyoshi, *Japanese Patent Application No. H1-125741* (1994).

12) Y. Inoue, *Japan Fats and Oils Monthly*, Published by Miyuki Syobo, Tokyo, 1993, **46**, 45 to 61.

13) L.M. Breyer & C.E. Walker, *Journal of Food Science*, 1983, **48**, 955 to 987.

14) K. Hosomi, M. Uozumi, K. Nishio & H. Matsumoto, *Journal of Japanese Society for Food Science and Technology*, 1992, **39,** 806 to 812.

15) H. Kanematsu, T. Maruyama, R. Niiya, T. Matsumoto, *J. Japan Oil Chemists' Soc.*, 1982, **31**, 273 to 276.

16) S. Nakamura, "Food Emulsifiers and Emulsifying Technology", published by Kogyo Gijutsu-Kai, Tokyo, 1995, p.124 to 153.

17) M. Oishi, K. Onishi, M. Nishijima, K. Nakagome, H. Nakazawa, *Journal of the Pharmaceutical Soc. of Japan, Hyg. Chem. Sec.*, 1990, **16**, 69 to 73.

Use of Mixtures of Alkyl Alkoxylates and Alkyl Glucosides in Strong Electrolytes and Highly Alkaline Systems

Ingegärd Johansson*, Christine Strandberg, Bo Karlsson, Gunvor Karlsson and Karin Hammarstrand

AKZO NOBEL SURFACE CHEMISTRY AB, 44485 STENUNGSUND, SWEDEN

INTRODUCTION

Nonionic surfactants used to be more or less synonymous with alkyl alkoxylates. Today a whole range of polyhydroxyl surfactants of nonionic type has been developed from natural raw materials due to environmental considerations and is currently being used in augmenting amounts. Application areas are as wide as for the conventional nonionics including both the use in general cleaning and specific uses as wetting and penetrating agents in agro formulations. The most widely investigated polyhydroxyl based surfactants are the alkyl polyglucosides. The name covers a mixture of oligomers. In this paper they will be referred to as glucosides and will be the focus of the investigation.

The hydroxyl functions that provide hydrophilicity to the glucosides, substituting the ether linkages of the alkoxylates, give potentially different properties to these products. The differences are currently being looked at in many groups both within the universities and the industries.[1-8]

This paper deals with combinations of alkyl alkoxylates and alkyl glucosides in water as well as in highly concentrated electrolyte solutions and highly alkaline systems, covering properties like solubility, wetting, cleaning ability and foaming. These systems can be used in for instance detergents, institutional and industrial cleaning and textile treatments including mercerization and scouring.

The most obvious differences between alkoxylates and glucosides are found in the temperature dependence of the phase behaviour, alkoxylates having cloud points and glucosides sometimes said scarcely to have any. Combinations will influence the temperature dependent solubility in varying ways.

Also the electrolyte effect differs, alkoxylates usually not being very soluble in alkali and glucosides sometimes being more soluble in alkali than in pure water.

1. Synergistic effects between alkoxylates and glucosides

Synergistic effects for alkyl glucosides in mixtures with many other surfactant types have been claimed as an argument for their commercial interest. They are said to have effect on viscosity, cleaning, foaming and mildness of certain combinations with above all anionic products.[9-11] However, it should be interesting to investigate the possible synergies in mixtures between glucosides and ethoxylates as well since this should give non charged product combinations with low sensibility to for instance cationic additives.

1.1. Pure compounds. A few publications in this field exist discussing the behaviour of dodecyl ß-maltoside,[12,13] dodecyl ß-glucoside, decyl ß-glucoside and decyl ß-maltoside[14] in terms of influences of the ratio of mixtures with the ethoxylates C12(EO)8 or C12(EO)7 on CMC and surface tension. However, in neither of these cases any appreciable deviation from ideal mixing was seen. A surprisingly strong interaction was found between the glucoside and the maltoside thus hinting at an explanation of the differences between pure monoglucosides and commercial mixtures which contain an immense number of alkyl oligoglucosides due to the Fischer process with which they are produced.[15]

1.2. Commercial products. One study[16] has been made comparing synergies between one alkyl polyglucoside, C12-14 with an average of 1.3 glucoside units (=APG), and lauryl benzene sulphonate (=LAS), C12-14 ether (2EO) sulphate (=AES), or C12-14 (EO)10 (=AE). The investigation was made using a simulation of the whole surface tension versus concentration isotherm and not only the CMC. This technique shows which ratio that gives the highest synergy in each case. Surprisingly LAS showed very little synergy with APG, while both AES and AE gave a high synergistic interaction parameter. The results were then confirmed by cleaning tests which show typical synergistic enhancement of the efficiency in specific mixtures e.g. for a molar ratio around 0.8 for APG + AE.

Different results for the pure compounds and the commercial mixtures may be due to the different qualities of the products but this is as yet only speculations. In both ref. 15 and 16 true synergies, i.e. the mixture leading to results that are more pronounced than what either of the ingredients show, are evaluated.

2. Use of alkyl glucosides as hydrotropes

Already 1979 alkyl glucosides were recommended for use as hydrotropes e.g. to solubilize defoamers or low foaming alkoxylates of blockpolymer type into highly alkaline solutions intended for use in metal cleaning.[17]

2.1. The hydrotrope concept. Discussions have been going on in the literature[18,19] in articles by Friberg and co-workers concerning what influence a hydrotrope has in a water solution, being a solubilizer and/or destabilizing the liquid crystalline phases in the solution. Friberg has shown that the action of a hydrotrope as a disturbance in the ordered phases may well be the way it enlarges the solubilization of hydrophobic elements like oils or non-water-soluble surfactants. However, there may be other mechanisms in specific cases.

Usually the hydrotropes are rather water soluble molecules with bulky hydrophobes but also more surface active types like ethoxylated quaternary alkyl ammonium compounds are being used with a function more similar to a co-surfactant. The ammonium compounds mentioned are very efficient but also not totally benign to the environment. Thus there is a need for greener products such as the alkyl glucosides.

2.2. Alkyl glucosides as hydrotropes. We have chosen to study one specific model formulation with 5% of a hydrophobic alkyl ethoxylate of narrow range type, C10(EO)4, in different amounts of NaOH and glucoside as solubilizer with the aim of producing highly concentrated alkaline cleaners. The minimum amounts of hydrotrope to get a clear solution at room temperature are shown in figure 1.

The products chosen for comparison are glucosides with different chain lengths ranging from C4 to C8 with straight as well as branched hydrophobe structures and most of them with the same degree of glucosidation, DP=1.4-1.6. The C4-glucoside is Simulsol SL4 (DP=1) from SEPPIC, the C8 straight and C8 branched glucosides are AG6201 and AG6202 respectively from Akzo Nobel Surface Chemistry AB. The C5, C6 and C7 are lab products made with the Fischer process technique[15] at Akzo Nobel Surface Chemistry AB from glucose and isoamyl alcohol, hexanol (Sigma) and Exxal-7 alcohol (Exxon). The ethoxylate is a commercial product from Akzo Nobel Surface Chemistry AB. Two conventional hydrotropes, sodium cumene sulfonate (Hüls) and octyl imino dipropionate (Ampholak YJH-40 from Akzo Nobel Surface Chemistry AB), are put in as references. NaOH (Merck) as well as the surfactant mixtures are dissolved in water of medium hardness, 3.8°dH.

The results show an unexpectedly high efficiency for the hexyl glucoside being the only one functioning in 30% and 40% NaOH at reasonable amounts. Also within the glucoside family the hexyl glucoside is found to be more efficient than glucosides with shorter as well as longer hydrocarbon chain lengths.

A similar investigation has been made by M. Svensson and co-workers.[20] They have investigated the behaviour of a series of glucosides as hydrotropes in combination with some nonionics in a typical builder mixture, 5% tetrapotassium pyrophosphate and 5% sodium metasilicate as well as the phase behaviour for a mixture with SDS and pentanol. Their findings are interpreted as medium chained (C8 and C10) glucosides being most effective as solubilizers of nonionics (raising the cloud point) while the short chain one (C4) destabilises the crystalline phases most efficiently. Thus the judgement of hydrotrope efficiency is very much dependent on the test method used. We have not looked at the destabilisation mechanism but regard the hexyl glucoside efficiency as the result of its intermediate size and as the most suitable one in this special formulation.

However, a more detailed analysis of the solubilization in different media with different salts was needed and can be found below under Cloud points and Electrolyte effects.

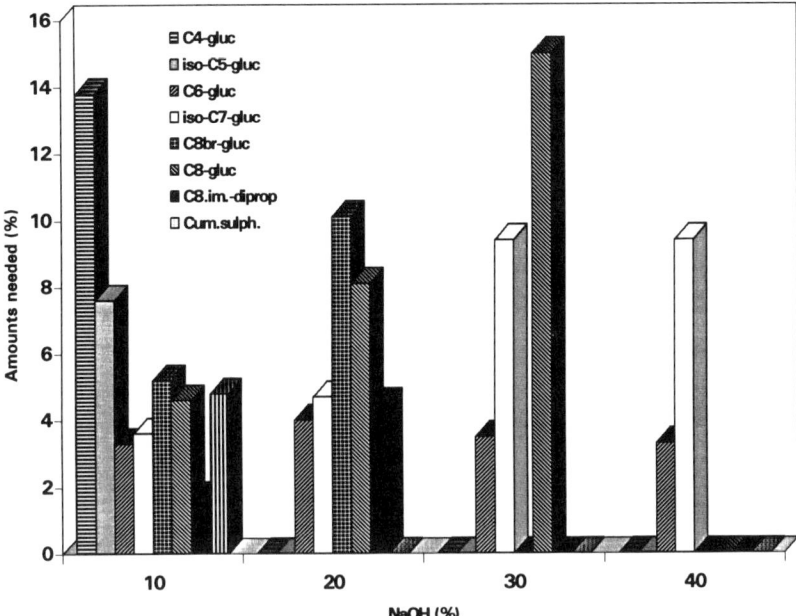

Figure 1 Hydrotrope efficiency. The formulations contain 5% C10(EO)4, different NaOH concentrations and the minimum amount hydrotrope needed to get a transparent solution at room temperature. No columns indicate that it is not possible to formulate a clear solution.

3. Microemulsion systems

Microemulsions consist of oil, water and surfactants in balanced proportions to create thermodynamically stable systems which are translucent to transparent in their appearance. They are used for cleaning purposes when a high degreasing effect is wanted and are less environmentally risky than the solvent based systems. In many areas with a need to introduce oil soluble substances in water based formulations microemulsions are useful. Examples of such situations can be found within cosmetics, pharmaceuticals and agro formulations. Today well balanced mixtures of fatty alcohol ethoxylates are used as surfactants. However, their skin compatibility may not be sufficient which could be improved by adding a glucoside, known to reduce skin irritation.

Another problem with microemulsions based on alkyl ethoxylates is their temperature sensitivity. Since the glucosides do not show this dominating temperature dependence it has been a tempting idea to use them alone or tuned with different additives to get microemulsion systems with a broad temperature stability. This idea has been the starting point for a whole range of investigations published during the last four years.

3.1. Using co-surfactants. To adjust the spontaneous curvature of the interphase to a planar type resulting in low interfacial tensions and potential microemulsions short hydrophobic nonions such as ethylhexyl glycerol ether,[21] short alkanols or the less toxic alkyldiols,[22] or larger hydrophobic molecules like sorbitan esters,[23] decanol and geraniol[24] have been used. Especially Kahlweit and co-workers[22] give practical hints on how to tune the systems also for the usually more difficult natural oils, triglycerides, e.g. rape seed oil methyl ester as well as isopropyl myristate.

The common feature for all these investigations seem to be the insolubility of the glucosides in hydrocarbon which means that it is very difficult to achieve the whole sequence of changing a system from oil in water to microemulsion to water in oil. The ethoxylates change their solubility with temperature being more and more hydrophobic at higher temperatures. With glucosides this tuning then must be done by adding co-surfactants instead. This has been most thoroughly discussed by Kaler and co-workers[25] in their investigation of mixtures of alkyl ethoxylates and alkyl glucosides. Common for all the practical solutions seem to be changing the nature of the oil phase by adding a co-surfactant which partly acts as a co-solvent and makes the unpolar oil more polar resulting in a better solubility of the alkyl glucoside therein.

3.2. Changing glucoside structure. Good results have also been achieved by Wades group[26] by changing the structure of the glucosides using branched hydrophobes like Guerbet alcohols and then combining with suitable nonionics, usually with long hydrophobes like C16 and C14. They work with the ratio between ethylene oxide units (EO) and the hydroxyl units (OH) both in the mixtures but also within the molecules using ethoxylated alcohols as hydrophobes for the glucosides. In one example they end up with an EO/OH ratio of 0.51 mol/mol to get a temperature independent microemulsion with decane as oil phase. This technique is developed for use in soil remediation.

The same area has initiated the investigation by Clemens.[27] The resulting system in his case consists of C10-12 glucoside with a DP=1.3 mixed with a very hydrophobic nonionic C16(EO)2 and rape seed oil as oil phase.

4. Cloud points

4.1. Temperature dependent phase behaviour for glucosides and ethoxylates. Due to the very steep temperature dependence of the border lines in the phase diagrams[1-8] the glucosides often have been referred to as not having cloud points. However, as has been shown by Rohm and Haas[17] and Balzer[28] there is a clouding phenomena that appears at very specifically balanced glucosides. A change in aglycon chain length of one carbon can change the cloud point from being above 100°C to being below 0°C.[8] A typical phase diagram for the lower concentration range is shown in figure 2 (from ref. 28) for a glucoside, C12-14(G)1.8, together with that of an alkoxylate, C12-14(EO)7. Here the typical very strong concentration dependence of the cloud point for the glucoside is in contrast to the shallow curve of the ethoxylate.

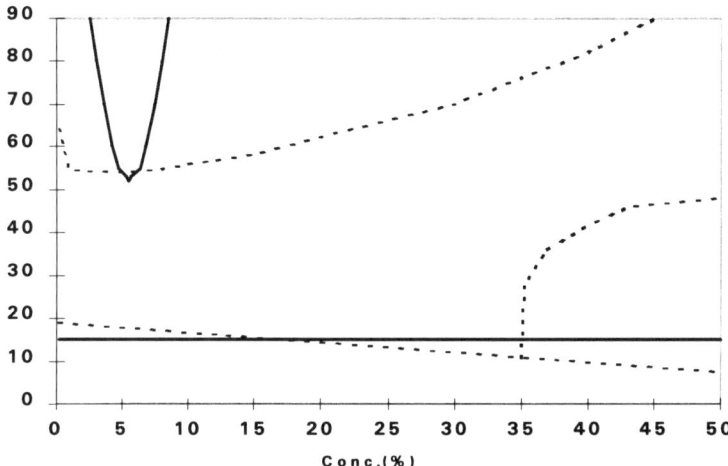

Figure 2 *Phase diagrams (low concentrations) of C12-14 APG 1.8 (full line) and C12-4(EO)7 (dotted line) from ref. 28.*

In the industrial use nonionics often are characterized by their "cloud point". This value is measured according to a procedure given in the norm ISO1065 in which five ways of measuring are described using different media for nonionics which are extremely water soluble, water soluble, sparingly water soluble and not at all water soluble. In each case only one concentration is recommended (even if it is important to state which concentration that has been used), a procedure which relies on the fact that the clouding of a nonionic is rather insensitive to the concentration as can be seen in figure 2. Since the concentration dependence of the clouding of the glucosides is extremely concentration dependent, the minimum of the cloud point curve can be given as a characterisation of the glucoside.

4.2. What will happen when two products with as different characteristics as these are mixed? Since they often are used together in cleaning of hard surfaces it is important to know what to expect. We have measured the cloud point at different total concentrations of the surfactant mixture (glucoside+ethoxylate) keeping the ratio constant. In all investigations deionized water has been used. In figures 3, 4 and 5 the results for three different short chain glucosides mixed with a C10(EO)4 narrow range ethoxylate are given. The glucosides all are fully water soluble and have a cloud point above 100°C. The alkoxylate is not water soluble at low concentrations and has a cloud point below 0°C. Three different weight ratios glucoside/alkoxylate have been tested for each of the three glucosides. Mixing the two types gives a clouding behaviour that for high amounts of glucoside is similar to what is shown as typical for a balanced glucoside mixture (figure 2, C12-14G1.8) and for a higher nonionic content becomes more like the ethoxylate clouding. What is seen in all three cases is the dip at low concentrations. If then the formulation being balanced in the practical application only is tested and characterized at one concentration, for instance at 1%, the answer can be very misleading.

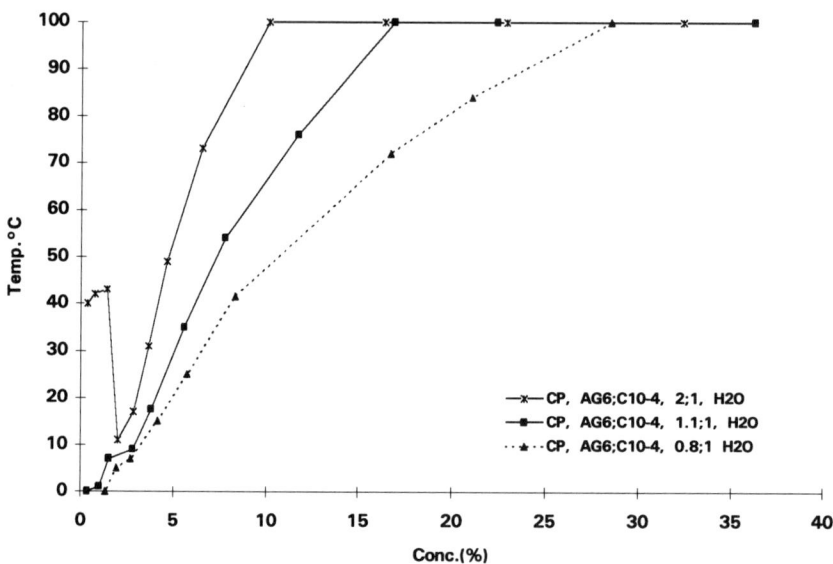

Figure 3 *Cloud point C6-glucoside and C10(EO)4 in water*

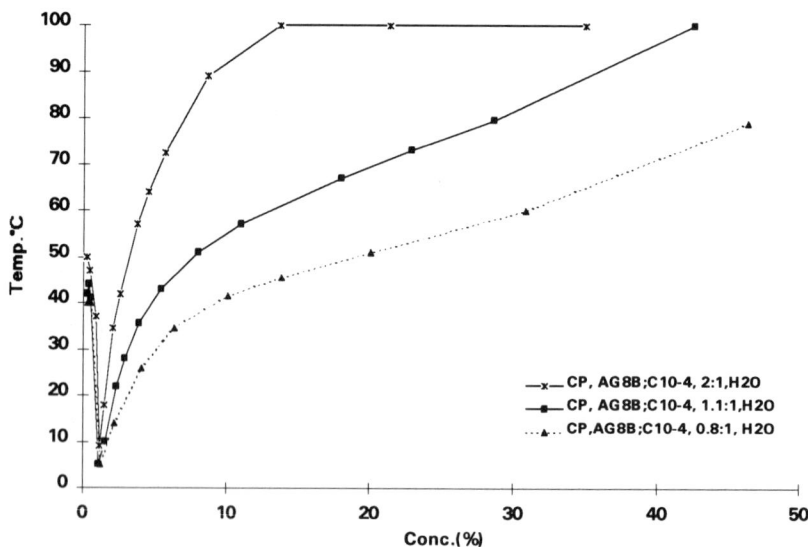

Figure 4 *Cloud point C8-branched glucoside and C10(EO)4 in water.*

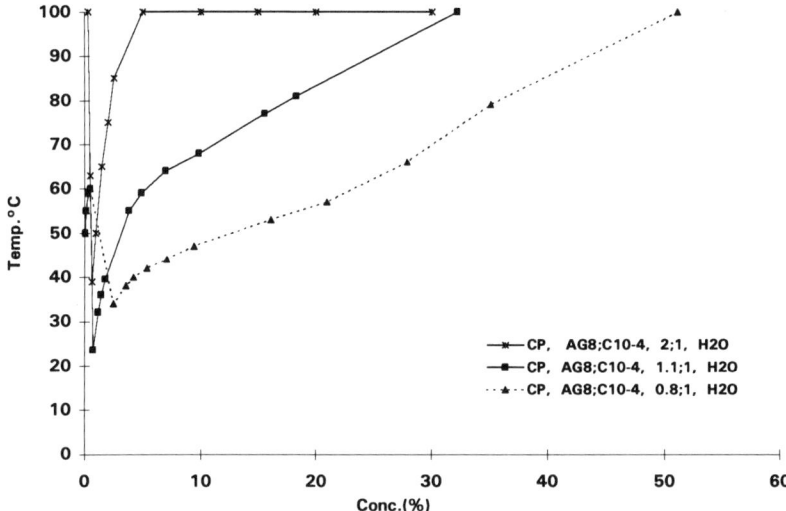

Figure 5 *Cloud point C8-glucoside and C10(EO)4 in water.*

5. Electrolyte effects

5.1. Glucosides or alkoxylates in highly concentrated electrolytes. In practical applications the formulations very often contain electrolytes, as thickening aid (e.g. NaCl in personal care or house hold cleaning), as complexing agent (detergents and industrial cleaning agents), to regulate pH (highly alkaline cleaning formulations, textile treatment additives) or to get specific effects (agro chemical additives).

An extensive investigation of electrolyte effects on cloud points in glucoside solutions as compared to alkyl ethoxylates has been made by Balzer.[28] He has pointed out the higher sensitivity to electrolytes in the case of the glucosides. The lowering of their cloud points is more pronounced than those of alkyl ethoxylates. It is explained by the somewhat anionic character that is ascribed to the commercial alkyl glucosides. This charge helps to keep the micelles apart. When small amounts of salt are added this effect is screened and the cloud point is lowered. However, as is remarked by Kahlweit[22] and Somasundaran[29] the alkyl glucosides over all are rather insensitive to electrolytes. This can also be seen in the salt dependence curve shown by Balzer since only the very first small amounts of salt lowers the cloud point. To decrease it further high amounts of salt have to be added. This is for instance the case when the cloud point for a hydrophilic alkyl glucoside like octyl monoglucoside is decreased below 100°C.[22] The same thing happens with mixed micelles consisting of an alkyl ethoxylate and minute additions of an anionic surfactant. The cloud points of these mixtures are drastically lowered in the presence of electrolytes at concentrations that are considerably smaller than those affecting the cloud point of the nonionic surfactant on its own.[30]

An exception is the sensitivity to alkali. An addition of e.g. NaOH results in an increase of the cloud point temperature for the glucoside and a decrease for the ethoxylate. The specific effect on the glucosides may be due to a deprotonation of weakly acidic OH groups on the glucose unit (pKa=12.35 for glucose and 12.51 for sucrose).[31]

A parallel can be seen in the use of sodium glucoheptanoate as a complexing agent in highly alkaline solutions. It is said to function only at extremely high pH, which should be consistent with deprotonation resulting in negative charges also from the weakly acidic hydroxyl groups and a larger tendency to coordinate to positive ions.[31]

5.2. What is then the effect on mixtures of glucosides and alkoxylates of electrolytes? The tendency towards highly concentrated formulations for environmental reasons is strong and well known for instance in the detergent industry. We have for that reason tested the behaviour of mixtures between C6-, C8-, or C8-branched glucosides and a narrow range C10(EO)4 alkoxylate in extremely high concentrations of different builders as well as alkali over a concentration range from 30-40% total surfactant down to around 0.5%. Three different weight ratios of glucoside/ethoxylate have been studied for each surfactant combination, 2:1, 1.1:1 and 0.8:1.

5.2.1. NTA. In figures 6 and 7 the decreasing of the cloud point curve by NTA (nitrilotriacetic acid, a complexing agent) in rather high amounts is given. Figure 6 shows the behaviour of the two C8-glucosides which both seem to be salted out together with the ethoxylate. The C6-glucoside (figure 7), however, can stand not only 20% NTA but also 30 and 35%. In this case the choice par preference should be the short C6-glucoside.

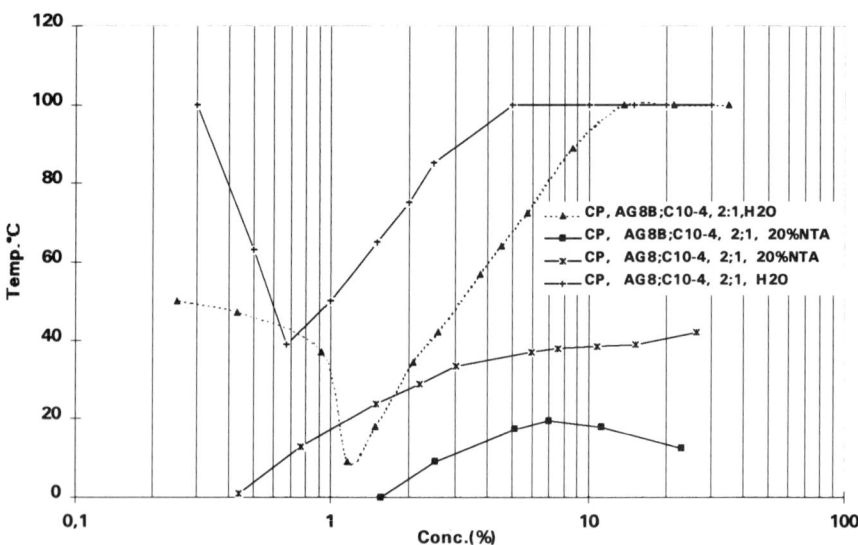

Figure 6 *Cloud point C8-glucoside, branched and straight, C10(EO)4 in water or 20% NTA.*

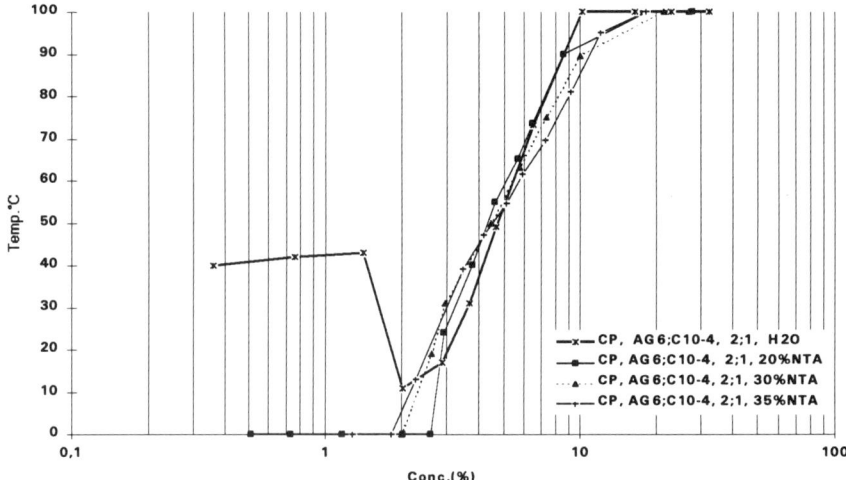

***Figure* 7** *Cloud point C6-glucoside and C10(EO)4 in water or different NTA concentrations.*

5.2.2. NaOH. A very different picture is seen when glucoside/alkoxylate mixtures are dissolved in highly alkaline solutions.

- Figure 8 shows the results for C8 and C6 glucosides in 20% NaOH. The behaviour differs in various ways. C6 gives rather similar results for all three ratios but C8 is more sensitive to the relative amounts giving cloud points above 100°C for 2:1 around 65°C for 1.1:1 and below 0 for 0.8:1. For C8-branched only the ratio 2:1 was rich enough in glucosides to give a clear solution.

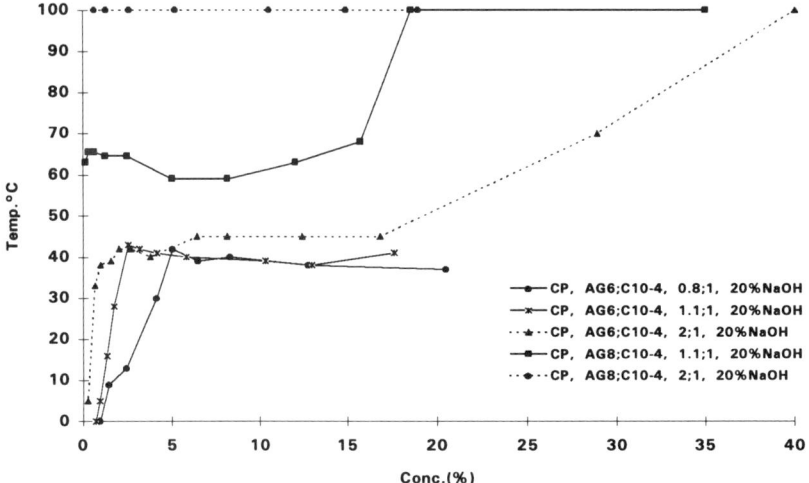

***Figure* 8** *Cloud point C8 and C6-glucosides and C10(EO)4 with 20% NaOH (CP C8 glucoside 0.8:1, <0°C).*

- In figures 9, 10 and 11 the cloud point curves for hexyl glucosides, octyl glucosides and branched octyl glucosides respectively, in mixtures with a decyl alkoxylate (C10EO4) dissolved in 5, 10, 15 and 20% NaOH are shown. For comparison the behaviour in water in each case is given. At lower concentrations the mixtures are more soluble in alkali than in water and often also more dissolvable in higher concentrations of alkali than in lower. Typically the dip in the cloud point for the surfactant in water is cut-off and the solution is stable also at temperatures up to around 30-40°C. At higher surfactant concentrations the cloud point often is depressed by higher alkali quantities. When scrutinized in detail the differences in behaviour are obvious, each glucoside showing optimal properties in certain regions.

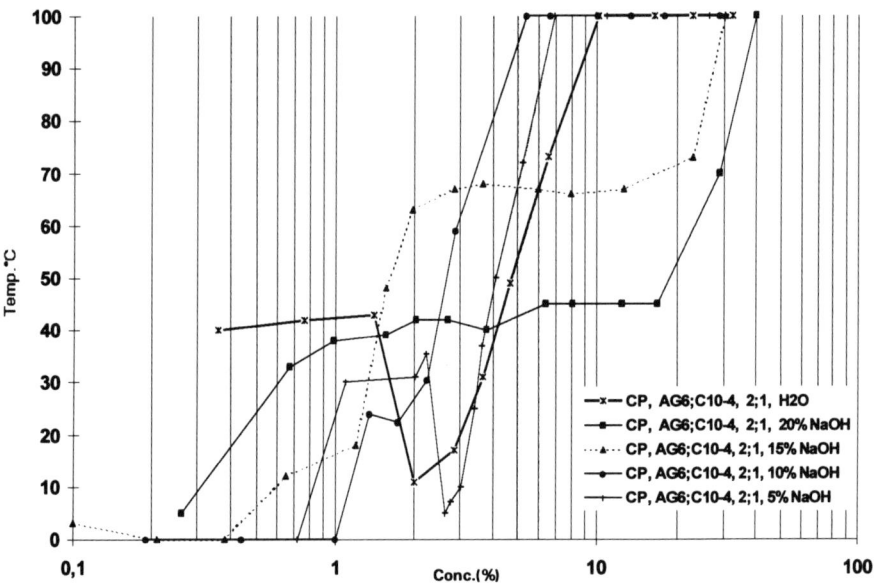

Figure 9 *Cloud point C6-glucoside and C10(EO)4 in water or different NaOH concentrations.*

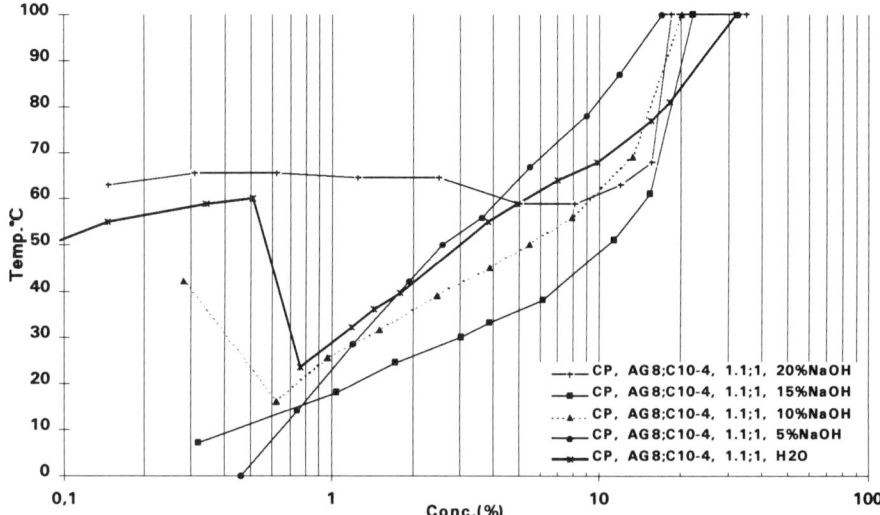

Figure 10 *Cloud point C8-glucoside and C10(EO)4 in water or different NaOH concentrations.*

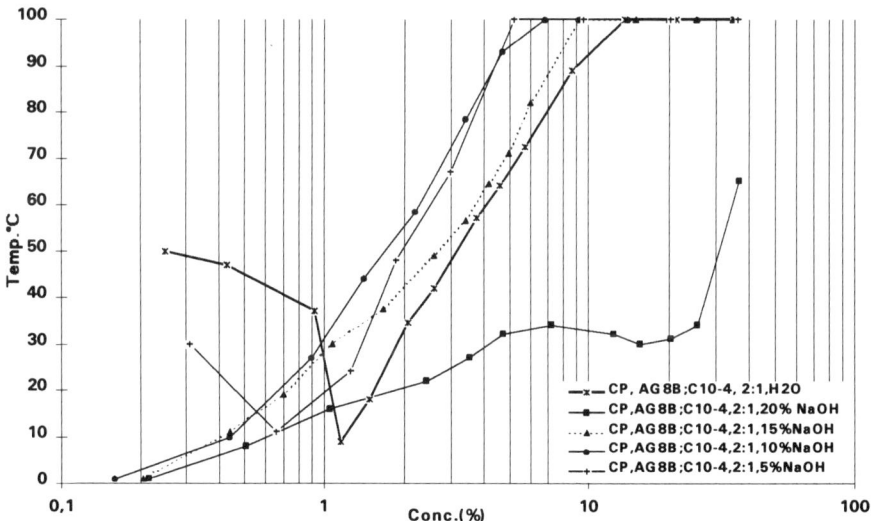

Figure 11 *Cloud point, C8-branched-glucoside and C10(EO)4 in water or different NaOH concentrations.*

- Of special interest is the behaviour of the hexyl glucoside, figure 12, where in each case the mixture at low concentrations is more soluble in 20% NaOH than in water.

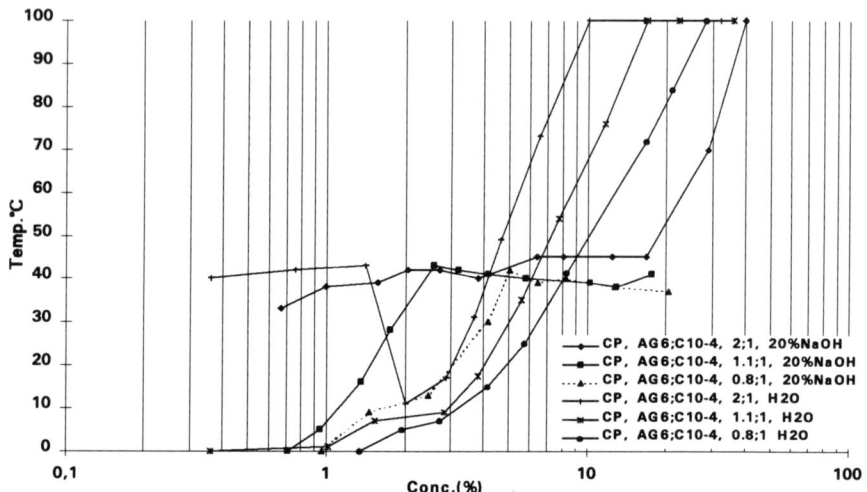

Figure 12 *Cloud point C6-glucoside and C10(EO)4 in water or 20% NaOH.*

5.2.3. TPPP and sodium metasilicate. A formulation with a commonly used concentration of conventional builders giving an elevated pH around 13 as compared to 11 for NTA and 14 for the NaOH solutions has been tested as medium for the same mixtures as above. The results are shown in figures 13-15. The same common feature with hexyl glucoside being most efficient overall and the octyl glucosides having a somewhat higher cloud point at low concentrations emanates. If the aim is to find a formulation which is transparent in all dilutions a mixture of the two types could be a possibility.

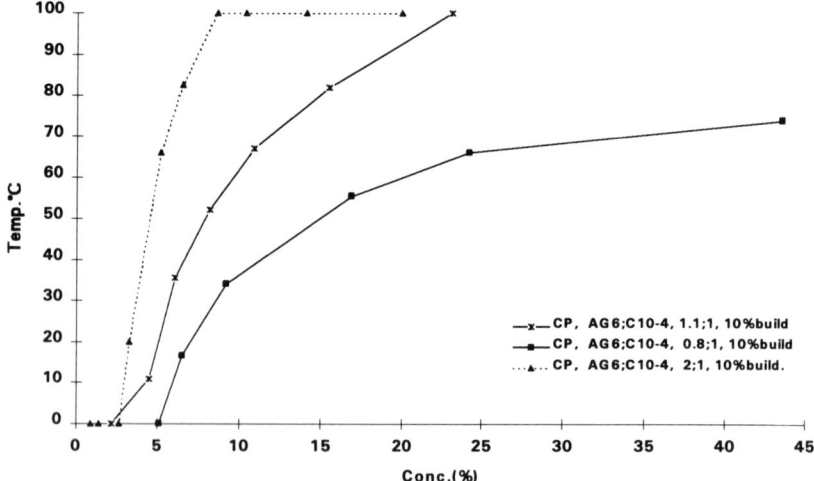

Figure 13 *Cloud point C6-glucoside and C10(EO)4 with 6% TPPP and 4% Metasilicate.*

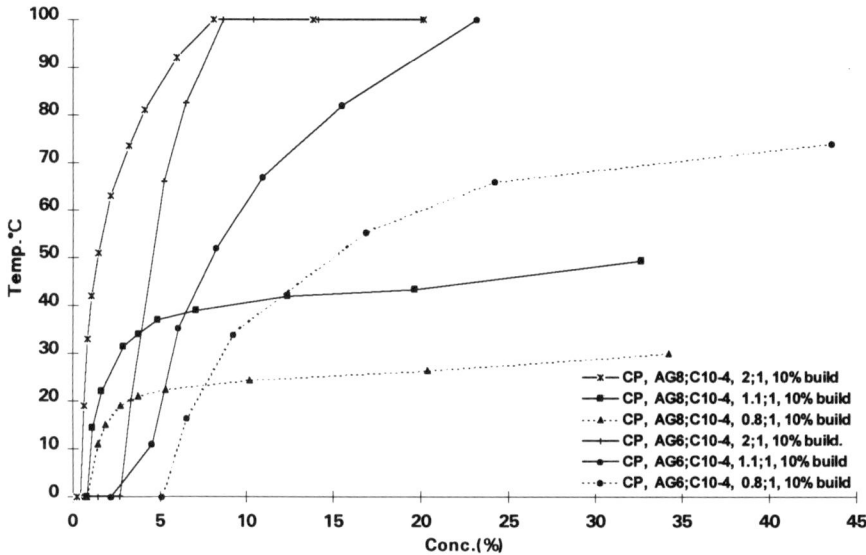

Figure 14 *Cloud point C8-glucoside or C6 glucoside and C10(EO) with 6% TPPP and 4% Metasilicate.*

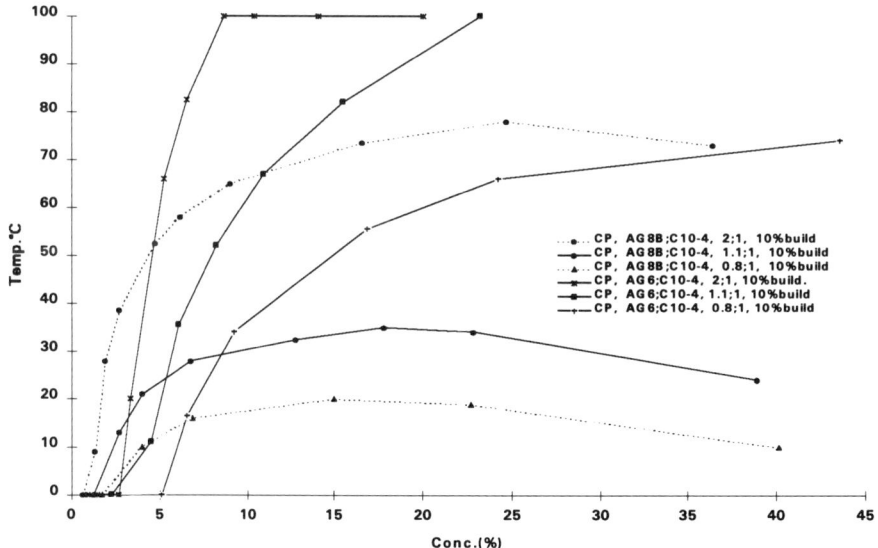

Figure 15 *Cloud point C8-branched-glucoside and C10(EO)4 with 6% TPPP and 4% Metasilicate.*

However, as will be discussed in the next paragraph, using combinations that result in working solutions being very close to a cloud point border can be beneficial from an application point of view.

Comparing the results in all three electrolytes gives the impression that two effects are operating, one due to the high pH and one to a more common salt effect. The octyl glucosides seem more sensitive to the "salting-out" phenomenon than the hexyl glucoside. This could be understood in terms of the longer hydrocarbon chains. The effect at very high pH points toward a change in character of the glucosides being more and more anionic. All the OH-groups of the glucose moiety are said to be weakly acidic.[31]

6. Application results

The balance in the solution is important for its function. The complex process of cleaning, be it of fabrics[32] or of hard surfaces,[33] has been studied and analyzed in terms of the packing of the surface active agents in the cleaning solution or when adsorbing to the oily soil changing its contact angle. Different mechanisms are related to different aspects of surface activity, but one dominating feature is the vicinity to zero spontaneous curvature (packing parameter close to one) which also gives minimum in interfacial tensions to the oil phase involved which is beneficial for spontaneous emulsification and for creation of a microemulsion state. Traditionally proximity to a cloud point border line also indicates growing aggregates in the solution with less surface curvature.

6.1. Cleaning. We have performed some tests to relate cleaning performance to our findings concerning the rough phase diagrams shown in the preceding paragraph.

The cleaning efficiency was evaluated according to the following method. White painted plates are smeared with a model dirt consisting of an oil-soot mixture obtained from diesel engines. The cleaning solutions are then poured onto the top of the dirty plates and left there for one minute. The plates are rinsed off with a rich flow of water. All solutions and the water are kept at a temperature of about 15-20°C. The solutions to be compared are tested at the same plate. The reflectance of the plate is measured with a Minolta Chroma Meter CR-200 before and after cleaning to get the % soil washed away.

A cleaning test was performed with a series of formulations containing hexyl glucoside, C10(EO)4 and TPPP + metasilicate where the ratio between the surfactants was varied slightly but the total content was kept constant at 10%. This concentrate also contained 6% TPPP and 4% metasilicate. The test was made at a 1:20 dilution, i.e. 0.5% surfactant in the working solution. The result can be seen in figure 16 which also shows the foam of the same solutions. Obviously the fluid has to be balanced carefully. Too much co-surfactant destroys the cleaning result and too little gives more foam in this case.

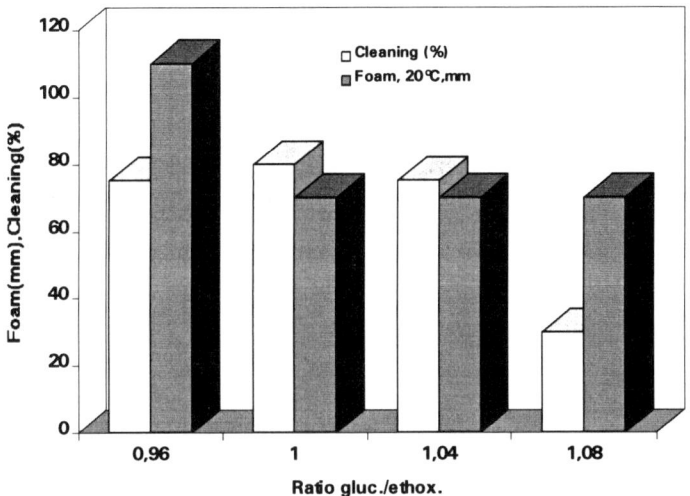

Figure 16 *Cleaning and foam, C6-glucoside and C10(EO)4 with TPPP + Metasilicate; 0.5% tot. surfactant.*

6.2. Foam. The amount of foam is also dependent on the type of solution. The picture is rather complicated, however, and I will only give some examples of what factors that may influence the result.

The foam is measured as mm foam produced in a 500 ml measuring cylinder (with 49 mm inner diameter) from 200 ml surfactant solution when the cylinder is turned around 40 times in one minute. The test can be made at different temperatures and the foam height is registered directly and after 1 min.

In the vicinity of phase borders the foam may be suppressed as for example is seen in the temperature effect on the foam in the alkaline solutions of C6-glucoside and C10(EO)4, illustrated in figure 17. The foam volume grows with the concentration both of surfactant and NaOH. As has been pointed out earlier the glucoside may be more anionic in the stronger NaOH medium and therefore may pack with the uncharged ethoxylate in a similar way as anionic surfactants do with fatty alcohols giving rise to very stable foams.

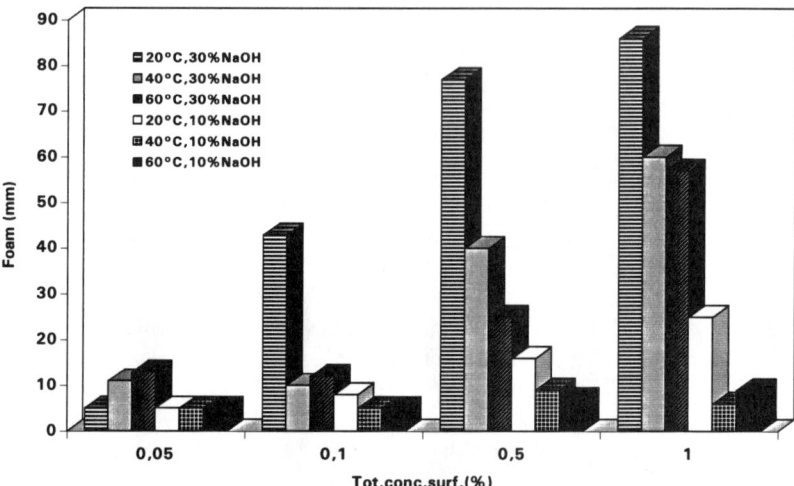

Figure 17 *Temperature dependence of foam, C6-glucoside and C10(EO)4, (1;1) in 10 or 30% NaOH.*

The different foaming behaviour of solutions of optimized mixtures of C6- C8- or C8-branched glucosides and C10(EO)4 is shown in figure 18. The concentrates contain 5% C10(EO)4, 6% TPPP, 4 % metasilicate and glucoside. The added amounts of this co-surfactant are minimized to get the same temperature interval of stability, i.e. transparency, for all the concentrates. The measurement is made on dilutions 1:20. The amount of foam is rather high especially in the C8- or C8B-glucoside mixtures at 20°C but disappears eventually at 60°C where the fluids probably have passed the cloud point borderlines.

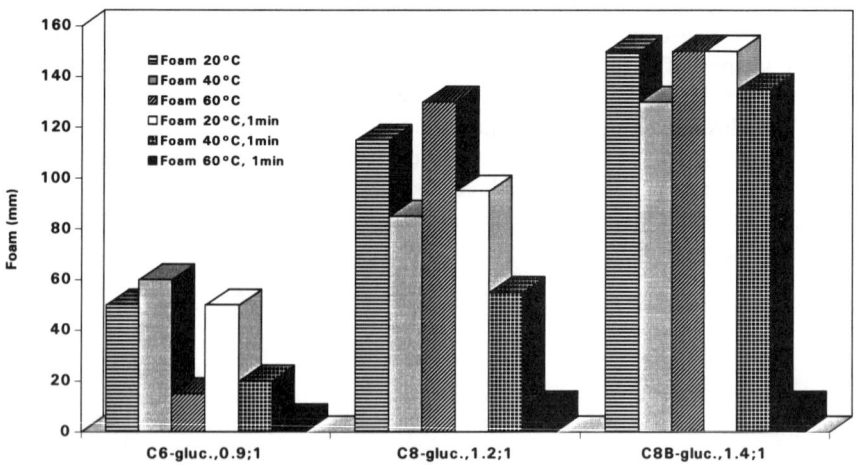

Figure 18 *Foam, glucosides and 5% C10(EO)4, 6% TPPP and 4% Metasilicate diluted 1:20 at different temperatures.*

It is also seen that the level of the foam overall is lower in the hexyl glucoside mixtures. An explanation could be the effect of the shorter chain glucosides on the lamellar states (discussed in paragraph 2) which may be operating also in the foam lamellas thus destabilizing the foam.

6.3. Wetting in NaOH solutions. The surface tension of a NaOH solution is very high. Since highly alkaline fluids are used in many areas to modify various surfaces such as grease covered walls in the food industry, natural or synthetic fibres, cellulose before derivatization to water soluble polymers etc., the low penetration ability due to the high surface tension is a problem. Consequently there is a need for efficient wetting additives that are soluble in these media. The glucosides have been used as such but are not always efficient enough. We have used the contact angle of a droplet of the solution at a hydrophobic surface as a measure of the wetting ability.

In figure 19 a comparison between the wetting effectivity of the C6-glucoside/ C10(EO)4 mixture (ratio1:1) and three different glucosides dissolved in 40% NaOH is shown. The contact angle has been measured with a goniometer 1 min after application of the fluid. The fluids were diluted and the effect on the contact angle investigated. As can be seen the wetting with the ethoxylate containing mixture is more efficient than that of the glucosides alone.

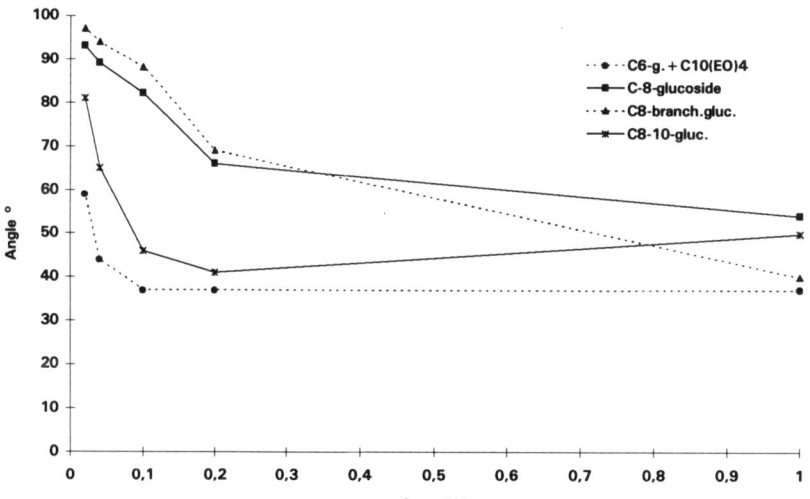

Figure 19 *Contact angle, C6-glucoside + C10(EO)4 and glucosides only in 40% NaOH.*

The effect of this increased wetting on a mercerization process is shown in figure 20. The technique here is to measure the shrinkage rate of a normalized yarn in a highly alkaline medium with and without wetting agent (ISO 6836-1983, Hintzman). Even deceptively small increases of the penetration rate, which equals shrinkage rate, are important in the extremely high speed equipments that are used in the modern textile industry. Two ethoxylate/C6-glucoside mixtures (1:1) are tested against a fully formulated commercial product Mercerol QW and give similar results.

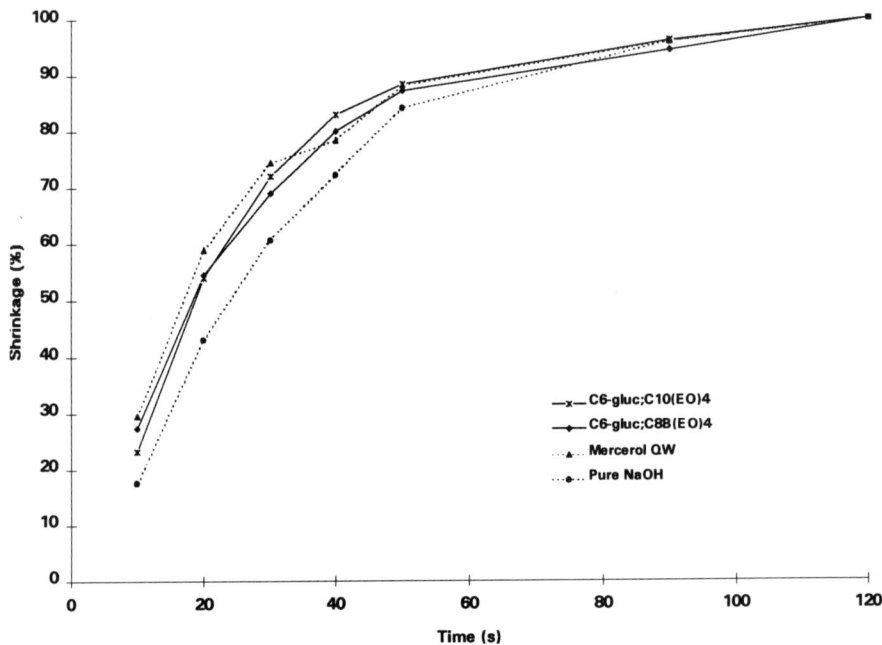

Figure 20 *Shrinkage rate at 30% NaOH, method ISO 6836-1983.*

CONCLUSION

The limited mapping that has been accounted for here includes a restricted set of substances. Much more has to be done to get a clear picture of how the glucosides and the etoxylates interact in water solutions and at interphases. However, our impression is that these mixtures offer new and exciting areas for research and development resulting in as yet unforeseen possibilities.

REFERENCES

1.   W. v. Rybinskij, In Curr. Op. Coll. Int. Sci., 1996, **1**, 587.
2.   K. Shinoda, A. Carlsson and B. Lindman, Adv. Coll. Int. Sci., 1996, **64**, 253.
3.   D. Balzer, Tenside Surf. Det., 1996, **33**, 2.
4.   F. Nilsson, INFORM, 1996, **7(5)**, 490.
5.   F. Nilsson, O. Söderman and I. Johansson, Langmuir, 1996, **12**, 902.
6.   F. Nilsson, O. Söderman and I. Johansson, Langmuir, 1997,**13**, 3349.
7.   F. Nilsson, O. Söderman and I. Johansson, J. Coll. Int. Sci., accepted.
8.   F. Nilsson, O. Söderman, P. Hansson and I. Johansson, Langmuir, submitted.

9. C. Nieendick and K.-H. Schmid, Performance chemicals, 1995, **Aug.**, 34
10. C. Nieendick and K.-H. Schmid, SÖFW-Journal, 1995, **6**, 412.
11. H.I. Leidreiter and U. Maczkiewitz, SÖFW-Journal, 1996, **10**, 674.
12. C.J. Drummond, G.G. Warr, F. Grieser, B.W. Ninham and D.F. Evans, J. Phys. Chem., 1985, **89**, 2103.
13. C.J. Drummond, G.G. Warr, F. Grieser, B.W. Ninham and D.F. Evans, J. Phys. Chem., 1986, **90**, 4581.
14. M.L. Sierra and M. Svensson, Langmuir, submitted.
15. J. Thiem and T. Böcker in "Industrial Applications of Surfactants III" (Ed. D. R. Karsa), Royal Society of Chemistry, Cambridge, 1992, 123.
16. R. Wüstneck and G. Wasow, Tenside Surf. Det., 1996, **33**;2, 130
17. Product broschure for Triton BG-10 printed by Rohm and Haas.
   T.J. Kaniecki, U. S. Pat 4,147,652, 1979.
18. S.E. Friberg, C. Brancewics and D.S. Morrison, Langmuir, 1994, **10**, 2945.
   R. Guo, M.E. Compo and S.E. Friberg, J. Disp. Sci. Tech.,1996, **17(5)**, 493.
19. S. Friberg in "Industrial Applications of Surfactants III" (Ed. D.R. Karsa), Royal Society of Chemistry, Cambridge, 1992, 227.
20. A. Matero, Å. Mattsson and M. Svensson, J. Surf. Det., submitted.
21. K. Fukuda, O. Söderman, B. Lindman and K. Shinoda, Langmuir, 1993, **9**, 2921.
22. M. Kahlweit, G. Busse and B. Faulhaber, Langmuir, 1995, **11**, 3382.
   M. Kahlweit, G. Busse and B. Faulhaber, Langmuir, 1996, **12**, 861.
   M. Kahlweit, G. Busse and B. Faulhaber, Langmuir, 1997, **13**, 5249.
   H. Kahl, K. Kirmse and K. Quitzsch, Tenside Surf. Det., 1996, **1**, 26.
23. T. Förster, B. Guckenbiehl, H. Hensen and W. von Rybinski, Progr. Colloid. Polym. Sci., 1996, **101**, 105.
24. C. Stubenrauch, E.-M. Kutschmann, B. Paeplow and G.H. Findenegg, Tenside Surf. Det., 1996, **33**, 237
   C. Stubenrauch, B. Paeplow and G. H. Findenegg, Langmuir, 1997, **13**, 3652.
25. L.d. Ryan, K.-V. Schubert and E.W. Kaler, Langmuir, 1997, **13**, 1510.
   L.d. Ryan and E.W. Kaler, Langmuir, 1997, **13**, 5222.
26. Kyoung-Hee Oh, J.R. Baran, W.H. Wade and V. Weerasooriya, J. Disp. Sci. Tech., 1995, **16(2)**, 165.
27. W.D. Clemens, Dissertation, Entwicklung von Mikroemulsionen aus biologisch abbaubaren Komponenten und ihre Anwendung zur Remobilisierung polyzyklischer aromatischer Kohlenwasserstoffe aus Böden, Institut für Angewandte Physikalische Chemie, Forschungszentrum Jühlich GmbH, 1994.
28. D. Balzer, Langmuir, 1993, **9**, 3375.
29. L. Zhang, P. Somasundaran and C. Maltesh, Langmuir, 1996, **12**, 2371.
30. L. Marszall, Langmuir, 1988, **4**, 90.
31. J.A. Rendleman in "Advances in Chemistry series, #117", American Chemical Society, Washington DC, 1973, p.51.
32. L. Thompson, J. Coll. Int. Sci., 1994, **163**, 61.
33. M. Malmsten and B. Lindman, Langmuir, 1989, **5**, 1105.

A New Surfactant Made from Kelp Seaweed

M. E. Levey[1], E. B. Revell[1], R. Dahm[2] and V. Machowski[3]

[1] NATURAL TECHNOLOGIES INTERNATIONAL LIMITED, HONILEY, WARWICKSHIRE CV8 1NP, UK
[2] DEPARTMENT OF CHEMISTRY, SCHOOL OF APPLIED SCIENCES, DE MONTFORT UNIVERSITY, LEICESTER LE1 9BH, UK
[3] MANRO PERFORMANCE CHEMICALS LIMITED, STALYBRIDGE, CHESHIRE SK15 1PH, UK

1. INTRODUCTION

Surfactants are manufactured from many materials of plant, animal and petrochemical origin and combinations of these raw materials, however the authors are unaware of any other surfactants made from seaweeds. Certain esters of alginates are noted to exhibit a degree of surface wetting capability and soda ash, obtained by burning certain species of seaweed, has been used to saponify animal and vegetable fats.

The rationale for the development of more natural surfactants stems from the growing public awareness of environmental issues and the effects of chemical pollution. This awareness and desire to protect the planet was encapsulated in the Agenda 21 protocol adopted by world leaders at the 1992 Earth Summit in Rio de Janeiro, giving a commitment towards pollution reduction and sustainable development. Natural alternatives to replace chemicals must not only be cost effective in their application, but must also be energy efficient to manufacture and competitively priced.

There are many species of seaweeds in the world's oceans, with over four hundred and fifty identified in the coastal waters of the United Kingdom, alone. They contain varying amounts of different polysaccharides, amino acids and trace elements. Seaweeds have been used for centuries as sources of sodium carbonate and iodine and more recently as sources of polysaccharides such as alginates and carregeenan for food, pharmaceutical and industrial use. The other principal uses of seaweeds are as fertilisers, processed to extract biologically active compounds to stimulate plant growth and in animal and human nutrition.

2. MANUFACTURE

A process to manufacture a powerful surface active agent from brown kelp seaweed (*Laminaria sp.*) has been developed from work originally carried out to extract biologically active compounds (plant growth stimulants) from the kelp. This process uses food grade materials as feedstock and no hazardous chemicals are employed. Its details are proprietary and cannot be described here. However, methods that are not used in this manufacturing process include bacterial fermentation of seaweeds to produce volatile fatty acids, which may be alkoxylated or sulphonated, used to produce other esters or reacted with sugars at high temperatures.

3. CHARACTERISTICS

The kelp surfactant was originally developed for application to plants and is ecologically sound and has not shown any detrimental effect to the environment (they have proven to be useful tools in bio-remediation of land). It is a complex entity and laboratory work with solvent extraction techniques has proven that it is possible to extract the surfactant molecules, but this has proven not to be necessary for the applications described in this paper below. The components of the kelp surfactant and its characteristics are summarised Table 1 and Table 2 below.

Table 1 Kelp surfactant components

Polysaccharides
Proteins, amino acids
Natural oils
Trace elements
Betaines

Table 2 Kelp surfactant characteristics

Classification	Anionic
pH	4.0 to 4.5
Freezing point	Start 0°C Final −6°C
Boiling point	Start 98°C Final 104°C
Density at 20°C	1.01 gcm^{-3}
Solubility in water	Miscible in all proportions

The kelp surfactant performs as well as many other surfactants at much lower concentrations in terms of wetting and emulsification. Its surface tension reduction is constant at approximately 31 mNm^{-1} over a considerable concentration range, from 0.1% down to 0.0001% active molecules in water, this is shown in Figure 1.

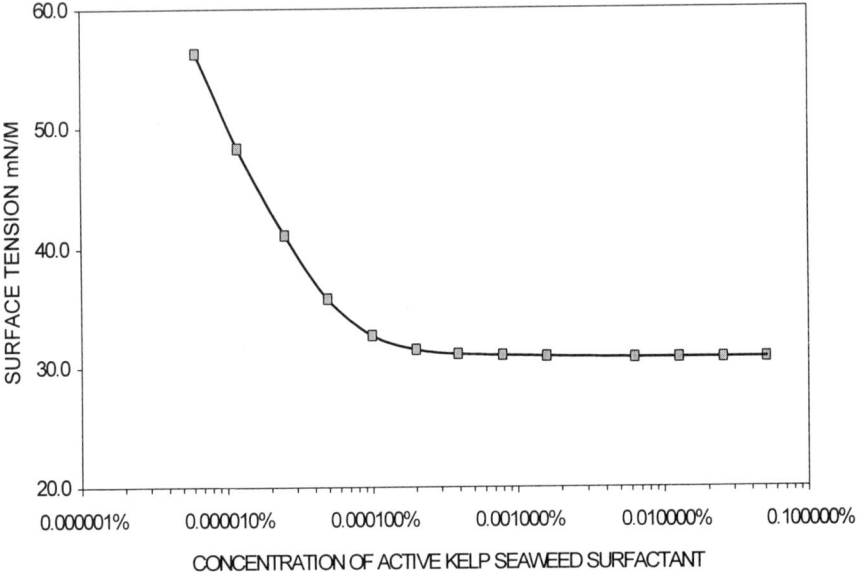

Figure 1 Surface tension against concentration in water of kelp surfactant.

The kelp surfactant is compatible with non-ionic, anionic, and amphoteric surfactants. However, its inclusion in a cationic system, also requires the inclusion of an amphoteric surfactant.

A comparison of the surface tension of the kelp surfactant with other surfactants is shown in Figure 2 at varying concentrations. It can be seen from this that the kelp surfactant can be an important addition to formulations to reduce surface tension at very low inclusion rates.

3.1. Preservation

The kelp surfactant is rapidly degraded, once dilutions are contaminated by bacteria. Trials have shown that the surfactant is compatible with most commercial preservative systems, however an effective and almost innocuous preservative system has been adopted for most applications potassium sorbate (<0.5%) plus silver chloride on titanium dioxide substrate (20 mg/Kg). This is compatible with all investigated applications of the surfactant, including skin contact.

3.2. Safety

The kelp surfactant has not shown any toxicity and shows good skin tolerance. There are no health and safety implications from its use, however the conditions under which the kelp surfactant and formulations that include it, may require protective clothing to be worn.

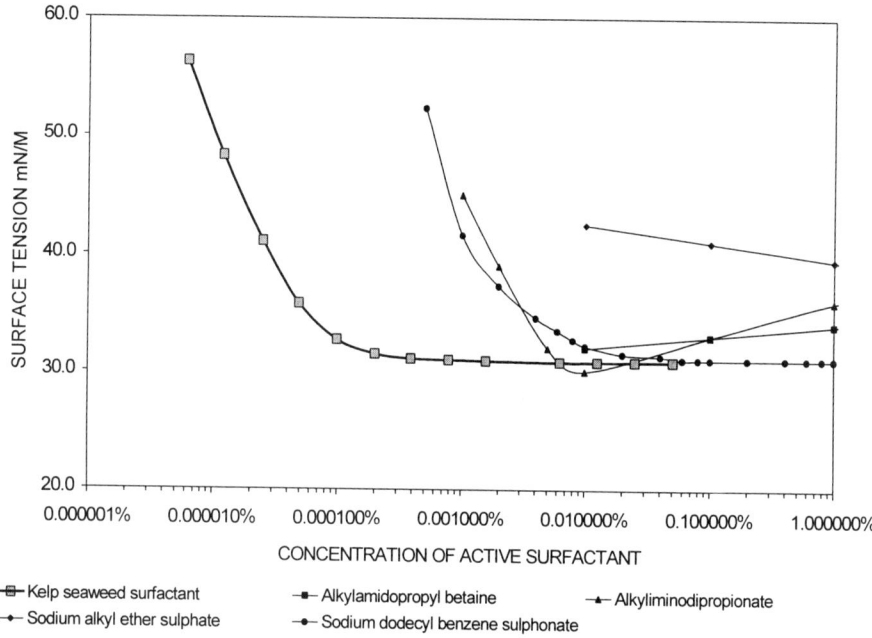

Figure 2 Surface tension against concentration of kelp surfactant in water.

4. APPLICATIONS OF THE KELP SURFACTANT

The kelp surfactant may be used

- alone as a replacement for other surfactant systems: aqueous degreasing, cleaning hard surfaces, fabrics, as a wetting agent in combination with plant nutrients, etc.
- as a formulation component: aqueous degreasers, cleaners, toiletries and cosmetics, etc.
- in a number of unique applications, due to its nature: plant growth stimulation, soil conditioning, land remediation and oil spill clean-up, etc.

Extensive trials have been carried out to determine the performance of the kelp surfactant in the applications summarised in Table 3, under varying conditions. The kelp surfactant has proven to be highly effective under most conditions, however it ceases to function at extremes of pH.

Table 3 Applications for kelp surfactants

Industrial cleaners and de-greasers
Domestic cleaning products
Toiletries and cosmetics
Waste water and effluent treatment
Land remediation and oil spill clean-up
Agriculture, horticulture and amenity

4.1.   Cleaning and degreasing

Surfactant formulations for cleaners and degreasers have been extensively researched and many are available commercially. Trials have been carried out to determine the performance of the kelp surfactant in hard surface cleaning and de-greasing applications, both by itself and in formulation with other surfactants.

Tests have shown that although the kelp surfactant is highly effective, when used alone it is more costly than most other formulations for cleaners and degreasers. However, the addition of a small amount of the kelp surfactant to other tested formulations improved their cleaning and de-greasing performance, this included combinations of sodium lauryl ether sulphate, amphoterics such as alkylamidopropyl betaines and alkylpolyglycosides. The following observations were made about the kelp surfactant in the cleaning and de-greasing applications, it offered:

− good cleaning and grease removal efficiency
− minimal amounts of transparent residues
− excellent solubilising, including body fats
− stable against acids and alkalis (pH 12 or less)
− improvement of low temperature properties of surfactant combinations
− good skin tolerance
− reduction in total amounts of surfactant used, therefore less discharge into drains

Comparisons between concentrations of the individual components of typical general purpose cleaners and ready to use glass cleaners, using standard ingredients and the kelp surfactant are given in Table 4 and Table 5 below. It can be seen that lower concentrations of other surfactants are required, where the kelp surfactant is included and the lowest total surfactant concentration occurs when the kelp surfactant is used alone.

Products made using the kelp surfactant meet consumer demands for ecologically sound cleaning products. Unfortunately, the surfactant's cost renders its inappropriate for all but the most environmentally committed of consumers. The kelp surfactant is an ideal addition to plant derived surfactant formulations to enable higher performance products to be developed for this market.

Table 4 Typical cleaner formulations

Ingredients		Concentrated general cleaner (w/w%)	Combination cleaner (w/w%)	Kelp surfactant cleaner (w/w%)
Kelp surfactant		0	0.002 to 0.005	0.1
Surfactant	Alkyl sulphate	10 to 20	5 to 10	0
	Alkyl ether sulphate			
	Alkylamidopropyl betaine			
	Fatty acid alkanolamide			
	Soap			
	Alkyl polyglycoside			
Builders	Citrate	1 to 4	1 to 2	0
	Carbonate + bicarbonate			
Solvents	Alcohol	0 to 5	0 to 5	0
	D-limonene			
Additives	Fragrances	<2	<1	<1
	Dyes			
	Preservatives			
Water		Balance	Balance	Balance

Table 5 Typical ready to use glass cleaner formulations

Ingredients		Ready to use Glass Cleaner (w/w%)	Combination cleaner (w/w%)	Kelp surfactant cleaner (w/w%)
Kelp surfactant		0	0.001	0.005
Surfactant	Alkyl sulphate	0.025 to 0.25	0.025 to 0.16	0
	Alkyl ether sulphate			
	Alkylamidopropyl betaine			
	Alkyl polyglycoside			
Solvents	Alcohol	5 to 25	0 to 5	0
	Acetic acid			
Additives	Fragrances	<1	<1	<1
	Dyes			
	Preservatives			
Water		Balance	Balance	Balance

4.2. Industrial de-greasers

Industry uses large quantities of de-greasers, both volatile chlorinated organic solvents and aqueous, mainly alkali systems. The kelp surfactant has proven to be an excellent aqueous de-greaser for cleaning metal and plastic components by itself and it will remove oil and grease from concrete without damaging the structure. The inclusion of the kelp surfactant in other aqueous formulations, enables lower performance, plant derived surfactants to be used in formulations, where the higher performance petrochemical surfactants would otherwise be used. A comparison of formulations for aqueous de-greasers that demonstrates this is given in Table 6 below.

Table 6 Typical de-greaser formulations (using natural derived components)

Ingredients		De-greaser (w/w%)	Combination de-greaser (w/w%)	Kelp surfactant de-greaser (w/w%)
Kelp surfactant		0	0.005 to 0.01	0.1
Surfactant	Alkyl sulphate	20 to 40	5 to 10	0
	Alkyl ether sulphate			
	Alkylamidopropyl betaine			
	Soap			
	Alkyl polyglycoside			
Builders	Carbonate + bicarbonate	2 to 7	2 to 4	0
	Sodium glucoheptonate			
Solvents	Alcohol	0 to 5	0 to 5	0
	D-limonene			
Additives	Preservatives	<1	<1	<1
Water		Balance	Balance	Balance

Following extensive trials, a major car manufacturer in southern Europe has selected a de-greaser formulation based upon the kelp surfactant, with sodium lauryl ether sulphate, cocoamidopropyl betaine and a potassium soap to reduce foaming. The chosen formulation was determined to work as effectively at 45°C at pH 9, as the manufacturer's previous preferred de-greaser. This was a surfactant/caustic/ metasilicate system used at a higher temperature of 80°C at pH greater than 12. The use of the new formulation will give substantial energy savings.

4.3. Toiletries and cosmetics

The market for cosmetics and toiletries is growing and there is increasing consumer demand for naturally based products with effective performance. This is placing demands upon formulation chemists, who are forced to be more innovative and to search out new and effective natural and naturally derived materials.

Seaweeds and extracts have long been used in toiletries and cosmetics. There is considerable potential for a product that is a new natural surfactant from plants, which shows good skin tolerance. Currently the concentrated kelp surfactant is a dark brown in colour and on dilution produces solutions that are pale yellow to golden, depending upon concentration. The dark brown colour of the concentrate has proven difficult to remove.

In shampoo formulations, the kelp surfactant has been observed to leave a surface charge on the hair strands, which helps the binding of cationic conditioners.

The kelp surfactant has been tested in many formulations including skin cleansers, shampoos, bath and shower gels. Commercial products are under development by several manufacturers that use it as a formulation ingredient.

4.4. Agriculture, horticulture and amenity

Seaweeds have been applied to crops for centuries and in the last fifty years extracts, particularly alkaline ones have been used. Benefits have often been noted, however it is only in the last decade that biochemists have begun to identify the active compounds and their mechanisms. The surfactant contains these important biologically active compounds at a few parts per million, trimethyl glycine (glycine betaine) and γ-aminobutyric acid betaine (they are present in many seaweed extracts and would normally be included in the total amino acid/protein content). These betaines are used in mechanisms that help plants to resist environmental stress.

The excellent wetting and penetrating properties of the kelp surfactant, its resistance to freezing, together with the biologically active compounds makes it an ideal adjuvant for the addition of crop nutrients and crop protection products, either by itself or in combination with amphoteric surfactants, e.g. alkylamidobetaines. Work to prove the compatibility and benefits with natural and inorganic crop nutrients, in the field has been carried out and results appear promising.

Compatibility with a number of other crop protection products has been proven in the laboratory, but not in general use. All products used as components of pesticide formulations in the UK must be approved by the Pesticides Safety Directorate and this surfactant has not yet been registered.

4.5. Waste water and effluent treatment

The betaines, combined with the surfactancy, make the product effective in waste-water treatment, for example removing fat and grease deposits from drains and grease traps. The introduction of small quantities of the kelp surfactant into a grease trap helps to effectively wet and emulsify the fats and greases, to make them available as food to the bacteria and the betaines stimulate colonies of aerobic bacteria present to multiply and use the available food source.

The effects of the use of the kelp surfactant in the grease trap at a fast food restaurant in London are shown in Figure 3. A metered amount was automatically dosed into the drain lines and grease trap each night, after the restaurant had closed. The decline in the amount of oil and grease in the grease trap over a twenty week period is summarised 3 below.

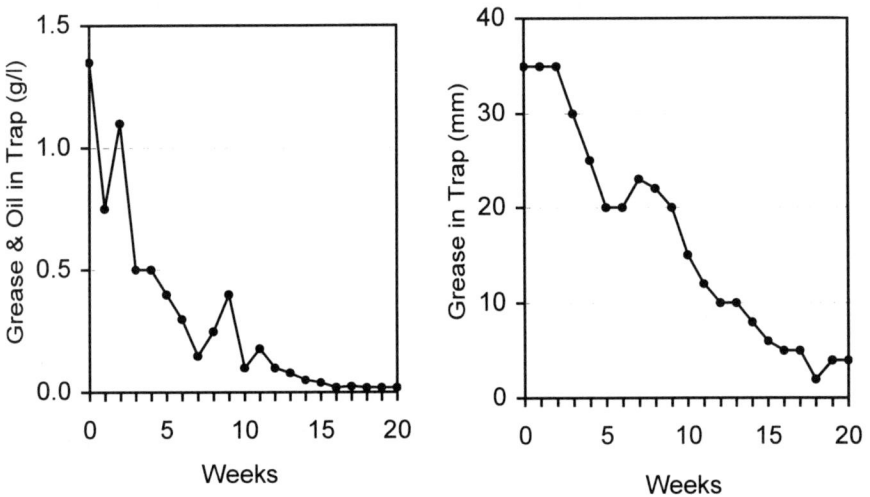

Figure 3 Levels of grease and oil in grease trap.

4.6.   Land remediation and oil spill clean-up

The kelp surfactant is an ideal product to use to clean-up hydrocarbon contaminated land and to disperse oil spills. A dilute solution of the kelp surfactant is applied by spray to the contamination (several applications one week apart for land and a single, more concentrated application for water). The hydrocarbons will be emulsified by the kelp surfactant and the betaines will stimulate naturally occurring aerobic bacteria to multiply and to break them down.

For example, the kelp surfactant has been used to clean-up heavily contaminated rail track ballast at a station in London. Three treatments with a dilute solution one week a part each, then left for three months, reduced a high total petroleum hydrocarbon content to a negligible amount at the base of the ballast, which averaged one metre thick.

4.7.   Summary

The kelp seaweed surfactant has many applications as a formulation ingredient to reduce the total active surfactant concentration in use and as a new solution to many problems.

Starch-derived Products in Detergents

Roland Beck

APPLICATION CENTRE PHARMA & CHEMICAL, CERESTAR, HAVENSTRAAT 84, 1800 VILVOORDE, BELGIUM

1 INTRODUCTION

The use of starch-based raw materials in surfactant synthesis has gained considerable interest over the past 5 years. Whereas the production capacity of starch-based surfactants was approximately 5000 t/a in 1992, today's production capacities are in the range of 60-70.000 t/a in Western Europe. With this figure starch-based surfactants have catapulted themselves from a Cinderella existence into the class of established speciality surfactants. This growth is mainly attributed to the use of these surfactants in detergent products such as dishwashing liquids and personal care items such as shampoos or pH-neutral liquid soaps.

The purpose of this paper is to shortly present the established generation of starch-based surfactants and to describe their strength and weaknesses. Based on this consideration a concept for a new generation of APG-type surfactants is developed, a process for their production is presented, and the basic application profile is compared with that of established starch-based surfactants.

The recent developments, although today mostly academic, in the field of glucamine derived surfactants is also presented.

2 ESTABLISHED STARCH-BASED SURFACTANTS

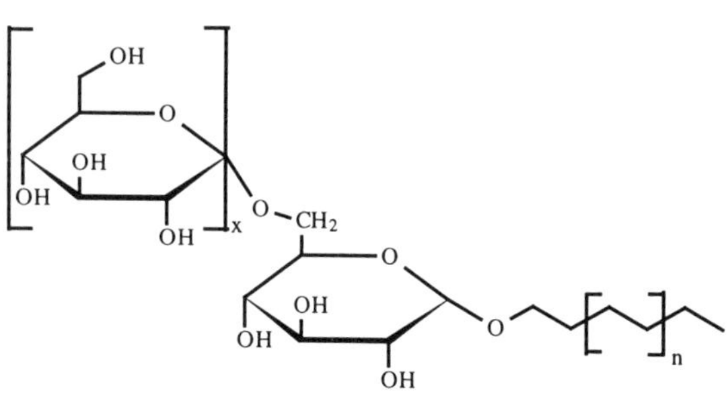

Figure 1 *Basic structure of Dodecyl polyglucoside*

Two different types of starch-based surfactants are responsible for this enormous growth, i.e. alkyl polyglucosides, known under their acronym APG, and glucamides also sometimes erroneously referred to as glucosamides. Figure 1 shows one of the many possible structures of dodecyl polyglucoside.

In Figure 2 the most common glucamide, i.e. the C_{12}-glucamide, is shown.

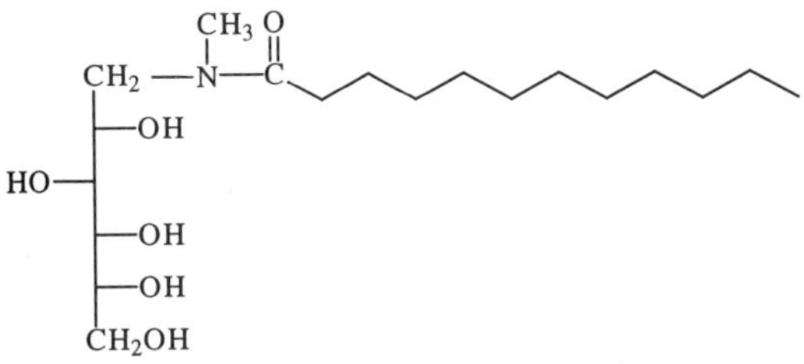

Figure 2 *Structure of C_{12}-glucamide*

The basic profile of these two types of surfactants has been extensively described in the literature. The most important are:

- Alkyl polyglucosides and glucamides belong formally to the class of non-ionic surfactants
- Particular detergency profile, e.g. towards removal of fatty stains
- Synergistic effects with common anionic surfactants
- Readily biodegradable
- Low irritation potential

The hydrophilicity of the hydrophilic head group, i.e. the carbohydrate part, is determined by the number of carbohydrate residues bound to the hydrophobic tail. In the case of alkyl polyglucosides the hydrophilic head group is determined by the average degree of polymerisation and is essentially fixed in the range of 1.3 to 1.4, as explained later in detail. For glucamides a chemically defined structure, that of a sorbityl residue, of course unequivocally determines the hydrophilic character.

3 STRUCTURAL FLEXIBILITY OF ETHOXYLATE SURFACTANTS

Alkyl alkoxylate surfactants, the classical non-ionic surfactants, possess in contrast to the starch-based surfactants an almost unlimited flexibility. In Figure 3 the basic reaction scheme for alkyl ethoxylate surfactants, the most widely used non-ionic detergent surfactants, is given. Both the hydrophobic moiety, determined by the chain-length of the fatty alcohol represented by n, and the hydrophilic moiety, effected by the number of ethoxylate residues m, can be chosen according to the desired HLB value, solubility, and surface activity.

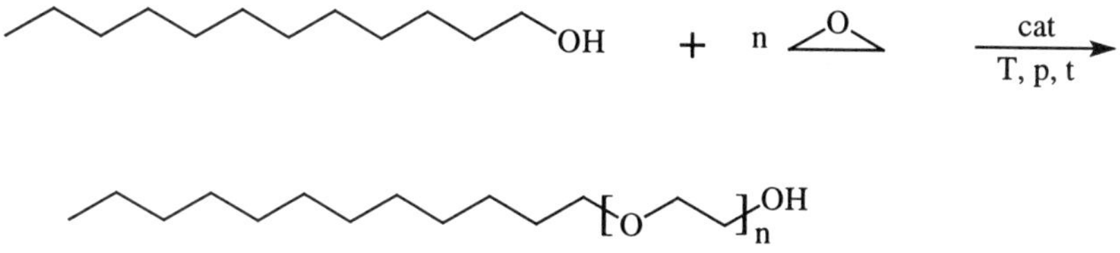

Figure 3 *Basic reaction scheme for the synthesis of alkyl ethoxylate surfactants*

In the case of today's starch-based surfactants their structural flexibility is restricted to variation of the hydrophobic moiety, the hydrophilic appears to be fixed. In this respect starch-based surfactants resemble anionic surfactants, in which also only the hydrophobic tail can be varied.

If starch-based surfactants could be as versatile as classical non-ionic surfactants the application profile and thus their market volume would certainly increase. A wider range of alcohols to be used, an adapted solubility, a tailored HLB value, and a specific surface activity must simply add to the versatility of starch-based surfactants.

4 THE BASIC CHEMISTRY OF ALKYL POLYGLUCOSIDES

The acid-catalysed formation of alkyl glucosides is known since about a century as 'Fischer glycosidation', whereby a monosaccharide is reacted with an excess of an alcohol. The main reaction products are the two isomeric alkyl monoglycosides, with smaller amounts of di-, tri-, and higher oligomeric glycosides. Alkyl polyglycosides are essentially 'Schulz-Flory' distributed, and the average degree of polymerisation (DP) is used for their characterisation. (Figure 4).

In the case of alkyl polyglucosides the two reactants, fatty alcohol and crystalline dextrose, are not miscible, a homogeneous state is only reached at the end of the reaction, when all the glucose has reacted to alkyl (poly)glucosides.

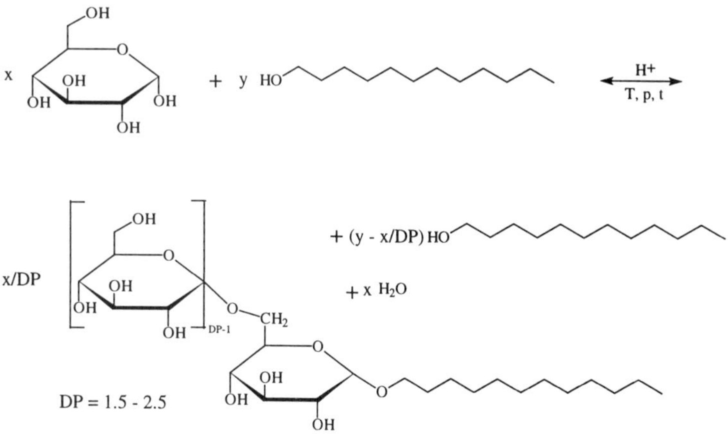

Figure 4 *Simplified Reaction Scheme of Alkyl Polyglucoside Synthesis*

4.1 High DP APGs

For the production of higher DP alkyl polyglucosides three principle strategies are possible:

➤ Change the ratio of alcohol to glucose
➤ Use a higher oligomeric carbohydrate as reactant, e.g. maltose
➤ Try to shift the equilibrium

Since no high DP alkyl polyglucosides are on the market it was obvious, that the solution to this problem was not trivial.

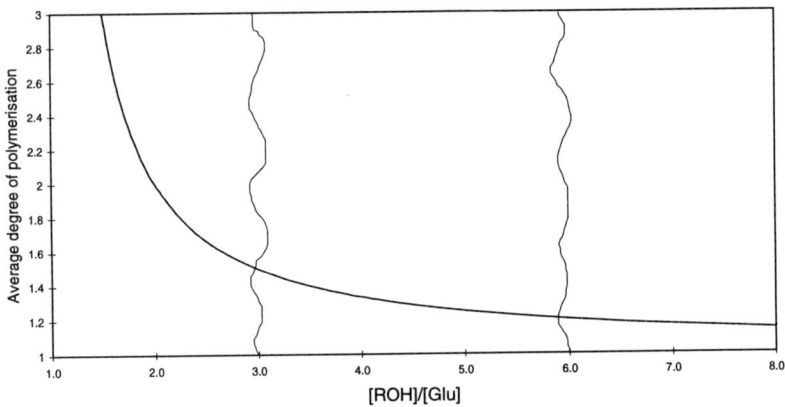

Figure 5 *Average degree of polymerisation of an alkyl polyglucoside in relation to the molar ratio fatty alcohol to glucose used during synthesis*

Very quickly it became clear that lowering the ratio fatty alcohol to glucose did not work as anticipated. As shown in Figure 5 the ratio fatty alcohol to glucose should be well below 3:1 to give the desired high DP alkyl polyglucosides. At this reaction conditions, however, the formation of by-products are strongly enhanced, and a standard alkyl polyglucoside with a DP of about 1.40-1.50 with a lot of 'polyglucose' was the result.

Using a disaccharide as starting material did as well not yield the wanted products. Either a reaction called 'alcoholysis' split the disaccharide when applying severe reaction conditions, or the disaccharide remained unreacted or simply transformed into 'polyglucose'.

The only remaining approach left was thus to try to shift the equilibrium somehow. As, however, explained the first approach, simply changing the ratio of the two starting reactants, was not successful. The operation of the equilibrium shift had to be tackled at a state of reaction, when the glucose had not got the possibility of self-polymerisation anymore.

We decided therefore to shift the equilibrium at the end of the reaction. This means that the some of the excess fatty alcohol, in which the alkyl polyglucoside is dissolved in, has to be removed. Figure 6 shows a simplified block diagram of the alkyl polyglucoside process (US Pat. 5,756,072).

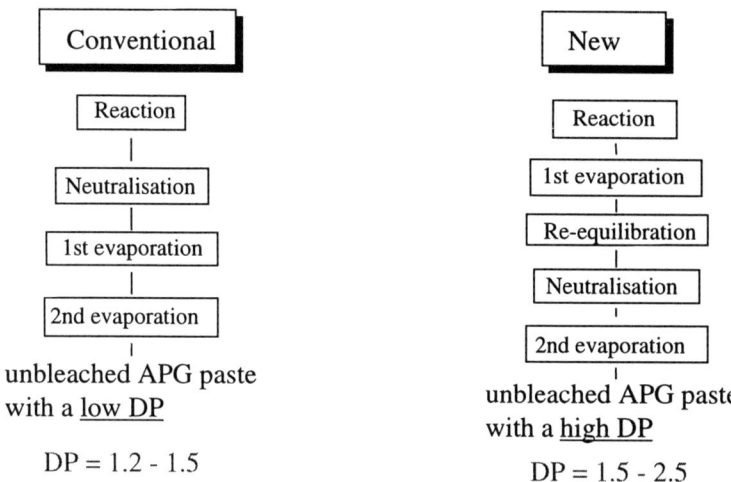

Figure 6 *Simplified Block Diagram of the Alkyl Polyglucoside Process*

In the conventional process for the production of alkyl polyglucosides the reaction mixture is neutralised and then the excess fatty alcohol is removed by single or double-pass evaporation.

Removing part of the excess fatty alcohol prior to neutralisation, in principle retains the possibility for a re-equilibration of the alkyl polyglucoside. In Figure 7 the average degrees of polymerisation (DP) of an octyl polyglucoside in function of the re-equilibration time are plotted for three different amounts of remaining fatty alcohols.

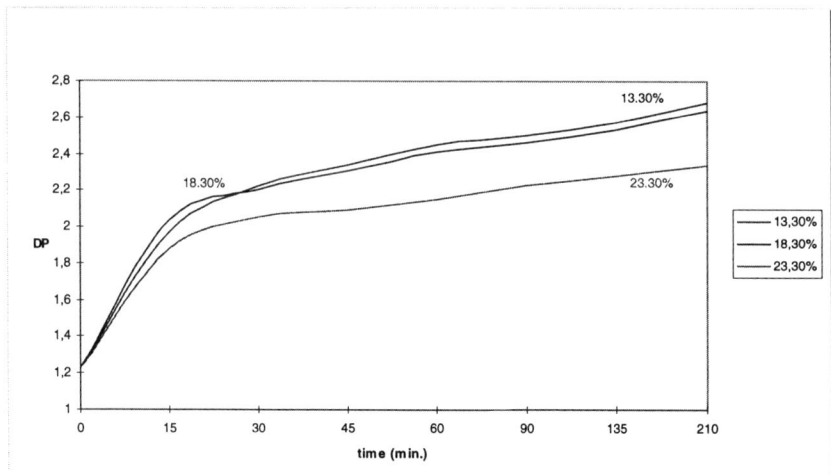

Figure 7 *Influence of the Re-equilibration Time of an Octyl Polyglucoside in function of the remaining fatty alcohol content*

After a fast initial re-equilibration the reaction speed slows down and reaches a plateau. Average DP values between 2.0 and 2.5 can be reached by employing this technique of re-equilibration of an unneutralised alkyl polyglucoside by evaporating the excess reaction alcohol.

Figure 8 shows the polymer distribution according to 'Schulz-Flory' of an alkyl polyglucoside in function of its average degree of polymerisation. As can be seen clearly the strongest effect of an increase in the average degree polymerisation is in the reduction of the monoglucoside amount and the enlargement in the DP3$^+$ region.

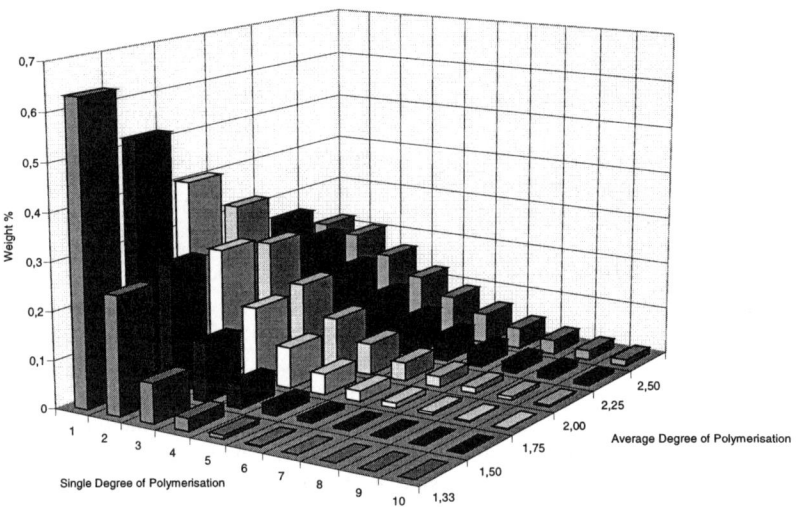

Figure 8 *Distribution of Alkyl Polyglucosides in function of their Average Degree of Polymerisation*

4.2 Basic Application Profile of a High DP Octyl Polyglucoside in comparison with a standard Octyl Polyglucoside

To verify the basic assumption that increasing the average degree of polymerisation of an alkyl polyglucoside would change the application profile, we have compared two octyl polyglucoside, a standard with a DP of 1.40 and a high DP octyl polyglucoside with a DP of 2.30.

The polymer distribution of these two octyl polyglucosides is given in Figure 9. The monoglucoside amount is approximately halved, and the weight percentage of the DP3$^+$ region is strongly increased.

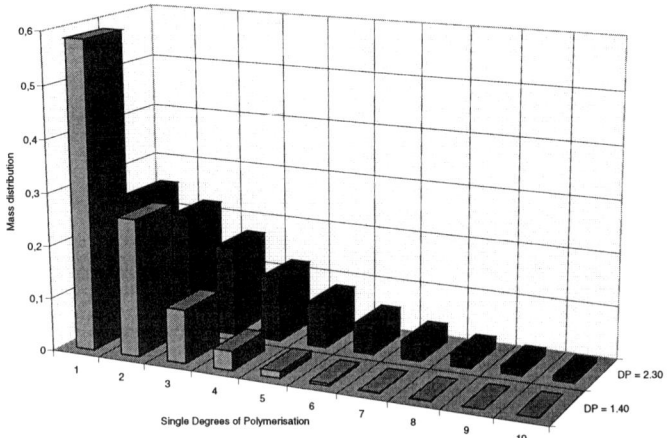

Figure 9 *Distribution of the two Octyl Polyglucosides used for testing the basic Application Profile*

The first test performed with any kind of surfactants is the determination of the surface activity. Figure 10 shows the surface tension in function of the surfactant concentration. Being a relatively hydrophilic surfactant, C_8- based surfactants are at the lower boarder, we did expect a worsening of the surfactant characteristics, that means an increase in the critical micelle concentration. However, as the first surprise in the application testing, the high DP octyl glucoside gave lower surface tensions at the same concentration compared with the standard octyl glucoside.

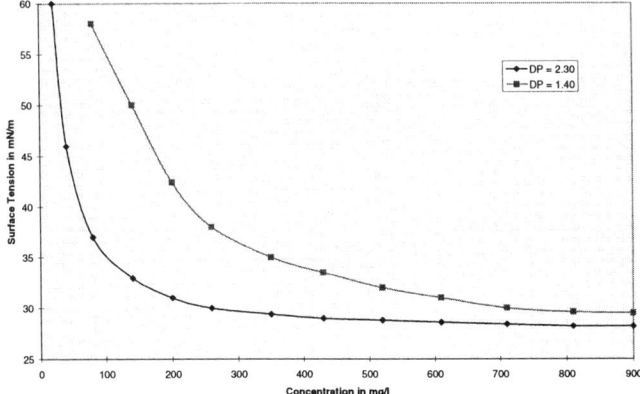

Figure 10 *Distribution of Alkyl Polyglucosides in function of their Average Degree of Polymerisation*

An other important property of surfactants is the wetting characteristics. The wetting characteristics at different temperatures and pH values were determined and are shown in Figures 11 and 12.

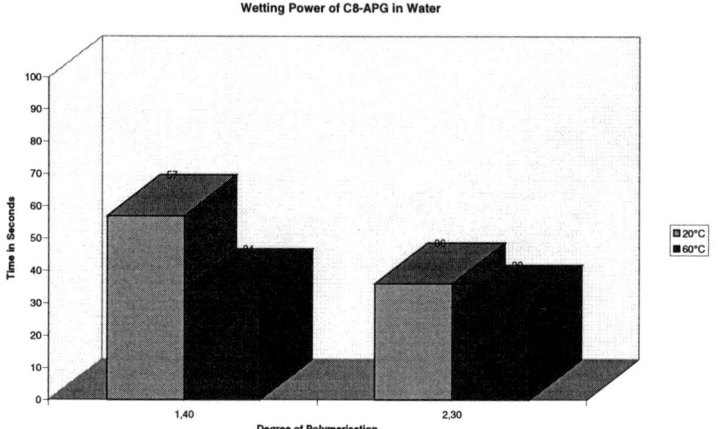

Figure 11 *Wetting Behaviour at neutral pH in function of the Average Degree of Polymerisation*

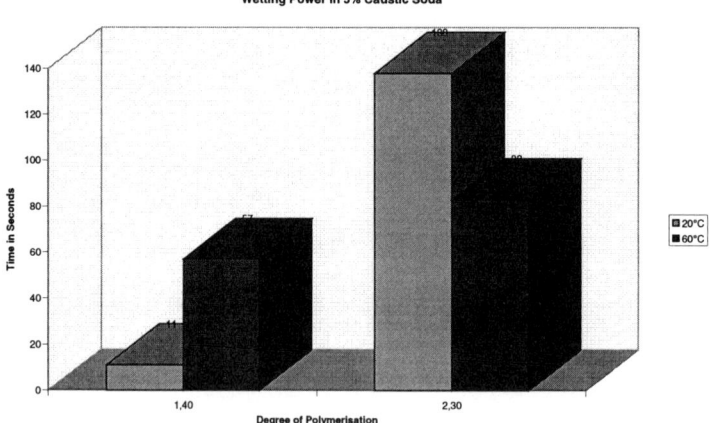

Figure 12 *Wetting Behaviour at alkaline pH in function of the Average Degree of Polymerisation*

At neutral pH the high DP octyl polyglucoside shows a somewhat faster wetting than the standard octyl polyglucoside, whereas in 5% caustic soda the picture is reversed. There the standard octyl polyglucoside wets faster, in particular at lower temperatures. A reverse trend in dependency of the temperature is observed between the two octyl polyglucosides.

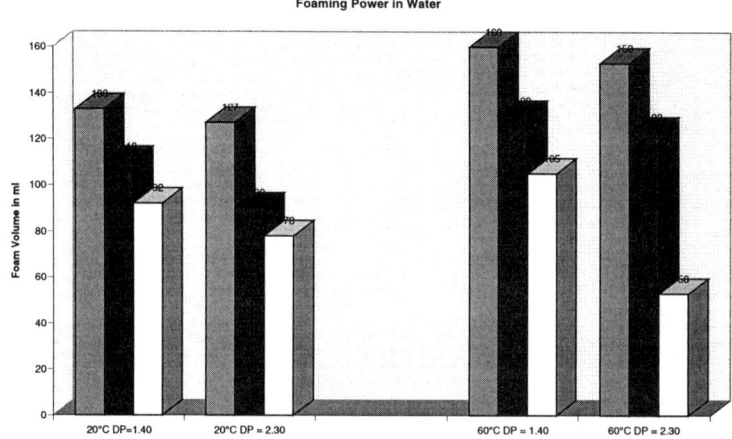

Figure 13 *Foaming Characteristics at neutral pH in function of the Average Degree of Polymerisation*

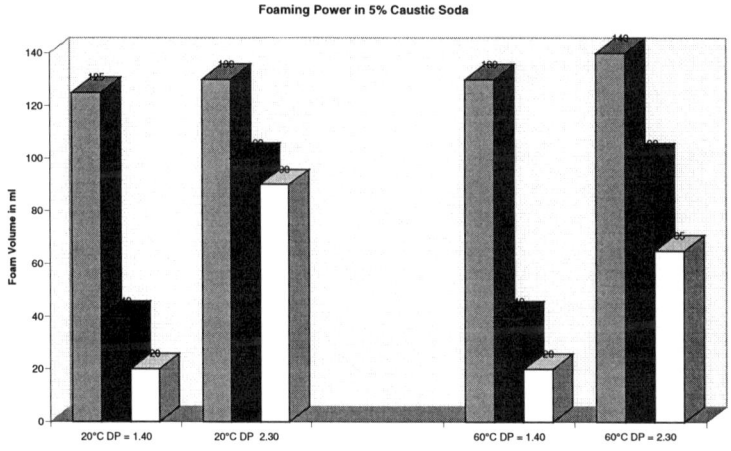

Figure 14 *Foaming Characteristics at alkaline pH in function of the Average Degree of Polymerisation*

The foaming characteristics, both at neutral pH and in the presence of 5% caustic soda at two different temperatures are shown in Figures 13 and 14. At neutral pH the standard octyl polyglucoside foams somewhat more than the high DP octyl polyglucoside. At alkaline pH however the foaming power of the high DP octyl polyglucoside is significantly higher than that of the standard octyl polyglucoside.

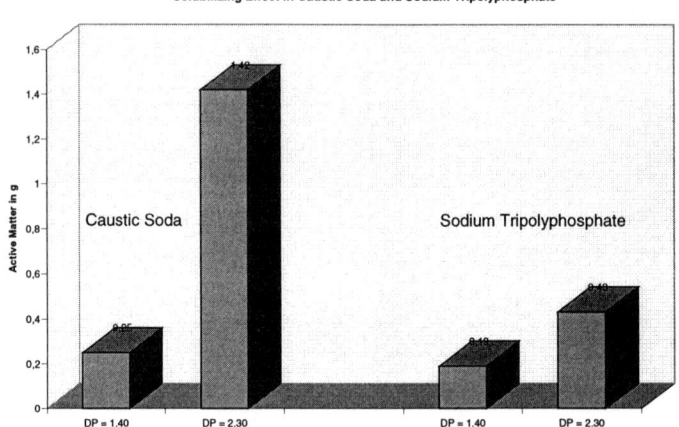

Figure 15 *Solubilising effect in presence of caustic soda and sodium tripolyphosphate in function of the Average Degree of Polymerisation*

Whereas the solubilising power of the standard octyl polyglucoside is hardly at all effected by the type of alkali, i.e. caustic soda and sodium tripolyphosphate, respectively, that of the high DP octyl glucoside is strongly effected (Figure 15).

5 GLUCAMINES AND GLUCAMIDES

Glucose can be functionalised by introduction of an amine group. This is most conveniently done by reductive amination, also referred to as reductive alkylation. Glucose is reacted with ammonia or a primary amine under hydrogenation conditions to give glucamines (Figure 16). Reaction product obtained with ammonia is glucamine or 1-amino-1-deoxy sorbitol, reaction products obtained with primary amines are the secondary amines N-alkyl glucamines or N-alkyl-1-amino-1-deoxy sorbitols (EP 0 536 939).

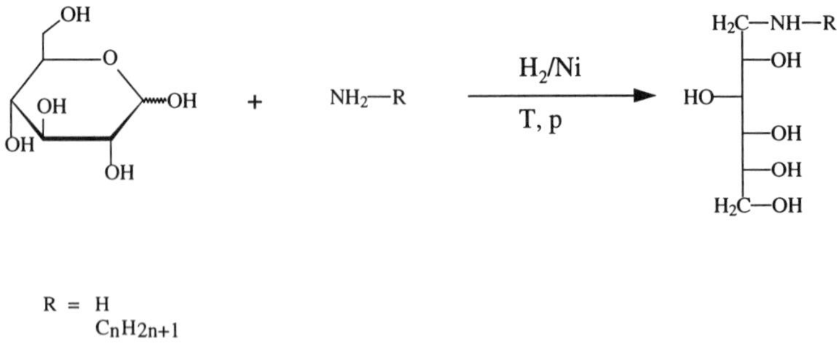

R = H
  C_nH_{2n+1}

Figure 16 *Aminopolyols - Glucamines*

5.1 Glucamides

Glucamides have gained considerable importance in the past few years as naturally based surfactants (Figure 17). Glucamides, like APGs, show synergistic effects with other types of surfactants and, due to their polyhydroxy structure, have a low irritation potential. Glucamides are thus used as secondary surfactants, and also because their solubility in water, in the absence of other surfactants, is low.

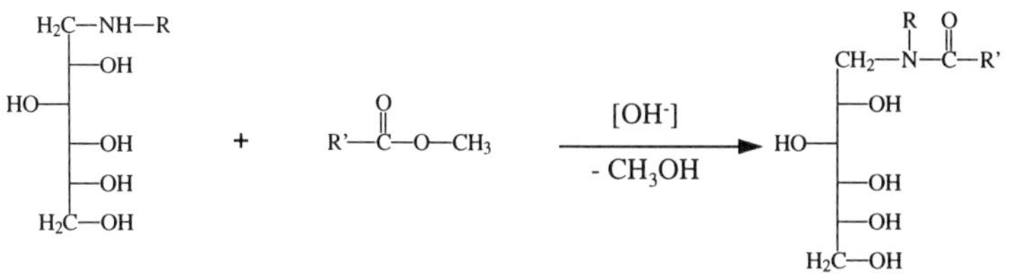

$R = C_nH_{2n+1}$
$R' = C_nH_{2n+1}$

Figure 17 *Glucamides*

5.2 Other Types of Glucamine Surfactants

To demonstrate the synthetic potential of glucamines an overview of the recent developments in glucamine derived surfactants is presented here. Some of the molecules have been reported in the scientific or patent literature and some were prepared and characterised in our own laboratories. An earlier extensive overview on glucamine based surfactants is given by R. Beck ("Application of Starch derived Products in Detergents", in "Industrial water soluble polymers", ed. C.A. Finch, RSC, 1996, p 76-91).
Figure 18 shows the synthetic potential of N-methyl glucamine for the production of amphoteric, cationic or non-ionic surfactants.

More novel types of glucamine based surfactants are shown in Figure 19, i.e. block-type surfactants, triangle shaped and gemini surfactants. Although the application potential of these compounds is today not known, it demonstrates the activity in research and development both in industry and academia.

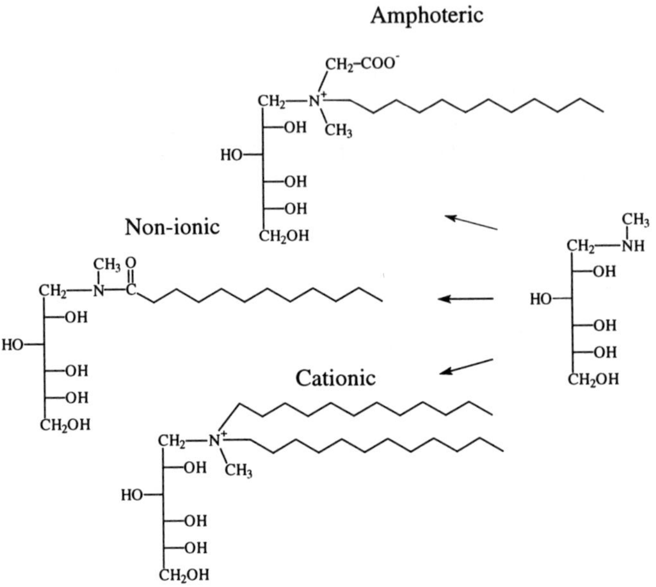

Amphoteric

Figure 18 *N-methyl glucamine based surfactants*

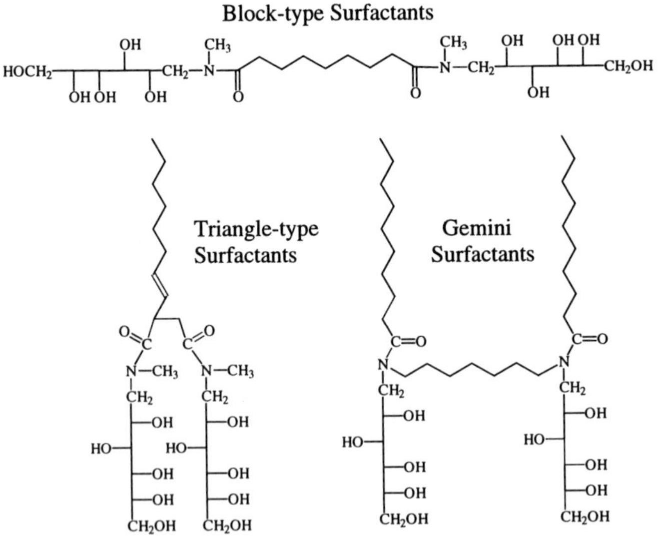

Figure 19 *Block-, Triangle- and Gemini Surfactants based on N-methyl glucamine*

6 CONCLUSION

- Carbohydrate-based surfactants, APGs and glucamides, have been the fastest growing type of surfactant of the nineties
- Within just one decade the application area has shifted from low volume to high volume applications, like detergents (Figure 20).

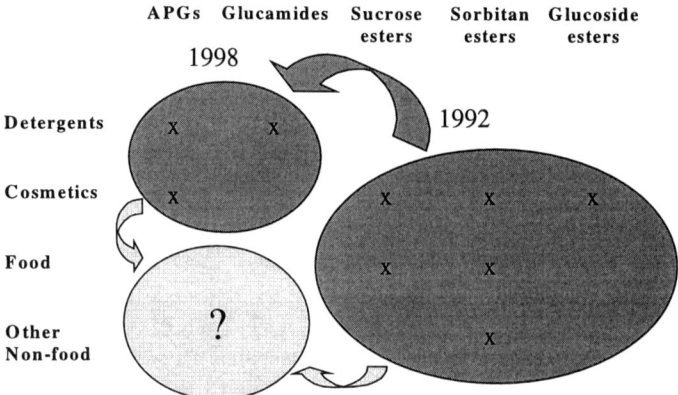

Figure 20 *Application areas of carbohydrate based surfactants*

- Synthetic potential of carbohydrates as hydrophilic building blocks for surfactants is not yet fully utilised
- The accessible DP has been widened from 1.3-1.4 to up to about 2.5, increasing the HLB value by about 3 points, thus increasing the flexibility of carbohydrate-based surfactants.
- Scientific and patent literature is still growing for new carbohydrate-based surfactants as well as for new applications and formulations
- The original marketing argument 'green product' has been replaced by performance characteristics and mildness
- Carbohydrate-based surfactants took their place as secondary surfactants in formulations
- May-be their future will be as 'sesquinary' surfactants

The Properties and Industrial Applications of *N*-Acyl ED3A Chelating Surfactants

Joseph J. Crudden

HAMPSHIRE CHEMICAL CORPORATION (A SUBSIDIARY OF DOW CHEMICAL CORPORATION), 2 EAST SPIT BROOK ROAD, NASHUA, NEW HAMPSHIRE 03060, USA

1 INTRODUCTION

Ideal surfactants for industrial use are expected to exhibit low toxicity to mammals, aquatic species and microorganisms, be stable under use conditions, exhibit high efficacy for the purpose employed, and to be compatible with other ingredients in the formulation. A recently developed class of surfactants, N-acyl ED3A chelating surfactants, meet these criteria for many applications and since they are multifunctional, acting as surfactants, chelates, hydrotropes and corrosion inhibitors, may provide preferred solutions to many applications problems.

2 KEY FEATURES OF THE PRODUCTS

The N-acyl ED3A chelating surfactants are strong and gentle surfactants which provide optimum surface tension reduction between pH 5 and 8. Because the products retain the ethylenediaminetriacetate structure characteristic of EDTA, they can chelate multivalent cations such as Mg^{++} and Ca^{++}. However, since the products are surface active, they act differently than conventional chelates in many important respects. Unlike conventional chelates, which tend to corrode metals, these products act as corrosion inhibitors. Conventional chelates such as EDTA tend to exhibit calcium chelation values which are not concentration dependent. The chelation of calcium by Lauroyl ED3A is strongly dependent on concentration. Chelation is strong above the CMC but much weaker at low concentration. The products have been found to exhibit extremely low toxicity and irritancy, to biodegrade quite rapidly and to exhibit very low toxicity to microorganisms. Conventional anionic surfactants exhibit low tolerance to water hardness and are rapidly de-lathered and precipitated by it. These products, by contrast, exhibit synergistic lather enhancement in the presence of salinity and hardness. Anionic surfactants such as sodium lauryl sulfate rapidly denature proteinaceous substances such as enzymes and are co-precipitated by cationic surfactants. The N-acyl ED3A chelating surfactants are compatible with enzymes and cationic surfactants. The products have also been found to act as extremely efficient hydrotropes even in highly alkaline systems.

3 SYNTHESIS AND STRUCTURES

The structure of N-lauroyl ethylenediametriacetate, LED3A, chelating surfactant is presented in Figure 1(c). The product is an analog of the conventional chelate ethylenediamenetetraacetate, EDTA, Figure 1(a). where one of the acetate groups, which usually remains pendant in solution and is not essential for chelation, is replaced by a lauroyl group. The synthesis, developed by Parker[1-6], is effected by stabilization and subsequent acylation of the intermediate Ethylenediamine triacetate, ED3A, Figure 1(b).

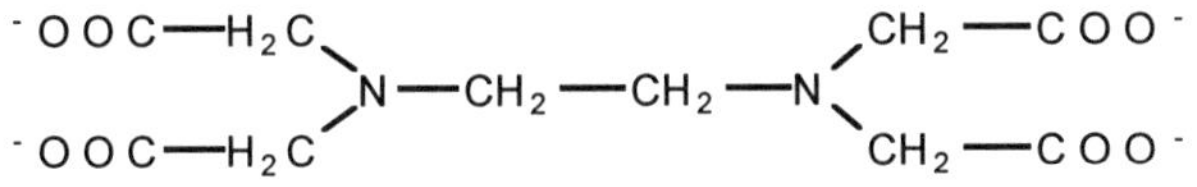

(a)

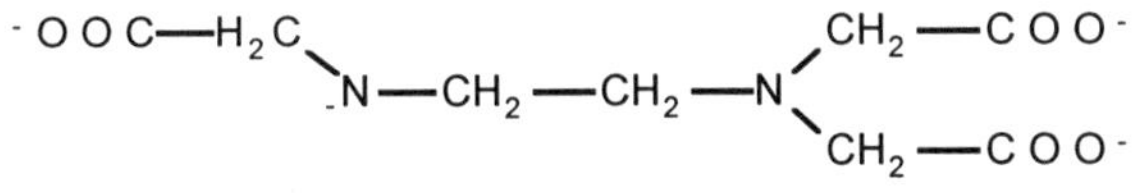

(b)

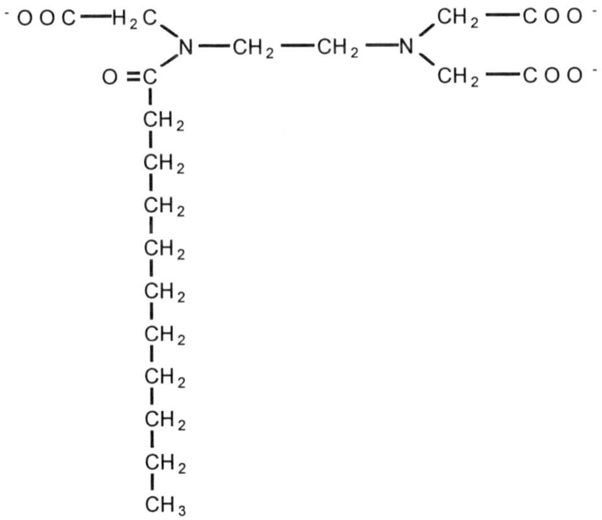

(c)

Figure 1   *(a) Ethylenediaminetetraacetate, (b) Ethylenediaminetriacetate,*
(c) Lauroyl ethylenediaminetriacetate

The acylation, a fatty acid chloride condensation, is carried out under carefully controlled conditions and produces a pure product in high yield. Upon acidification of the condensate, LED3A acid, a white crystalline water insoluble powder, is precipitated from solution. Neutralization of two of the acid groups with base forms highly surface active water soluble acetate salts. By varying the fatty acid used to produce the fatty acid chloride a wide range of structures can be produced. Furthermore, any base can be used to neutralize the acid. It is therefore possible to produce a wide array of surfactant structures with diverse properties. A range of easily accessible structures is presented in Table 1. Longer acyl groups and more hydrophobic counterions, such as TEA, will produce products with lower HLB numbers whereas shorter chains and more hydrophilic counterions, such as sodium, will produce products with higher HLBs.

Table 1 *Easily Accessible Structures of N-acyl ED3A*

Counterion	Acyl Group							
	C_8	C_{10}	C_{12}	C_{14}	C_{16}	C_{18}	$C_{18=}$	Cocoyl
Sodium	*	*	•	*	*	*	*	*
Potassium	*	*	•	*	*	*	*	*
Ammonium	*	*	•	*	*	*	*	*
Monoethanolamine	*	*	•	*	*	*	*	*
Diethanolamine	*	*	•	*	*	*	*	*
Triethanolamine	*	*	•	*	*	*	*	*
N-Propylamine	*	*	•	*	*	*	*	*
Isopropylamine	*	*	•	*	*	*	*	*
2-Amino-1-butanol	*	*	•	*	*	*	*	*
2-Amino-2-methyl-1,3-propane diol	*	*	•	*	*	*	*	*
2-Amino-2-methyl-1-propanol	*	*	•	*	*	*	*	*
2-Amino-2-ethyl-1,3-propane diol	*	*	•	*	*	*	*	*
Tris(hydroxymethyl) aminomethane	*	*	•	*	*	*	*	*

* Structures for which TSCA Registration is being sought.

• TSCA Registration filed.

4 PRODUCT PURITY

Unlike betaines, which typically contain significant quantities of salt, N-acyl ED3A surfactants contain very little residual salt. Upon acidification of the crude condensate the N-acyl ED3A acid is precipitated as water insoluble crystals while water soluble impurities such as sodium sulfate remain dissolved in the aqueous phase. The solid acid can be rinsed free of residual fatty acid to a purity of greater than 99.%.

5 USEFUL pH RANGE

The titration of lauroyl ED3A with sodium hydroxide is presented in Figure 2. Just above pH 5 two of the acetate groups become neutralized and the solubility of the product rapidly increases. At about pH 8 the third acetate group begins to react. Surface tension reduction is optimum within this pH range. However LED3A is quite soluble in up to 20% sodium hydroxide solution. The titration of LED3A with triethanolamine is presented in Figure 3. The profile is similar to that for sodium hydroxide.

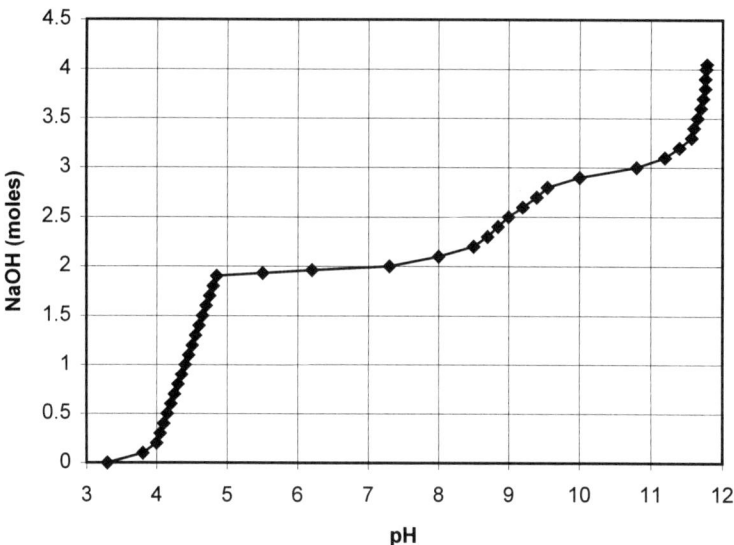

Figure 2 *Titration of LED3A with NaOH*

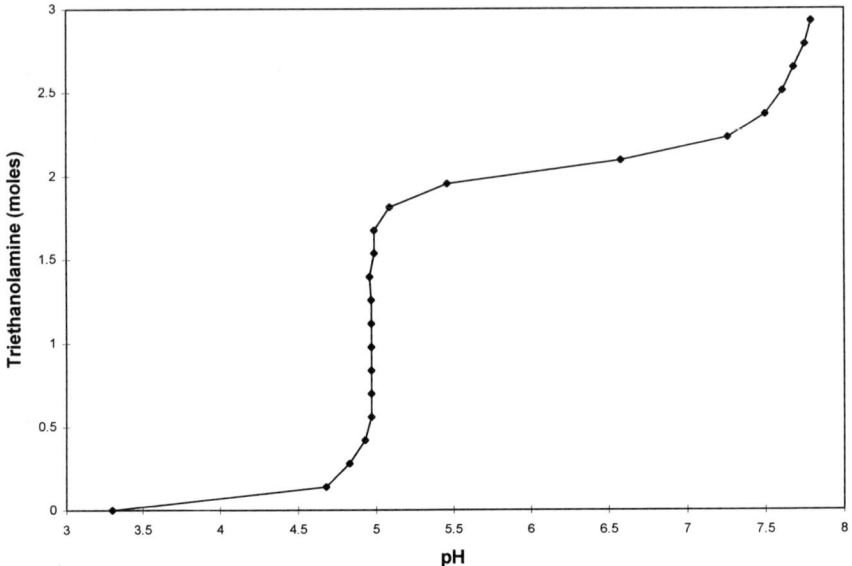

Figure 3 *Titration of LED3A with TEA*

6 SURFACE ACTIVITY

The minimum surface tension and CMC for a range of N-acyl ED3A sodium salts, at pH 7, are presented in Table 2. Sodium Myristoyl ED3A exhibits both the lowest minimum surface tension and CMC. All these structures act as strong surfactants, providing 5 to 10 dynes/cm greater surface tension reduction than sodium lauryl sulfate, at this pH.

Table 2 *CMC and Minimum Surface tension of Sodium N-acyl ED3A*

	CMC *(w/w)% x 10^{-1}*	CMC *Moles L^{-1} x 10^{-3}*	Minimum Surface *Tension Dynes cm^{-1}*
Na Lauroyl ED3A	1.7	4.0	25.0
Na Cocoyl ED3A	1.7	4.0	24.7
Na Myristoyl ED3A	0.27	0.6	21.5
Na Oleoyl ED3A	0.99	2.0	28.0
Sodium Lauryl Sulfate	2.4	---	33.0

7 pH DEPENDENCE OF SURFACE TENSION

The pH dependence of surface tension for the sodium, TEA and ammonium salts of LED3A is presented in Figure 4. Because the ionization of the headgroup is pH dependent, the surface activity of the molecule is influenced by the pH of the system. At high pH, above 9, where the headgroup is fully neutralized, surface activity is not very well developed. However, as the pH is reduced below 8, towards 7 and below, the reduction in surface tension becomes much more pronounced. As the pH is dropped below 5 the acid form of the product begins to precipitate from solution and the surface tension again rises steeply.

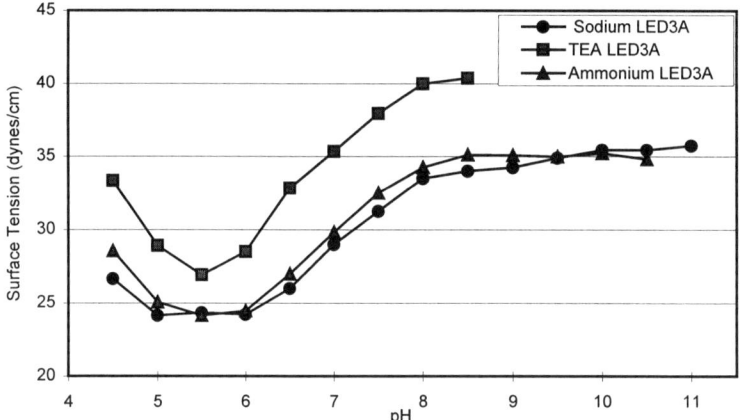

Figure 4 *Surface Tension vs. pH (LED3A 1% Solution)*

8 LATHER STABILITY

The lather stability of surfactant solutions can be determined by the method of Hart and DeGeorge[7]. A Waring blender is used to uniformly agitate and lather 200 ml of the surfactant solution under test. The contents of the blender are poured rapidly into a plastic Nalgene PF150 funnel, which rests on a 200 mesh sieve. The time for the foam to subside and drain through the funnel, to a marker wire near the neck, is taken as the lather drainage time. A lather drainage time of 10 seconds or less would indicate that the system does not produce very stable lather; whereas, a drainage time of 40 seconds or better would be expected from a strong lathering agent.

8.1 Influence of Sodium Chloride

Sodium chloride is often used to build viscosity in surfactant systems or may be present in a system because only sea water is available. Some surfactants are salted out of solution, and may be severely delathered, by the presence of sufficient salt. The dependence of lather stability of a 1% solution of NaLED3A on the presence of added sodium chloride is presented in Figure 5. Results for sodium lauryl sulfate are included for comparison. The lather drainage time for LED3A is significantly enhanced by the addition of sodium chloride whereas the addition of sodium chloride strongly reduces the propensity of sodium lauryl sulfate to form stable lather.

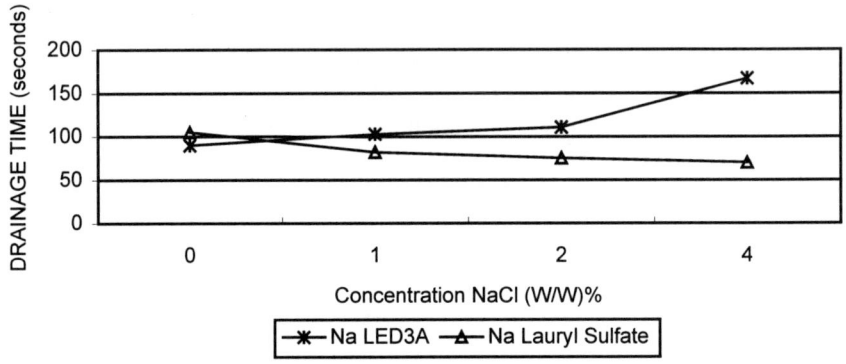

Figure 5 *Lather Drainage Time vs. Salinity*

8.2 Influence of Calcium

Water hardness ions such as calcium and magnesium have a very strong tendency to precipitate and de-lather soaps and most anionic surfactants. The unsightly deposits formed in bathrooms, where high levels of hardness are present in the groundwater, are insoluble calcium salts of fatty acid soap. In laundry products, sequestering agents such as zeolites or polyphosphates must be added to deactivate the hardness ions.

       The influence of water hardness on the lather drainage time of solutions of NaLED3A was determined using the method of Hart and DeGeorge, Figure 6. Surprisingly, it was found that the product is not only very tolerant to high levels of hardness but that the addition of water hardness ions significantly enhances the stability of lather on the surfactant solution.

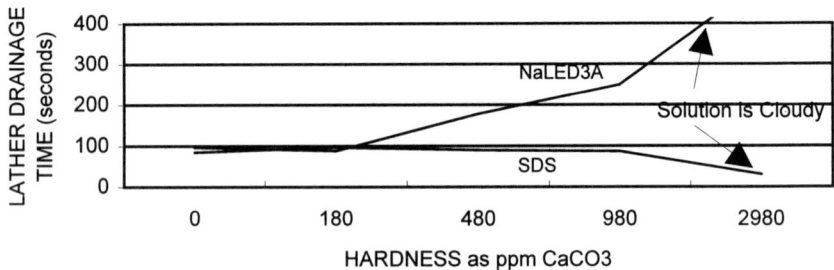

Figure 6 *Lather Drainage Time vs. Water Hardness*

It may be that the divalent counter-ions interlink adjacent surfactant headgroups in the surface film, thereby increasing surface viscosity and lather stability. Since the surfactant headgroup is trivalent it is unlikely to be precipitated by a divalent counterion.

9 DE-LATHERING AND DE-FOAMING

Silicone defoamers effectively delather NaLED3A in soft water. However, in the presence of water hardness they are less effective.

Sodium Pelargonoyl ED3A produces very little lather in the absence or presence of water hardness.

10 DRAVES WETTING TIME

The Draves wetting time for a range of fiber types by a range of surfactant systems are presented in Table 3.

Table 3

Surfactant System 1%, 23°C	Fabric	Draves Wetting Time Seconds
NaLED3A	Cotton	570
NaLED3A	Nylon	20
NaLED3A	Worsted Wool	80
NaLED3A:SDS (1:1)	Cotton	230
NaLED3A:NaLS (1:1)	Cotton	75
NaLED3A:LDAO (1:1)	Cotton	10

LED3A is not very effective at wetting cotton but mixtures with lauryl dimethyl amine oxide are very effective.

11 COMPATIBILITY WITH ENZYMES

Enzymes are achieving increasing importance in many surfactant and detergent systems. Many detergent systems now incorporate enzymes of varying types in order to cope with problem soils. Linear alkyl benzene sulfonate has been shown to deactivate protease enzyme whereas salts of Lauroyl ED3A have been shown to be compatible with and enhance the performance of protease enzyme[8].

12 COMPATIBILITY WITH CATIONIC SURFACTANTS

N-acyl ED3A surfactants have been found to be compatible with cationic surfactants, presumably because the headgroup is large and multivalent.

13 SYNERGY IN MIXED SURFACTANT SYSTEMS

Although the products do not produce the strong synergy in surface tension reduction on mixed surfactants which is characteristic of the N-acyl sarcosinates, they do tend to exhibit strong synergy in lather enhancement on mixed systems, particularly in the presence of water hardness ions. The lather stability on blends of NaLED3A and sodium lauryl ether sulfate, SLES, at a constant concentration of 1%, is presented in Figure 7. Synergistic lather enhancement is evident above 20% NaLED3A and the effect is even more pronounced in the presence of 400 ppm water hardness. Similar lather enhancement was found with lauryl dimethyl amine oxide and sodium lauroamphoglycinate[9].

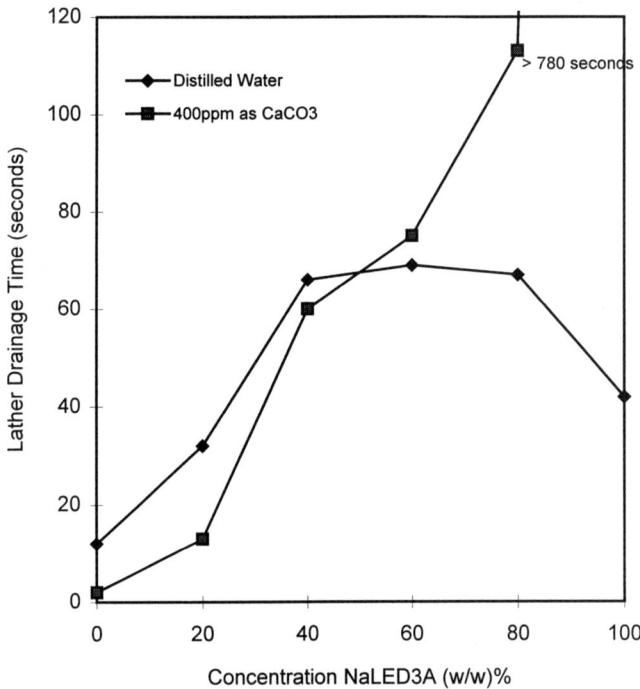

Figure 7 *Lather Drainage Time of Sodium Laureth Sulfate:Sodium LED3A*

The ability of NaLED3A to enhance the lathering potential of a commercial body wash which contains cocamidopropyl betaine, sodium cocoyl isethionate and sodium laureth sulfate as primary surfactants was assessed by determining lather drainage time, as before. The lather drainage times are presented in Table 4. The body wash was found to contain about 20% solids, therefore 20% Na LED3A was used in order to maintain constant activity.

Table 4

Test System	Water	LDT (Secs.)
Body Wash	Soft	12
Body Wash	Hard	0
Body Wash: NaLED3A 4:1	Hard	40

Hard = 300 ppm as $CaCO_3$

The body wash is strongly compromised by the presence of high levels of water hardness but performance can be recovered to a level which is exceeds the product's best performance, by a factor of three, by replacement of 20% of the product with NaLED3A at an equivalent concentration.

14 HYDROTROPY

A hydrotrope is a chemical which has the property of increasing the aqueous solubility of various slightly soluble organic compounds. They are commonly employed in the formulation of liquid detergents, especially where high alkalinity can salt out the surfactants. Sodium and ammonium xylene sulfonate are commonly used as hydrotropes in liquid detergents. NaLED3A was found to be five times more efficient at clearing a solution of sodium dodecyl benzene sulfonate in 5% sodium hydroxide than ammonium xylene sulfonate.

15 CHELATION OF METALS

The order of chelation of metals by LED3A, determined by UV-visible spectroscopy, is:

$$Mg^{2+} < Cd^{2+} < Ni^{2+} = Cu^{2+} < Pb^{2+} < Fe^{3+}.$$

The calcium chelation value of a chelate, expressed as grams of calcium chelated per gram of chelate, can be determined by a calcium oxalate titration. In this method[10], calcium ions are titrated into a solution of calcium oxalate and the chelate being tested. When all the chelate present is depleted by chelation of added calcium, the excess calcium produces a precipitate of insoluble calcium oxalate. In order to be effective, the titration must be carried out at pH 11. Conventional chelates such as EDTA exhibit concentration independent calcium chelation values. One mole of chelate complexes one mole of calcium no matter how much chelate is present. The calcium chelation value for EDTA is 260 mg $CaCO_3$/gm.

The calcium chelation value for NaLED3A was found to be dependent on the concentration of the surfactant. The dependence of calcium chelation value on NaLED3A concentration is presented in Figure 8. At low concentration the surfactant appears to act as quite a weak chelate, complexing less than 20 mg/gm.

However, as the concentration of the surfactant rises above the CMC, its chelating power increases, reaching almost 1:1 chelation at a concentration of 5%.

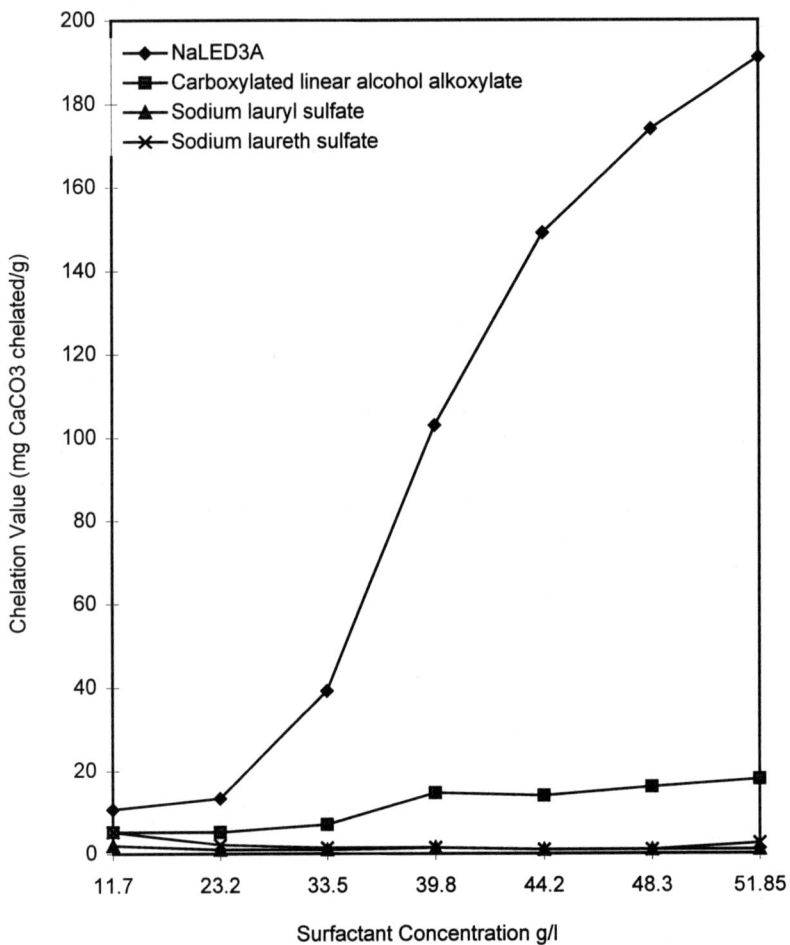

Figure 8 *Calcium Chelation Value*

16 CORROSION INHIBITION

The potential of LED3A to inhibit the corrosion of bright mild steel was evaluated. The steel coupons were immersed in 15% hydrochloric acid at 40°C, in the presence and absence of added LED3A. The rate of corrosion was determined by weight loss over a 96 hour period. The results are presented in Table 5. LED3A reduces the rate of corrosion by a factor of 10.

Table 5 *Corrosion Inhibition by LED3A*

	% Weight Loss	
Elapsed Time (Hours)	*Control*	*LED3A*
0	0	0
24	45	2
48	60	5
72	72	7
96	85	9

17 METAL PASSIVATION

Stainless steel is protected from corrosion under ambient conditions by a passivating oxide film which forms on the surface. High alkaline cleaners are often used to clean and decontaminate stainless steel vessels and reactors. These highly alkaline products strip away the passivating film. As a result, significant corrosion takes place during the cleaning process. Furthermore, significant levels of heavy metals such as nickel and chromium may be found in the rinse water. Sodium Lauroyl ED3A can deposit a highly substantive surface film on the surface of the metal which can reduce these effects by an order of magnitude.

18 ENVIRONMENTAL PROPERTIES

Surfactants for use in industrial applications must be evaluated to determine if they are safe for release into the environment and sufficiently innocuous to come into prolonged contact with humans and animals. N-acyl ED3A chelating surfactants were subjected to a battery of tests in order to determine if they were suitable for widespread use. The products proved to be surprisingly mild in terms of mammalian and aquatic toxicity and to be non-irritant in terms of skin and eye irritation. These studies have previously been reviewed in detail[11].

18.1 Biodegradation

An OECD closed bottle test for ready biodegradability revealed that NaLED3A is readily biodegradable at a concentration of 2 mg per liter.

18.2 Toxicity to Microorganisms

LED3A was found not to inhibit the growth of Pseudomonas Aeruginosa, Bacillus Cereus, Glicocladium Virens, Penicillium Funicolosum, Aspergillus Niger or Oscillatora Prolifera at concentrations up to 1,000 ppm. This indicates that the presence of the surfactant is unlikely to upset the stability of microbial ecosystems.

18.3 Mammalian Toxicity

LED3A was found to be nontoxic to rats at up to 5,000 mg/kg and can therefore be defined as nontoxic. Sodium lauryl sulfate, which is sufficiently innocuous to be used in toothpaste has an LD_{50} of 1,200 mg/kg.

18.4 Aquatic Toxicity

Since most surfactants, irrespective of application, will eventually be discharged to watercourses, it is important to establish that the aquatic toxicity is sufficiently low not to interfere with the viability of aquatic organisms. The US EPA requires that new products be evaluated on organisms along the food chain since significant toxicity at any level can lead to problems. Sodium LED3A was evaluated for toxicity to Rainbow Trout and Daphnia Magna by flow through tests and to Blue Green Algae in a static test. The results are presented in Table 6, with data for LAS, a surfactant widely used in laundry detergent, included for comparison.

Table 6

	Surfactant	
Species	LED3A	LAS
Algae	570 mg/L NOEC	~75 mg/L LC_{50}
Daphnia Magna	>110 mg/L NOEC	3.94 mg/L LC_{50}
Rainbow Trout	>320 mg/L NOEC	0.36 mg/L LC_{50}

NOEC = No observed effect concentration
LC_{50} = Concentration required to kill 50% in 96 hours

These results indicate that LED3A is extremely innocuous in terms of aquatic toxicity.

19 SKIN IRRITATION

In-vivo and in-vitro studies were carried out to determine the dermal irritation potential of LED3A.

Sodium LED3A was found to be a non-irritant when applied to the skin of New Zealand Rabbits, at a concentration of 1%. To assess the irritation potential of the surfactant at higher concentration, in-vitro methods were used. The "In-Vitro Skin Irritation Assay", which was developed by Advanced Tissue Sciences of La Jolla, California, quantifies the effect on viability of uniform specimens of live cultures of human tissue, caused by exposure to the test substance. Irritant or toxic substances will rapidly reduce the viability of the tissue, whereas milder substances will exert much less of an effect. The degree of reduction of viability of the tissue test specimens is determined by MTT Assay[12]. The in-vitro scoring classification is presented in Table 7.

Table 7 *In-Vitro Scoring Classification*

MTT - 50 (micro g/ml)	Classification
0 - 200	Severe
200 - 1,000	Moderate
1000 - 10,000	Mild
>10,000	Non-irritant

The degree of reduction of viability caused by exposure to many test substances has been shown to correlate well with known in-vivo results[13-14].

Test results for 10% solutions of sodium, potassium, TEA and ammonium salts of LED3A, along with results for sodium dodecyl sulfate, coco betaine, sodium lauroyl glutamate and sodium lauryl ether sulfate, are presented in Table 8.

Table 8 *In-Vitro Score and Classification for a Range of Surfactants*

Product	In-Vitro Score MTT - 50 (micro g/ml)	Classification
Sodium dodecyl sulfate	250	Moderate
Coco Betaine	400	Moderate
SLES	522	Moderate
Lauroyl Glutamate	4,000	Mild
NaLED3A	6,000	Mild
KLED3A	>10,000	Non-irritant
TEALED3A	>10.000	Non-irritant
NH$_4$LED3A	>10,000	Non-irritant

It is evident that N-acyl ED3A surfactants are much milder than currently available anionic surfactants such as coco betaine and SLES.

20 EYE IRRITATION

A 1% solution of Sodium LED3A caused no irritation when applied to the eyes of New Zealand White Rabbits[11].

Advanced Tissue Sciences have also developed test methods for assessing the potential of possibly irritating substances, such as surfactants, to cause eye irritation[15]. These methods use tissue cultures which contain stomal and epithelial components which act as in-vitro counterparts of the cornea and conjunctivae, structures in they eye which are important targets in ocular irritation. These tests indicate that the salts of LED3A are extremely mild in terms of eye irritation potential[9].

21 EYE STING

A solution of Tris Amino LED3A has been found to cause perceptibly less eye sting than a commercial baby shampoo at the same active concentration[9].

22 APPLICATIONS

22.1 Detergency

N-acyl ED3A chelating surfactants can act both as builders and surfactants so they could be expected to exhibit strong detergency. Sodium LED3A was evaluated for its ability to remove dust/sebum soil from cotton. The tests were carried out in a Tergotometer and the efficacy determined by brightness gain. The results with data for APG 600, sodium lauryl ether sulfate and linear alkylbenzene sulfonate, alone and in combination with LED3A included for comparison, are presented in Figure 9. The NaLED3A outperforms the other surfactants and can be seen to act synergistically with SLES and APG 600 to enhance soil removal.

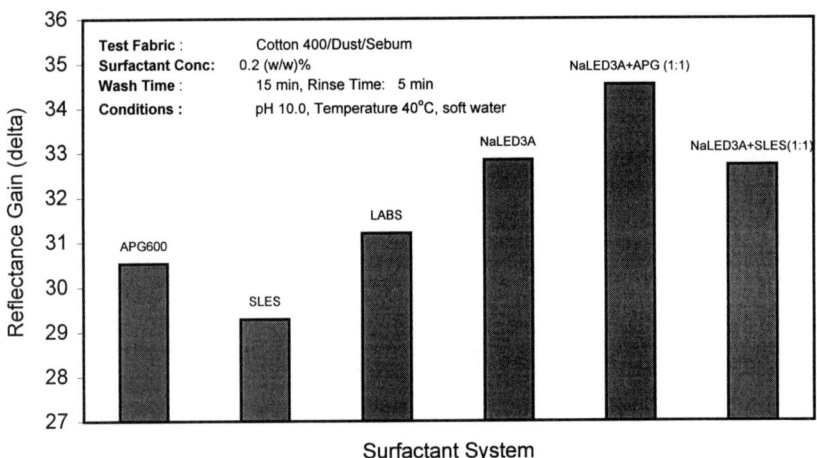

Figure 9 *Detergency of Surfactant Blends*

22.2 Heavy Duty Alkaline Cleaners

Heavy duty alkaline cleaners are used in industry to clean fabrics, reactors, vessels, vehicles, bottles, and even ovens and cages. Since N-acyl ED3A chelating surfactants are compatible with high levels of caustic, foam copiously in the presence of water hardness, act synergistically with co-surfactants, are compatible with enzymes, act as very efficient hydrotropes and are extremely mild in terms of toxicity and irritancy, they should be extremely useful in the formulation of alkaline and non-alkaline cleaners.

22.3 Ultra Mild Cleaners

The in-vitro and in-vivo dermal and ocular irritation tests carried out on N-acyl ED3A surfactants indicate that they should be very suitable for formulation of products even as mild as baby shampoos[8]. The strong surfactancy exhibited by the surfactants, even under adverse conditions, also indicates that they might confer enhanced performance to products of this type. In order to assess the potential of LED3A to enhance the performance of a mild baby shampoo the lather drainage time was determined, as before, for a 6% solution of the product, in the presence and absence of water hardness and in the presence and absence of LED3A. The results, presented in Figure 10, demonstrate that LED3A can dramatically enhance the performance of such products, particularly in the presence of water hardness.

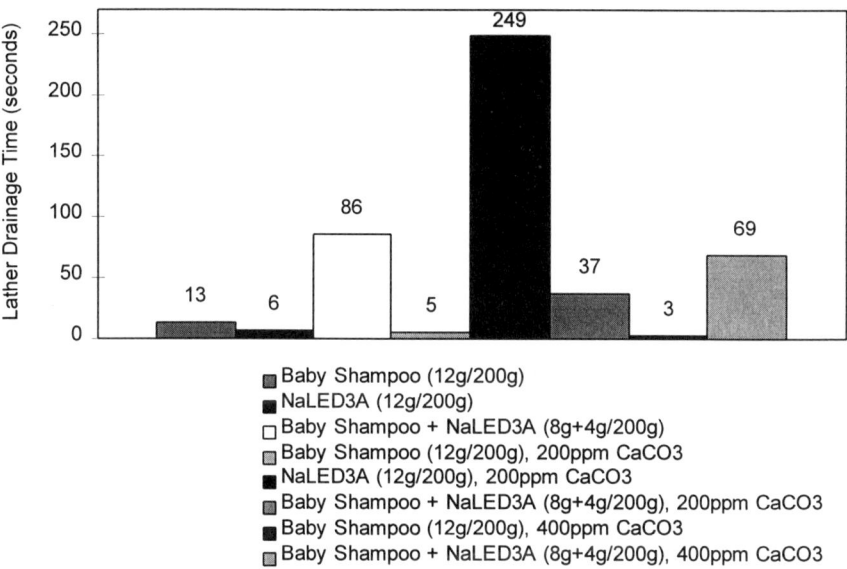

Figure 10 *Baby Shampoo Lather Stability*

    The irritation potential of the shampoo in the presence and absence of added LED3A was determined by the Advanced Tissue Sciences tissue toxicity test. The results, with comparative data for pure NaLED3A and SLES at equivalent activity, are presented in Table 9. It is evident that LED3A can be used to enhance both the mildness and lathering potential of such systems.

Table 9 *In-Vitro Skin Irritancy Screen*

	In-Vitro Score	
Product	*MTT - 50 (micro g/ml)*	*Classification*
Baby Shampoo	1,906	Mild
Baby Shampoo + NaLED3A	2,149	Mild
NaLED3A	>10,000	Non-irritant
Na Laureth(3) Sulfate	522	Moderate

22.4 Creams and Lotions

In a recent patent[16] assigned to Avon, Ptchelintsev discloses the use of N-acyl ED3A as active ingredients in preventative as well as therapeutic compositions to promote exfoliation and alleviate symptoms caused by abnormal keratinization. The patent claims that the products act as efficient exfolients without the irritation typically caused by alpha hydroxy acids. Sample formulations taken from the patent are presented in Tables 10 and 11.

Table 10 *Topical Lotion Formulation*

Component	*Wt. %*
N-acyl-N,N'N'-ethylenediaminetriacetic acid sodium salt	1.0
Alcohol	10.0
Glycerin	5.0
Polyoxyethylene (40M) Stearate	3.5
Ammonium Hydroxide	2.5
Octylmethoxycinnamate	2.0
Thickener	0.5
Fragrance	0.05

Final pH of approximately 7.0.

Table 11 *Topical Cream Formulation*

Component	*Wt. %*
N-acyl-N,N'N'-ethylenediaminetriacetic Acid	2.5
Cyclomethicone/Dimethicone Copolyol	11.5
Cyclomethicone Pentamer	10.0
Propylene Glycol	2.5
Mineral Oil	2.0
Sodium Chloride	1.0
Fragrance	0.15
Neutralizing Agent	---

Final pH of approximately 7.0.

22.5 Hard Surface Cleaners

Since these surfactants act as both chelates, surfactants and hydrotropes and are extremely mild, they should be very suitable for use in hard surface and soft surface cleaners.

22.6 Metal Cleaning and Degreasing

Since the products act as corrosion inhibitors, are substantive to surfaces and can form relatively oil soluble salts, they should be suitable for use in metal cleaning and degreasing formulations. Other properties of N-acyl ED3A, such as low toxicity and irritancy and the ability to act as hydrotropes and to tolerate water hardness, are also beneficial.

22.7 Metal Polishes

The products have the potential to dissolve tarnish and remove greasy soil.

22.8 Pesticide Adjuvants

The products can act as hydrotropes and should be capable of coupling pesticide-fertilizer combinations. They are very mild in terms of aquatic and mammalian toxicity and cause very little skin and eye irritation. They are compatible with many co-surfactants and should be sufficiently mild for use with biopesticides and enzyme containing systems. High salt tolerance should allow their use in products such as glyphosate and in regions of high water hardness. These surfactants can reach surface tensions sufficiently low, in the region of 25 dynes/cm, to allow the pesticide to penetrate plant leaves through the stomata.

22.9 Flotation of Minerals

The process of mineral flotation takes place by preferential adsorption of a flotation agent onto the surface of the component to be floated. The adsorbed flotation agent renders the surface of the particle sufficiently hydrophobic to be preferentially wetted by air. When the slurry is aerated the bubbles preferentially adhere to the hydrophobic particles thereby lifting and separating them from the contaminating components. Since N-acyl ED3A is strongly adsorbed onto many surfaces, and can be completely desolubilized by acidification, it should function well in flotation processes.

22.10 Lubricant Additives

Because N-acyl ED3A has a headgroup which is substantive to metal and also contains a straight chain acyl group, it is reasonable to assume that it might exhibit some lubricant properties. Since the product acts as a corrosion inhibitor and coupling agent it has the potential to act as a multifunctional lubricant additive.

The potential of a 10 % aqueous solution of an amine salt of LED3A to act as a lubricant was assessed by carrying out a Four Ball Wear Test with Coefficient of Friction.

The test was carried out by Petro-Lubricant Test Labs., Inc., 116 Sunset Inn Road, PO Box 300, Lafayette, NJ 07848. The test was run according to ASTM D-4172B (modified), 1200 RPM, 40 Kg Load, 23°C, 1 Hour.

The coefficient of friction rose from 0.075 to about 0.11 after 10 minutes and dropped back to about 0.09 after 30 minutes. Towards the end of the run the coefficient of friction had dropped below 0.08.

Wear Scar = 1.23 mm. Grand average 0.09. Y-0 intercept 0,103.

The control which was deionized water rapidly reached a coefficient of friction of 0.5 and the test had to be terminated after 10 seconds because of extreme stress on the apparatus.

The potential of a 0.1% addition of an amine salt of LED3A to enhance the lubricity of a synthetic motor lubricant, Mobil 1, 10W30, was also assessed using the Four Ball Wear Test With coefficient of friction (see Figure 11). The coefficient of friction rose to about 0.095 with the pure synthetic lubricant after about 5 minutes and remained flat for the remainder of the hour.

Wear Scar = 0.35 mm. Grand average = 0.093. Y-0 intercept = 0.90.

The test sample with added LED3A rose only to 0.08 after 5 minutes and dropped back to 0.07 at 35 minutes and remained at this level for the remainder of the test.

Wear Scar = 0.37. Grand average = 0.075. Y-0 intercept = 0.081.

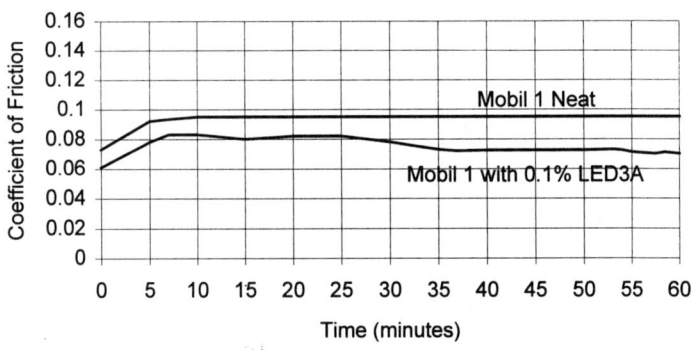

Figure 11 *Four Ball Wear With Coefficient of Friction*

It is evident that N-acyl ED3A salts may be useful for formulation of water based lubricants and can enhance the performance of hydrocarbon lubricants.

References

1. B. Parker., U. S. Patent 5,191,106, assigned to Hampshire Chemical Corp., 1993

2. B. Parker, B. Cullen, and R. Gaudette, U. S. Patent 5,250,728, assigned to Hampshire Chemical Corp., 1993.

3. B. Parker, U. S. Patent 5,449,822, assigned to Hampshire Chemical Corp.

4. B. Parker, U. S. Patent 5,177,243, assigned to Hampshire Chemical Corp.,1993.

5. B. Parker, U. S. Patent 5,191,081, assigned to Hampshire Chemical Corp. 1993.

6. B. Parker and B. Cullen, U. S. Patent 5,284,972, assigned to Hampshire Chemical Corp., 1995.

7. J. Hart and M. DeGeorge, "The Lathering Potential of Surfactants--A Simplified Approach to Measurement", J. Soc. Cosmet. Chem., September/October 1980, 31, 223-236.

8. J. Crudden, J. Lazzaro, and B. Parker, International Patent Application No. PCT/US97/04048.

9. J. J. Crudden, B. Parker, J. Lazzaro, and J. M. Crudden, International Patent Application No. PCT/US97/03961.

10. Hampshire Chemical Corporation Data Sheet, "Analytical Procedure for Calcium Chelation Value", HCT16.

11. J. Crudden and B. Parker, "4th World Surfactants Congress Proceedings", 1996, 3, 52-66.

12. J. Carmichael, W. Degraff, A. Gazdar, J. Minna, and J. Mitchell, "Evaluation of a Tetrazolium-based Semiautomated Colorimetric Assay: Assessment of Chemosensitivity Testing", Cancer Res. 1987,.47, 936-942.

13. M. Perkins, D. Roberts, and R. Osborne, "Human Skin Cell Cultures for In Vitro Skin and Eye Irritancy Assessments of Neat Test Materials", The Toxicologist, 1992, 12, 296.

14. J. Griffith, R. Nixon, R. Bruce, P. Reer, and E. Bannon, "Dose-Response Studies with Chemical Irritants in the Albino Rabbit Eye as a Basis for Selecting Optimum Testing Conditions for Predicting Hazard to the Human Eye", Toxicol. Appl. Pharmacol, 1980, 55, 501-513.

15.   R. Osborne, D. Roberts, M. Perkins, K. Wallace, G. Mun, and R. Curren, "Interlaboratory Validation of a New In Vitro Tissue Equivalent Assay (TEA) for Eye Irritation Assessments", Presented at Society of Toxicology, March 1993.

16.   D. Ptchelintsev, U. S. Patent 5,621,008, N-acyl-ethylene-triacetic acids, assigned to Avon Products, Inc., 1997.

Gemini Surfactants

Milton J. Rosen

SURFACTANT RESEARCH INSTITUTE, BROOKLYN COLLEGE, CITY UNIVERSITY OF NEW YORK, USA

1 INTRODUCTION

There is a considerable, recent patent literature on gemini surfactants, testifying to the intense interest of industrial organizations in these materials. However, since the products described in these patents generally are poorly characterized materials of indefinite composition from which it is impossible to draw valid structure/property relationships, this discussion will be confined to well-defined, well-characterized geminis reported in peer-reviewed scientific papers.

Gemini surfactants are surfactants that have two hydrophilic and two hydrophobic groups in the molecule, in contrast to conventional surfactants that have only one hydrophilic group and (usually) only one hydrophobic group. We first encountered these materials in the course of a project that we were working on, several years ago, for the Dow Chemical Company, in connection with their Dowfax surfactants. The types of chemical structures that are present in the Dow Chemical Company's Dowfax surfactants are shown in Fig. 1. Surfactants of these four types, well purified and well characterized, were synthesized for us in Dow's research laboratories. In investigating the surface properties of these materials,[1] we realized that the DADS (dialkyl diphenylether disulfonate) type of surfactant was much more surface active than we had expected. Its tendency to absorb at an interface and to form micelles was much greater than that of a conventional surfactant of comparable structure (i.e., having a similar [single] hydrophilic headgroup and an equivalent [single] hydrophobic group).

Just about the same time, we noticed that a group of Japanese investigators at Osaka University had started to publish data on surfactants having two hydrophilic and two hydrophobic groups in the molecule [2-6] and their molecules also were much more surface active than expected. Eventually, we joined forces and together published several papers on these surfactants.[7-10] Some diquaternary ammonium gemini surfactants had been synthesized by some other investigators [11-13] a few years earlier but their unusual surface properties appear not to have been commented upon.

We, and other investigators [14-24] have now studied several other homologous series of geminis (surfactants having two hydrophilic groups and two or three hydrophobic groups), anionic, cationic, and nonionic and have found that they indeed have some very unusual and unexpected properties.

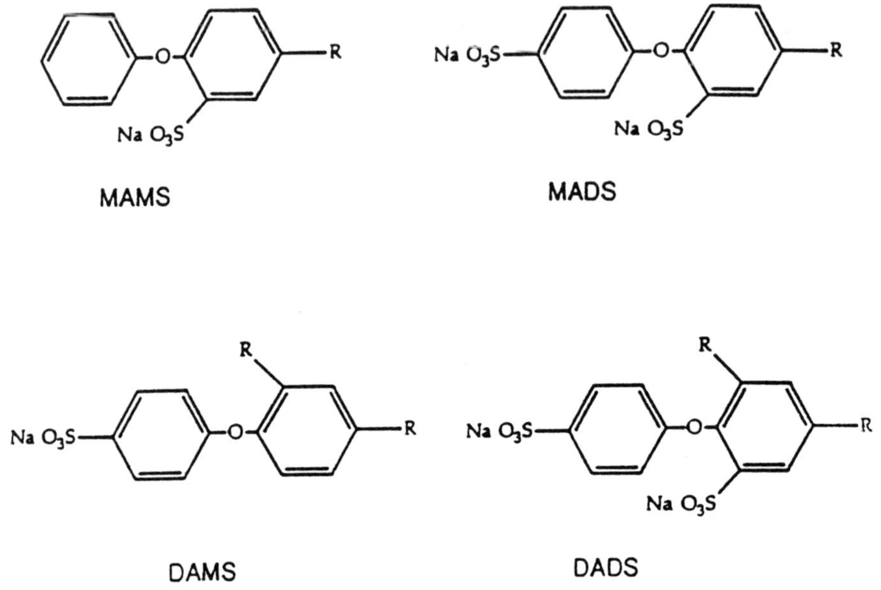

MAMS

MADS

DAMS

DADS

Figure 1 *Dowfax components*

2 PHYSICOCHEMICAL PROPERTIES

Table 1 shows some interfacial properties of some dianionic gemini surfactants, together with those of some conventional surfactants having similar (single) hydrophilic groups and roughly equivalent (single) hydrophobic groups in the molecule. It is apparent from the data that the geminis have cmc values 1 – 2 orders of magnitude smaller and C_{20} values (the molar surfactant concentration in the aqueous phase required to decrease the surface tension of the solvent by 20 dyn/cm, a measure of surfactant efficiency) 2 – 3 orders of magnitude smaller than the comparable conventional surfactants. Note that the surface tension at the cmc (γ_{cmc}) of the geminis is at least as low as, if not lower than, that of the conventional surfactants.

Table 2 shows data for some diquaternary ammonium cationic geminis, $[C_nH_{2n+1}N(CH_3)_2CH_2CHOH]_2^{2+}$. 2Br⁻, synthesized in our laboratory [23,24] or by other investigators, [14-22] together with those of comparable conventional surfactants. Their cmc values are again 1 – 2 orders of magnitude smaller than those of the latter, while their γ_{cmc} values are at least as low. Noteworthy is the effect of the structure of the linkage between the two hydrophilic groups. The cmc values are smallest when the linkage is flexible and hydrophilic (-CH₂CHOHCHOHCH₂-, -CH₂CH₂OCH₂CH₂-), larger when the linkage is flexible and hydrophobic (-CH₂CH₂CH₂CH₂-), and somewhat larger when it is rigid and hydrophobic (-CH₂C₆H₄CH₂-). Even with a rigid hydrophobic linkage, however, the cmc values are still 1 – 2 orders of magnitude smaller than those of the comparable conventional surfactants. For the cationic geminis with polymethylene linkages, it has been found that the cmc increases to a maximum when the number of polymethylene units is about five.[14] This is attributed to looping of the hydrophobic spacer into the interior of the micelle when it reaches sufficient length.

Table 1 *Properties of Some Anionic Gemini-Type Surfactants in Water at 25°C[a]*

Compound	Y	cmc (mM)	γcmc(dyn/cm)	$C_{20}(\pi M)$
A	-OCH$_2$CH$_2$O-	0.013	27.0	0.001
C$_{12}$H$_{25}$SO$_4$Na	---	8.2	39.5	3.1
B	-O-	0.033	28.0	0.008
B	-OCH$_2$CH$_2$O-	0.032	30.0	0.0065
B	-O(CH$_2$CH$_2$O)$_2$	0.060	36.0	0.01
C$_{12}$H$_{25}$SO$_3$Na	---	9.8	39.0	4.4

[a] Ref. 3, 4

Table 2 *Surface Properties of Some Cationic Geminis in Water at 25°C[a]*

Compound	cmc (mM)	γcmc(dyn/cm)	$C_{20}(\pi M)$
(C$_{10}$NMe$_2$CH$_2$CHOH)$_2$$^{2+}$. 2Br$^-$	3.7	35.5	0.58
(C$_{10}$NMe$_2$CH$_2$CH$_2$)$_2$$^{2+}$. 2Br$^-$ [b]	9.0	38.5	---
C$_{10}$NMe$_3$$^+Br^-$	68	---	---
(C$_{12}$NMe$_2$CH$_2$CH$_2$OH)$_2$$^{2+}$.2Br$^-$	0.7	35.4	0.13
(C$_{12}$NMe$_2$CH$_2$CH$_2$)$_2$ O^{2+}. 2Cl$^-$ [c]	0.5	39.2	0.25
(C$_{12}$NMe$_2$CH$_2$CH)$_2$$^{2+}$.2Br$^-$ [b]	1.2	---	---
[(C$_{12}$Nme$_2$CH$_2$)$_2$C$_6$H$_4$]$^{2+}$. 2Br$^-$ (50°C)	1.3	39.5	---
C$_{12}$NMe$_3$$^+Br^-$	16	---	8
(C$_{14}$NMe$_2$CH$_2$CHOH)$_2$$^{2+}$. 2Br$^-$	0.085	36.0	0.0028
C$_{14}$NMe$_3$$^+Br^-$	3.6	---	---

[a] Ref. 23, 24
[b] Ref. 14
[c] Ref. 22 (20°C)

We have found that the packing of the hydrophobic groups in the geminis at the aqueous solution / air interface (i.e., the surface area / chain), when the spacer between the hydrophobic groups is small or hydrophilic, is closer than that found in the comparable conventional surfactants, which may account for the observation that the surface tension at the cmc of the solution if often lower in the former than in the latter.

Why are the gemini surfactants so much more surface active than the comparable conventional surfactants? One would normally expect that when the number of alkyl chain carbon atoms per hydrophilic head group is the same in both conventional surfactants and in geminis, i.e., the hydrophile-lipophile balance in both compounds is the same, that their surface properties would be similar. The answer lies in the number of alkyl chain carbon atoms per molecule. It is well known that an increase in the number of carbon atoms in the alkyl chain of a (conventional) surfactant increases its surface activity, i.e., decreases its cmc and C_{20} values. This is due to the increased distortion of the water structure by the increased length of the alkyl chain, and its distortion of the solvent structure is the basis for the surface activity of the surfactant.[25] However, the increase in chain length also decreases the solubility of the surfactant in water and this limits the increase in the length of the chain (and hence, the increase in surface activity) of the conventional surfactant for use in aqueous systems. With two hydrophilic groups, however, the solubility of the gemini in water is increased greatly, and this permits the gemini molecule to contain many more alkyl chain carbon atoms and still remain water soluble, with the resulting great increase in surface activity.

What does this increased surface activity of the geminis mean for their practical utilization?

1. The higher surface activity means that less surfactant may be needed to perform a function for which surfactants are used, e.g., detergency, or emulsification. This means less raw material needed for the synthesis, less manufacturing by-product to be handled, and less environmental impact of the smaller surfactant quantity used to perform the particular function.

2. The much lower cmc of the gemini means that it may produce less skin irritation, because irritation is usually a function of the concentration of monomeric surfactant in the solution phase, and this decreases with a decrease in the cmc. A very low cmc value also means that such properties as solubilization of water-insoluble material can occur at much lower surfactant concentration with geminis than with comparable conventional surfactants, because solubilization occurs only when the cmc has been exceeded.

3. The tighter packing of the hydrophobic groups of gemini surfactants, compared to conventional surfactants, at the interface results in a more cohesive, more stable interfacial film. This means greater emulsion stability and greater foam stability. We have made stable emulsions with as little as 0.1% gemini emulsifying agent. Liposomes made with gemini surfactants have been found to be more stable than those made with comparable conventional surfactants.[26]

4. The double charge on an ionic gemini means that it should interact more strongly (attractively) with surfactants of neutral and opposite charge than do conventional surfactants. Because such attractive interaction between surfactants is the basis for synergy in their mixtures, this means that stronger synergy can be expected to be exhibited in mixtures of a gemini than in a comparable conventional surfactant with an oppositely charged or neutral surfactant.

Some surface tension data are shown in Figure 2. Values of the interaction parameters (β parameters) for various systems are shown in Table 3.   Then first noteworthy feature is that the mole fractions of approximately 0.5 of the two surfactants, both in the mixed monolayer (X) at the aqueous solution / air interface and in the mixed micelle (X^m) in the aqueous phase, indicate that the two surfactants are interacting in a 1:1 molar ratio, rather than in the 2:1 molar ratio that would be expected of a divalent gemini and a monovalent conventional surfactant. As a result of this 1:1 molar interaction, the complex formed is

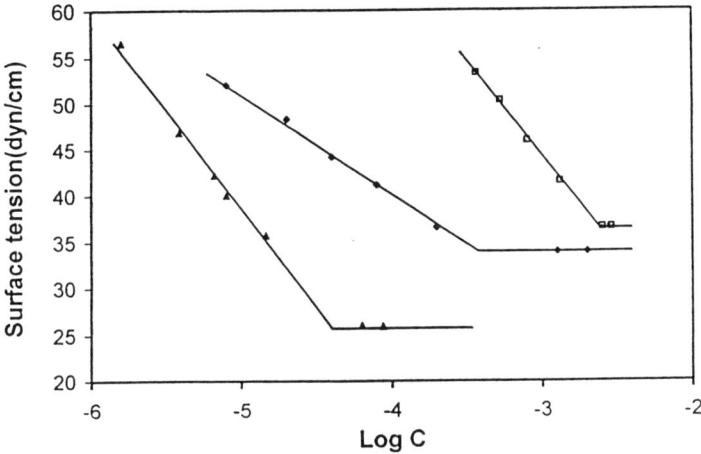

Figure 2 *Synergism in the system*

$(C_{10}NMe_2CH_2CHOH)_2{}^{2+}.2Br^--C_{12}H_{25}SO_3Na$ in 0.1 M NaBr at 25°C.[23] $\square$, $C_{12}H_{25}SO_3Na$;
$\blacklozenge$, $(C_{10}NMe_2CH_2CHOH)_2{}^{2+}.2Br$;
$\blacktriangle$, $(C_{10}NMe_2CH_2CHOH)_2{}^{2+}.2Br^--C_{12}H_{25}SO_3Na(\alpha = 0.67)$. Ref. 30.

Table 3 *Interaction Parameters for the Systems*
$(C_nNMe_2CH_2CHOH)_2{}^{2+}.2Br^--C_nSO_3{}^-Na^+$ at 25°C [a]

System	Medium 0.1 M	α	β^σ	β^m	X	X^m
$C_8Diq*-C_{10}SO_3Na$	NaBr	0.40	-26	-12	0.49	0.51
$C_8Diq*-C_{12}SO_3Na$	NaBr	0.50	-31	-14	0.45	0.45
$C_{10}Diq*-C_{10}SO^3Na$	NaBr	0.02	-34	-14	0.48	0.50
$C_{10}Diq*C_{12}SO_3Na$	NaBr	0.50	-34	-17	0.50	0.55
$C_{10}Diq*-C_{12}SO_3Na$	NaCl	0.22	-40	-19	0.44	0.48

water soluble, because there is a net charge on it. This is in contrast to what occurs in mixtures of conventional anionic and cationic surfactants, in which the 1:1 molar interaction product is a neutral complex that is often water insoluble and precipitates from solution.

The second point of interest in the data is the much weaker interaction between the gemini and the second surfactant in the mixed micelle (β^m), compared with their interaction in their mixed monolayer at the aqueous solution/air interface (β^σ). This is generally observed in gemini-containing mixtures and may be due to the greater difficulty of incorporating the two hydrophobic groups of the geminis into a convex micelle than of accommodating them at a planar interface.

A $-\beta^\sigma$ value larger than the $-\beta^m$ value, for a particular mixture of two surfactants, is one of the requirements for synergism in surface tension reduction effectiveness in that mixture (i.e., when the mixture reaches a lower surface tension value than attainable with either of the two surfactants comprising the mixture). Since geminis generally show much larger β^σ

values than β^m values when mixed with other surfactants, these mixtures often have surface tension values lower than those of the individual components.

3 PERFORMANCE PROPERTIES

Tables 4-6 show the properties of the gemini component (DADS) in a C_{10} linear alkylated diphenylether sulfonate (C_{10}DPE) Dowfax mixture relative to the other components (Fig. 1) and to other surfactants.

3.1 Hydrotropic Properties

The hydrotropic properties of surfactants stem from their ability to inhibit the formation of crystalline or liquid crystalline structures in the aqueous phase. Compounds that have a large hydrophilic portion relative to their hydrophobic portion are often good hydrotropes. Table 4 shows some data for the four types of alkylated diphenylether sulfonates (C_{10}DPE sulfonates) shown in Fig. 1, where R is a linear C_{10} alkyl chain (attached to the aromatic ring at different nonterminal carbon atoms). The data are for the addition of the different diphenylether sulfonates to a pasty, opaque, aqueous dispersion of linear alkylbenzene sulfonate (LAS). In each case, the added diphenylether sulfonate was 10% of the weight of the LAS (on a 100% active basis). As expected, the monoalkylated disulfonate (MADS), which has the largest ratio of hydrophilic:hydrophobic regions, is the best hydrotrope. However the gemini (DADS), with its two hydrophilic groups, is a better hydrotrope than the (conventional-structured) monoakylated monosulfonate (MAMS) or the dialkylated monosulfonate (DAMS).

3.2 Solubilization

To evaluate the solubilization properties of the C_{10} DADS gemini, relative to those of the other Dowfax components, the absorbance in the visible (at 600 nm) of mixtures of these materials with various water-insoluble surfactants at 1 g/l total surfactant concentration was measured. Data are shown in Table 5. In the 1-cm cell in which the absorbances were measured, in absorbance of 0.008 is given by an aqueous solution that is clear to the naked eye with only a trace of translucence; and 0.100 of more, by cloudy solutions.

It can be seen from the data that the DADS gemini, especially at 1% addition to the insoluble surfactant, is a better solubilizer than the other Dowfax components. This has been confirmed in our laboratory with other geminis, compared with comparable conventional surfactants, and is probably due to the very low cmc values of the geminis because, as mentioned above, solubilization commences only after the cmc has been exceeded. Cationic geminis have been shown [19] to solubilize several times more hydrocarbon than conventional surfactants.

3.3 Enhancement of Wetting

The solubilization of certain water-insoluble surfactants in aqueous medium often increases their wetting properties considerably.[27] Table 6 lists some data on this phenomenon. The data are all for 1 g/l total surfactant concentration in water at 25°C, with Draves skein wetting times (WOT) in seconds. Absorbances of the solutions at 600 Nm (in a 1-cm cell), equilibrium surface tension (γ_{eg}) in a dyn/cm and dynamic surface

Table 4 *Hydrotropic Properties of $C_{10}DPE$ Sulfonates* [a]

$C_{10}DPE$ sulfonate	Appearance[b]
None	Viscous, cloudy dispersion
MADS	Clear, flowable at 15°C
DADS	Clear, viscous at 25°C; cloudy at 15°C
MAMS	Viscous, cloudy dispersion
DAMS	Viscous, cloudy dispersion

[a] Ref. 1. Abbreviations as in Figure 1.
[b] Appearance of 23% LAS dispersion upon addition of 2.2% $C_{10}DPE$ sulfonate.

Table 5 *Absorbance at 600 nm of $C_{10}DPE$ Sulfonate Solutions of Water-insoluble Surfactants (conc. 1.0 g/L)* [a]

$C_{10}DPE$ Sulfonate (Conc., g/L)		Sil	H_2O – Insoluble Surfactant [b]	
			AE	APE
None		0.116	Insol.	insol.
DADS	(0.01)	0.020	0.008	0.014
MAMS	(0.01)	0.011	0.023	0.079
DAMS	(0.01)	0.007	0.044	0.145
MADS	(0.01)	---	---	---
DADS	(0.1)	0.008	0.008	0.006
MAMS	(0.1)	0.008	0.007	0.008
MADS	(0.1)	0.025	---	---
DADS	(0.2)	0.005	0.006	0.006
MAMS	(0.2)	0.007	0.007	0.007
MADS	(0.2)	0.017	0.069	0.107

[a] Ref. 1
[b] Sil = silicone-based surfactant; AE = alcohol ethoxylate; APE = alkylphenol ethoxylate.

tension at 1-surface age (γ_{ls}) in dyn/cm, are also listed. It is noteworthy that there is no correlation between the equilibrium surface tension value (γ_{eg}) and the wetting time (WOT). Even when γ_{eg} is very low (26-27 dyn/cm), WOT can be very high if the solution is cloudy. The main correlation is between the surface tension at 1-surface age (γ_{ls}) and WOT.[27]

In the $C_{12}EO_3$ (n-dodecylmonoether of trioxyethyleneglycol) – $C_{12}EO_8$ (n-dodecylmonoether of octaoxyethyleneglycol) system, the replacement of increasingly greater amounts of the water-insoluble $C_{12}EO_3$ by the water-soluble $C_{12}EO_8$ makes the aqueous phase somewhat clearer and decreases the wetting time considerably. However, all WOT values fall between those of the two individuals surfactants by themselves, indicating no synergy. The addition of a conventional sulfonate surfactant, $C_{12}H_{25}SO_3Na$ ($C_{12}SO_3Na$), produces somewhat similar effects, increasing the clarify of the solution somewhat and decreasing the WOT value of the $C_{12}EO_3$ considerably, but here the known stronger interaction of $C_{12}SO_3Na$ than $C_{12}EO_8$ with $C_{12}EO_3$ results in synergism in dynamic surface tension reduction (γ_{ls} for the mixtures is smaller than the γ_{ls} values for the individual surfactants by themselves) and this produces synergism in Draves skein wetting (WOT values for the mixtures are smaller than for the individual surfactants by themselves).

The addition of the gemini disulfonate ($C_{10}DADS$), which because of its double negative charge is expected to interact more strongly with $C_{12}EO_3$ than $C_{12}SO_3Na$, produces stronger synergy in dynamic surface tension (γ_{ls}) and in wetting (WOT) than the addition of the latter. Although the $C_{12}EO_3$ by itself has a WOT of 129 s and the $C_{10}DADS$ by itself has a WOT of 431 s, replacement of 20% of the $C_{12}EO_3$ by $C_{10}DADS$ yields a WOT of only 14.5 s.

Table 6 *Enhancement of Wetting of H_2O-Insoluble Surfactants (Total surfactant, 1g/l at 25°)* [a]

	Surfactant Ratio (w/w)			
System	*1:0*	*0.8 / 02*	*0.5 / 0.5*	*0:1*
$C_{12}EO_3 - C_{12}EO_8$				
Abs (600 nm)	Cloudy	0.082	0.045	Clear
γ_{eq}(dyn/cm)	27.1	27.7	28.1	34.8
γ_{ls}(dyn/cm)	45.9	43.9	35.9	37.3
WOT(s)	129	24.6	14.2	9
$C_{12}EO_3 - C_{12}SO_3Na$				
Abs (600 nm)	Cloudy	0.047	0.047	Clear
γ_{eq}(dyn/cm)	27.1	27.0	26.9	54.5
γ_{ls}(dyn/cm)	45.0	37.9	37.7	56.5
WOT(s)	129	19.3	16.8	28.0
$C_{12}EO_3 - C_{10}DADS$				
Abs (600 nm)	Cloudy	0.031	0.005	Clear
γ_{eq} (dyn/cm)	27.1	27.0	28.5	44.3
γ_{ls} (dyn/cm)	45.0	35.0	39.1	66.8
WOT(s)	129	14.5	17.5	431
Igepal CO-430 – $C_{10}DADS$				
Abs (600 nm)	Cloudy	0.024	0.006	Clear
WOT(s)	114	11.0	10.0	431
$C_{12}P - C_{10}DADS$				
Abs (600 nm)	0.465	0.007	0.006	Clear
γ_{eq}(dyn/cm)	26.6	26.8	29.7	44.3
γ_{ls} (dyn/cm)	68.6	33.1	39.0	66.8
WOT(s)	130	8.6	13.2	431

[a] Source: Ref. 1. Abbreviations: Abs, absorbance; WOT, wetting time. Other abbreviations as in Figure 1.

A similar dramatic decrease in wetting time is observed when $C_{10}DADS$ is added to the commercial nonylphenol ethoxylate (Igepal CO – 430), which is insoluble in water and has a WOT of 114 s.

Replacement of 20% of Igepal by $C_{10}DADS$ results in a wetting time of 11 s, even though the $C_{10}DADS$ by itself is a very poor wetting agent.

An even more dramatic effect is observed upon the replacement of 20% of n-dodecyl pyrrolidone ($C_{12}P$), a water-insoluble nonionic surfactant, by $C_{10}DADS$.   N-alkyl pyrrolidones have been shown [28] to be capable of accepting a proton and interacting as a cationic surfactant in the presence of an anionic surfactant. This interaction is stronger than the interaction of $C_{10}DADS$ with a nonionic surfactant incapable of accepting a proton. As a result of this stronger interaction, the replacement of 20% of the $C_{12}P$ by $C_{10}DADS$ produces a clear solution that has a WOT of 8-9 s.

3.4 Foaming Properties

The foaming properties of gemini surfactants depend, as in conventional surfactants, upon the chain length of the hydrophobic groups and upon the length and nature of the spacer between them. Kim *et al.* [21] have reported that, although conventional alkyl trimethylammonium cationic surfactants show very little foam at 0.1% concentration, C_{12} cationic geminis show very high foam, in some cases better even than that of sodium dodecyl sulfate.

3.5 Other Properties

The investigators at Osaka University have found that the anionic gemini surfactants that they have synthesized have much better lime soap dispensing ability [3-8] and much less sensitivity to Ca^{2+} and Mg^{2+} ions[9] than comparable conventional surfactants. Consequently, these geminis can be used in cleaning and other industrial processes where the presence of hard water or polyvalent ions would be expected to make conventional anionic surfactants ineffective.

Another useful, unique property of gemini surfactants is the viscoelasticity of their aqueous solutions at low concentrations, behaviour not shown by the comparable conventional surfactants. Zana and co-workers [18,21,22] have shown that cationic gemini surfactants with spacers of only two or three methylene groups produce entangled worm-like micelles and show this viscoelasticity at concentrations as low as 2%. Solutions of this type can be useful for building viscosity in formulations.

Diquaternary ammonium geminis prepared from N-dodecyl betaine [29] have been found to be more effective against nineteen gram-positive and gram-negative microorganisms than HTAB (hexadecyltrimethylammonium bromide). A nonionic gemini surfactant based upon dioctanoyl lysine, which was highly surface-active, was found [20] to be non-irritating and non-haemolytic, and consequently suitable for use in personal care and pharmaceutical formulations.

4 CONCLUSIONS REGARDING COMMERCIAL UTILIZATION OF GEMINIS

From the above data, there appear to be a number of situations where the commercial utilization of geminis would be reasonable. If the gemini can be produced at a price not much higher than that of a commercially utilized surfactant, then the gemini may have a cost effectiveness advantage, because less of the gemini may be needed to perform a particular function, due to its higher surface activity, foaming ability, or solubilization capacity. An additional benefit may be the lower skin irritation of the gemini. If the gemini can be produced only at a much higher cost than a commercially utilized surfactant, then the gemini may be used as a low-percentage additive to the conventional surfactant to enhance its properties, thus justifying the added cost.

Acknowledgements

This material is based on work published with my former collaborators, Drs. Xi Yuan Hua, Tao Gao, Letian Liu, Li Dong Song, and Mr Zhen Huo Zhu, and was supported by grants from the Colgate Palmolive Co., Dow Chemical Co., Reckitt and Colman, Rhone-Poulenc Surfactants and Specialties, Texaco Inc., Witco Corp., and the National Science Foundation.

References

1. M.J. Rosen, Z.H. Zhu and X.Y. Hua, *J. Am. Oil Chem. Soc.,* 1992, **69,** 30.
2. M. Okahara, A. Masuyama, Y. Sumida and Y.-P. Zhu, *J. Jpn. Oil Chem. Soc.,* (Yukagaku) 1988 **37,** 746.
3. Y-.P. Zhu, A. Masuyama, and M. Okahara, *J. Am. Oil Chem. Soc.,* 1990, **67,** 459.
4. Y-.P. Zhu, A. Masuyama, T. Nagata, and M. Okahara, *J. Jpn. Oil Chem. Soc.,* (Yukagaku) 1991, **40,** 473.
5. Y-.P. Zhu, A. Masuyama, and M. Okahara, *J. Am. Oil Chem. Soc.,* 1991, **68,** 268.
6. Y-.P. Zhu, A. Masuyama, Y.-I. Kirito, and M. Ohakara, *J. Am. Oil Chem. Soc.,* 1991, **68,** 539.
7. A. Masuyama, T. Hirono, Y.-P. Zhu, and M. Okahara, and M.J. Rosen, *J. Jpn. Oil Chem. Soc.,* 1992, **41,** 301.
8. Y-.P. Zhu, A. Masuyama, Y.-L. Kirito, M. Okahara, and M.J. Rosen, *J. Am. Oil Chem. Soc.,* 1992, **69,** 626.
9. Y-.P. Zhu, A. Masuyama, Y. Kobata, Y. Nakatsuji, M. Okahara, and M.J. Rosen, *J. Colloid Interface Sci.,* 1993, **158,** 40.
10. M.J. Rosen, T. Gao, Y. Nakatsuji, and A. Masuyama, *Colloids and Surfaces A,* 1994, **88,** 1.
11. H.C. Parreira, E.R. Lukenbach, and M.K.O. Lindemann, *J. Am. Oil Chem. Soc.,* 1979, **56,** 1015.
12. F. Devinsky, L. Masarova, and I. Lacko, *J. Colloid Interface Sci.,* 1985, **105,** 235.
13. F. Devinsky, I. Lacko, F. Bittererova, and L. Tomeckova, *J. Colloid Interface Sci.,* 1986, **114,** 314.
14. R. Zana, M. Benrraou, and R. Rueff, *Langmuir,* 1991, **7,** 1072.
15. F.M. Menger and C.A. Littau, *J. Am. Chem. Soc.,* 1991, **113,** 1451; *J. Am. Chem. Soc.,* 1993, **115,** 10083.
16. R. Zana and Y. Talmon, *Nature,* 1993, **362,** 228.
17. J. Seguer, C. Selve, and M.R. Infante, *J. Disp. Sci. Techn.,* 1994, **15,** 591.
18. D. Danino, Y. Talmon and R. Zana, *Langmuir,* 1995, **11,** 1448.
19. Th. Dam, J.B.F.N. Engberts, J. Karthauser, S. Karaborni and N.M. van Os, *Colloids and Surfs. A,* 1996, **118,** 41.

20. M. Macian, J. Seguer, M.R. Infante, C. Selve, and M.P. Vinardell, *Toxicology,* 1996, **106,** 1.
21. T.-S. Kim, T. Kida, Y. Nakatsuji, T. Hirao, and I. Ikeda, *J. Am. Oil Chem. Soc.,* 1996, **73,** 907.
22. T. Okano, N. Egawa, M. Fujiwara,and M. Fukuda, *J. Am. Oil Chem. Soc.,* 1996, **73,** 31.
23. M.J. Rosen, and L. Liu, *J. Am. Oil Chem. Soc.,* 1996, **73,** 885.
24. M.J. Rosen, and L.D. Song, *Langmuir,* 1996, **12,** 1149.
25. M.J. Rosen, "Surfactants and Interfacial Phenomena" 2nd Edn., Wiley, New York, 1989, p. 3.
26. K. Sakamoto, personal communication.
27. M.J. Rosen and Z.H. Zhu, *J. Am. Oil Chem. Soc.,* 1993, **70,** 65.
28. Z.H. Zhu, D. Yang and M.J. Rosen, *J. Am. Oil Chem. Soc.,* 1989, **66,** 998.
29. M. Diz, A. Manresa, A. Pinazo, P. Erra, and M.R. Infante, *J. Chem. Soc., Perkin Trans.,* 1994, **2,** 1871.
30. L. Liu and M.J. Rosen, *J. Colloid Interface Sci.,* 1996, **179,** 454.

Nonionic Surfactants: Achieving the Balance between Performance and Environmental Properties

H. R. Motson

ICI SURFACTANTS, WILTON, MIDDLESBROUGH, CLEVELAND TS90 8JE, UK

1 INTRODUCTION

Nonionic surfactants find use in many applications such as Household, Industrial and Institutional Cleaning, and wool scouring, from which they can be discharged unchanged into the effluent treatment system. Consequently the environmental impact of surfactants has been the focus of attention which has resulted in a number of regulatory and legislative constraints on their use. However the effects, such as wetting, detergency and foam control, which surfactants contribute to formulations are key to their effectiveness. A consequence of the regulatory and legislative pressure is a demand for effective surfactants with improved environmental properties such as biodegradability and aquatic toxicity.

Nonyl phenol ethoxylates and ethylene oxide/propylene oxide co-polymers are specific examples of surfactants which have performed effectively in a wide range of formulations for many years which are now considered to exhibit unsatisfactory environmental properties. As a result their use in a number of important applications, such as cleaning, has been restricted. This has created a demand for surfactants which can be used to replace them. Such replacement surfactants, however, must be at least as efficient as those which they are replacing whilst demonstrating more acceptable environmental properties.

Achieving the balance between performance and environmental properties can be difficult. However with the use of new techniques and a better understanding of structure effect relationships, it is possible to find satisfactory replacements for most applications.

This paper covers some of the issues and developments relating to alternatives for nonyl phenol ethoxylates and ethylene oxide/propylene oxide co-polymers. It describes how a better understanding of the structure/effect relationship and the consideration of parameters such as Phase Inversion Temperature (PIT) can contribute to the selection of more effective alternatives.

2 REPLACEMENT OF NONYL PHENOL ETHOXYLATES

Nonyl phenol ethoxylates are efficient, cost effective versatile products which have been used throughout the surfactants market for over forty years. They meet the primary biodegradability requirements of EC Directive 82/242. However the metabolites resulting from the degradation process do not readily degrade further and may have undesirable side effects on aquatic life. They are also believed to be weak oestrogens.

Nonyl Phenol Ethoxylates : Legislative Position

In 1974 the Commission responsible for the Convention for the prevention of marine pollution from land based sources (PARCOM) was established. In 1992 PARCOM made the following recommendations relating to nonyl phenol ethoxylates.

3 PARCOM RECOMMENDATION 92/8 ON NONYL PHENOL ETHOXYLATES
3.1.1 Considering that nonylphenol ethoxylates are widely used in large tonnages, giving rise to a high degree of exposure to the aquatic environment;

3.1.2 Considering that nonylphenol ethoxylates are degraded to persistent products such as nonylphenols which accumulate in biota;

3.1.3 Considering that the degradation products, nonylphenol and its mono-and diethoxylates, are toxic to aquatic organisms;

3.1.4 Considering, although these substances have been in use for some forty years, the consequences of continuing use are not well known. No environmentally negative effects have been observed, indicating that the situation may not be acute;

3.1.5 Considering satisfactory alternatives are available for the use as cleaning agents (domestic and industrial) and there are promising prospects for alternatives for the remaining uses or applications.

4 CONTRACTING PARTIES TO THE CONVENTION FOR THE PREVENTION OF MARINE POLLUTION FROM LAND-BASED SOURCES AGREE :
4.1.1 to study all uses of nonylphenol ethoxylates and similar substances, which lead to the discharge of these substances to sewer or to surface waters with a view to a reduction of such discharges;

4.1.2 that the use of nonylphenol ethoxylates as cleaning agents for domestic uses be phased out by the year 1995;

4.1.3 that the use of nonylphenol ethoxylates as cleaning agents (for industrial uses) be phased out by the year 2000;

4.1.4 that care shall be exercised to ensure that replacement materials for the current uses of nonylphenol ethoxylates are less damaging to the aquatic environment.

4.1.5 to report on the progress in the implementation of this Recommendation in 1994, 1997 and 2000 and to exchange information on acceptable substitutes.

This recommendation has been ratified by most European States and will be implemented through national or EC legislation.

In 1997 the recommendation was de drafted to and now refers to nonyl phenol and related products which should be taken to include octyl phenol ethoxylates.

5 OESTROGENIC EFFECTS

Some of the Nonyl phenol ethoxylates are believed to be weak oestrogens (several orders of magnitude less so than certain other substances showing endocrine effect). At a recent conference of Government Ministers of countries bordering on the North Sea a decision was made to restrict discharges of all substances showing endocrine effects.

This restriction will be implemented through PARCOM or its successor.

6 USES OF NONYL PHENOL ETHOXYLATES

The structure of nonyl phenol ethoxylates is such that it infers a unique set of properties on the series of products. Properties such as liquidity, alignment at the interface and solubility are all determined by the structure of the products.

This balance of properties is such that nonyl phenol ethoxylates are used in a wide range of applications as shown in the following table.

APPLICATION	EFFECT DELIVERED
Industrial laundry *	Detergent
Metal cleaning *	Wetting, detergency
Hard surface cleaning *	Wetting, detergency
Wool scouring *	Detergency
Traffic film removers *	Wetting
Hand cleaners *	Detergent, rheology modifier
Metal working fluids	Emulsifiers
Emulsion polymerisation	Emulsification
Paints and coatings	Dispersion
Agrochemical formulations	Wetting and emulsification.

* Covered by PARCOM recommendations

In many cases it is possible to find satisfactory replacements for nonyl phenol ethoxylates in such applications. In particular alcohol ethoxylates are finding wide use as replacements exhibiting improved environmental profile (when compared with nonyl phenol ethoxylates) and satisfactory performance properties.

A series of tests were carried out to compare the performance of alcohol ethoxylate and nonyl phenol ethoxylates in tests of particular importance to the I&I cleaning industry.

7 METAL DEGREASING

Products were evaluated at a concentration of 0.2% (w/w aq) on a stainless steel substrate at a temperature of 40 deg C using minimal agitation to minimise grease removal due to mechanical action. The soil used comprised

    15g Stearic acid
    15g Oleic acid
    30g Groundnut Oil
    25g Mineral Oil
    8g Octadecanol

Results are given in the following table

Moles EO	Nonyl phenol	C9-11 Alcohol	C10 alcohol	C13 Alcohol	C13-15 Alcohol	No surfactant
4		7		8		
6	10	96		50	2	0
8	61	96	99	96	97	
9	95	90	98	98	99	
12	92	94	99	94	96	

These data show that the alcohol ethoxylates evaluated demonstrate at least equivalent de-greasing performance to nonyl phenol ethoxylates.

8 HYDROTROPE REQUIREMENT

Nonyl phenol ethoxylates are used in many liquid formulations which can have relatively high electrolyte concentrations due to the use of builders, chelating agents etc in the formulation. In common with most nonionic surfactants nonyl phenol ethoxylates have limited solubility in such solutions and require the use of a hydrotrope to produce a single phase homogeneous formulation. Hydrotropes can have a significant bearing on the cost effectiveness of a formulation hence it is important to minimise the amount used.

Consequently any surfactants considered as replacements for nonyl phenol ethoxylates should not require the use of more hydrotrope for when substituted into a given formulation.

Hydrotrope requirement can be comparatively assessed using the following test.

9 TEST METHOD

Surfactant under test	2%
Electrolyte blend	8%
Water	90%
Hydrotrope	added to give required cloud point

Electrolyte blend is composed of

Sodium metasilicate	30%
Sodium tripolyphosphate	30%
Sodium carbonate	40%

The hydrotrope used in this experiment was sodium cumene sulphonate.

Surfactant, electrolyte blend and water are stirred at constant temperature, (25 deg C) and the hydrotrope titrated into the solution until a clear solution is obtained. The amount of hydrotope added is recorded.

Since it is desirable to minimise the use of hydrotrope less hydrotrope added to give to clear solution means a more cost effective formulation.

Nonyl phenol ethoxylates were compared with a number of alcohol ethoxylates using the method described. The results obtained are summarised in the following table.

Moles EO	Nonyl Phenol	C9-11 Alcohol	C10 Alcohol	C13 Alcohol	C13-15 Alcohol
4	43	>50		>50	>40
6	12.5	9.1		15	25
8	8	4.3	6.7	9.6	41
9	5.4	2.3	2.8	2	3.6
12	0.8	0	0	0	0

The results demonstrate that in general the alcohol ethoxylates tested require similar OR SMALLER AMOUNTS OF HYDROTROPE compared to nonyl phenol ethoxylates.

10   THE IMPORTANCE OF THE PHASE INVERSION TEMPERATURE IN EMULSION STABILITY

A key parameter in defining the emulsification properties of a surfactant is its HLB number. For conventional alcohol ethoxylates the approach of directly calculating the HLB number from the molecular structure has been a reliable guide to emulsifier choice for a given oil. However, for the new specialised alcoxylates currently being developed this method has been shown to be unreliable. To better characterise the emulsification properties of these new surfactant types a method following the pioneering studies of Shinoda[1] has been developed. A key finding in Shinoda's studies was that the phase inversion temperature (PIT) of an emulsion made with a given surfactant is directly related to the **effective** HLB number of the same surfactant. Following this discovery Shinoda suggested the use of the PIT or HLB temperature as an alternative to the calculated HLB number as a descriptor of surfactant emulsification behaviour.

The following example shows how the PIT may be used in this way. In this study three surfactants were compared. The first was a conventional nonyl-phenol ethoxylate (Synperonic NP8) whilst the other two were new specialised alkoxylates. The calculated HLB numbers of the three surfactants used are listed in the Table below

surfactant	calculated HLB number
Synperonic NP8	12.3
Development Product +7EO	12.2
Development Product +9.5AO	13.5

From a comparison of calculated HLB numbers one would expect that Development Product 7AO would be a good direct replacement for Synperonic NP8 whereas Development Product 9.5AO would be too hydrophilic. However, if the PITs of water in decane emulsions made using the Synperonic NP series, a family of linear hydrophobe alcohol ethoxylates or the development products are compared very different behaviour is observed as is shown in the Figure below.

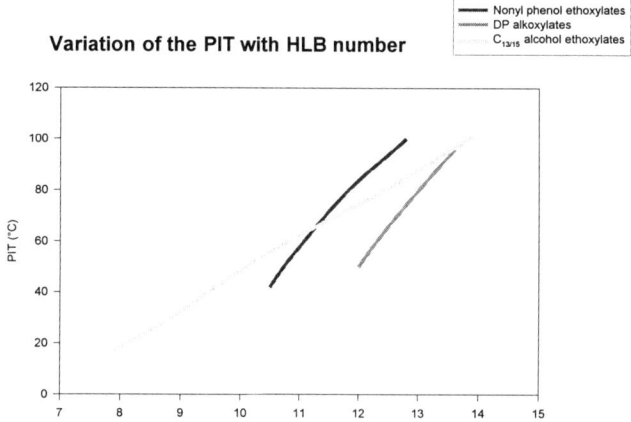

Variation of the PIT with HLB number

In the Figure the measured PITs are plotted against calculated HLB numbers for the surfactants considered. From this data it may be seen that the calculated HLB numbers for the Development Product overestimate the hydrophilic nature of the surfactant in comparison with a nonyl phenol ethoxylate and therefore cannot be used as helpful formulation guides. From the data presented in the Figure it may be seen that surprisingly the Development Product 9.5 AO is a better emulsification countertype to Synperonic NP8 than is Development Product 7AO. This conclusion is found to be true in practice as is shown in the Figure below.

Emulsion stability for decane in water emulsions

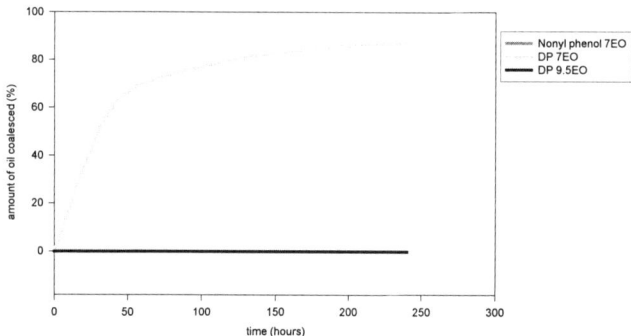

In the Figure the amount of coalesced oil which has separated out from a decane in water emulsion stabilised with the surfactants of interest is plotted against emulsion storage time. As may be seen the stability behaviour of the Synperonic NP8 and Development Product 9.5AO is similar whereas the emulsion made using Development Product 7AO is significantly less stable, in line with the PIT results discussed above.

11 DYNAMIC ADSORPTION RATES AND WETTING

An important physical parameter which determines wetting rates is the dynamic adsorption properties of the surfactant used in the formulation. A measurement which gives information directly about the dynamic adsorption properties of a surfactant is the maximum bubble pressure dynamic surface tension technique. The output from this experiment is a plot of surface tension against surface age from which the dynamic adsorption properties of different surfactants may be compared.

In the Figure below three surfactants of similar molecular weight are compared. This comparison shows how by simple variations in molecular structure significantly faster dynamic adsorption properties and hence better wetting performance may be built into a surfactant molecule.

Effect of molecular structure on DST performance
(surfactant concentration = 2g/l)

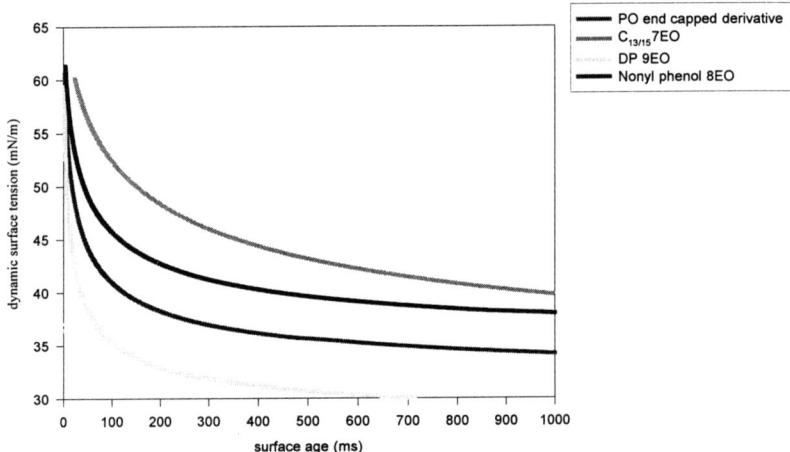

12 METHOD FOR DETERMINING THE PHASE INVERSION TEMPERATURE OF AN EMULSION

Surfactant was dissolved in water to give a 6wt% solution. To 25ml of this solution were added 25ml of decane and the mixture homogenised for two minutes using an Ultra Turrax homogeniser. The resulting emulsion was placed in a jacketed vessel containing a magnetic stirrer and 0.1g of sodium chloride added. A conductivity probe attached to a conductivity meter was then inserted into the emulsion and the temperature raised at a rate of ~2 °C/min. The conductivity was monitored during the course of the experiment as the temperature was increased to 95°C. At the phase inversion temperature a rapid decrease in the value of the emulsion conductivity was noted. The temperature at which this fall occurred was noted as the PIT.

13 PIT/DST EXAMPLES OF NEWER TECHNIQUES PROVIDE MORE EFFECTIVE PRODUCTS

The data presented, along with practical experience, demonstrate that in the majority of cases it is possible to effectively replace nonyl phenol ethoxylates with alcohol ethoxylates, particularly if the appropriate structure/effects relationships are considered.

However the classification of certain alcohol ethoxylates is an increasingly important factor which can influence the choice of alcohol ethoxylate for a given application.

14 CLASSIFICATION: DANGEROUS FOR THE ENVIRONMENT

The criteria for classification as Dangerous for the Environment are based on aquatic toxicity and biodegradability and are as shown below.

Hazard Classification Dangerous For the Environment

7th Amendment to 67/548/EEC (Dangerous Substances Directive)

Aquatic Toxicity LC 50	Ready Biodegradability	Risk Phrase
<1 mg/l	No	R50;R53
<1 mg/l	Yes	R50
1 mg/l< LC50 <10mg/l	No	R51;R53
10mg/l<LC50 <100mg/l	No	R52;R53

R50 Very toxic to aquatic organisms
R51 Toxic to aquatic organisms
R52 Harmful to aquatic organisms
R53 May cause long term adverse effects in the aquatic environment

Classification of Alcohol Ethoxylates

The classification of alcohol ethoxylates is carried out in accordance with published CESIO guidelines[2]. It has been shown that both the length of the alkyl chain and degree of ethoxylation influence aquatic toxicity and biodegradability.   Hence the CESIO guidelines take this into account as follows:

Alkyl chain length	Degree of ethoxylation	Risk Phrase
Linear C9-12	2 to 12	
Linear C12 -15	2 to 12	R 50
Linear C12-15	.>10	
Linear C16-18	all	
Branched C13	3-12	
Branched C13	>12	

The classification tales into account that toxicity and biodegradability are related to structure. However other properties of alcohol ethoxylates are also related to structure. In some cases the influence of structure on environmental properties and performance can be antagonistic. For example the efficiency of soil removal from fabric (detergency) is influenced by the alkyl chain length of the alcohol with optimum performance in the C 12 to C16 range as the following graph illustrates[3] :

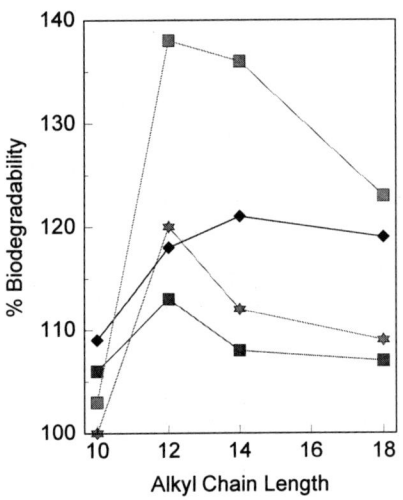

However the aquatic toxicity of alcohol ethoxylates increases with alkyl chain length. As a result the aquatic toxicity of products with the optimum chain length for detergency is such that they are classified as dangerous for the environment as the following data shows[4]

Species/Common Name	Surfactant	LC50(mg/L)	Test Duration Hrs
Salmo gairdneri	C9-11AE 2.5	5-7	96
Rainbow trout	C9-11 AE 5	8-9	96
	C12-15 AE3	1	96
	C12-15 AE9	1.2	96
	C14-15 AE7	0.9	96
	C16-18 AE 14	0.8	96

This classification precludes formulations containing these products from gaining approval under the various eco label schemes which have been introduced in Europe.

Hence there is a requirement for alcohol ethoxylates which have improved environmental performance but this must be achieved without compromising effectiveness.

By careful study of the relationship between structure and performance it some novel effects have been found which has lead to the development of alcohol ethoxylates which have achieved the balance.

The following table details the aquatic toxicity and ready biodegradability of a conventional C13-15 alcohol ethoxylate with that of a newly developed product.

	Daphnia Magna 48hr EC50 mg/l	Rainbow Trout 96hr EC50 mg/l	Algae ES 50 Growth mg/l	Ultimate Biodegradability	
				301F % BOD28/COD	301B % CO$_2$/ThCO$_2$
Development Product	1.1	3.9	1.3	*	*
Development Product	1.4	*	*	*	*
Development Product	18	2.4	1.3	66	99
Synthetic Alcohol Ethoxylates	0.75	1.8	*	57	80
"Natural" Alcohol Ethoxylates	0.75	*	*	58	*

* products not tested

The detergency performance of these products was compared using the following test protocol.

The surfactants under test were incorporated into a liquid detergent containing zeolite as a builder. The detergency of the formulations was determined using a Terg O Tometer at 40 C and 60 C in 100 and 300 ppm hardness water, on standard EMPA and Krefeld cotton and poly cotton cloths. The performance of each formulation was measured by determination of the increase in reflectance of the cloths.

The results are shown in the following graphs.

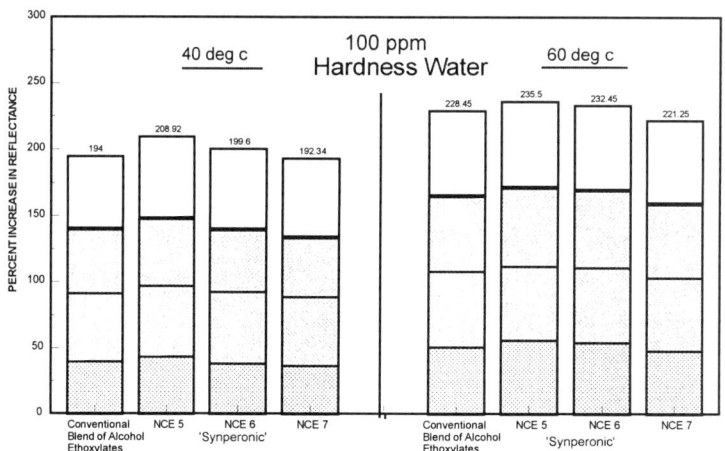

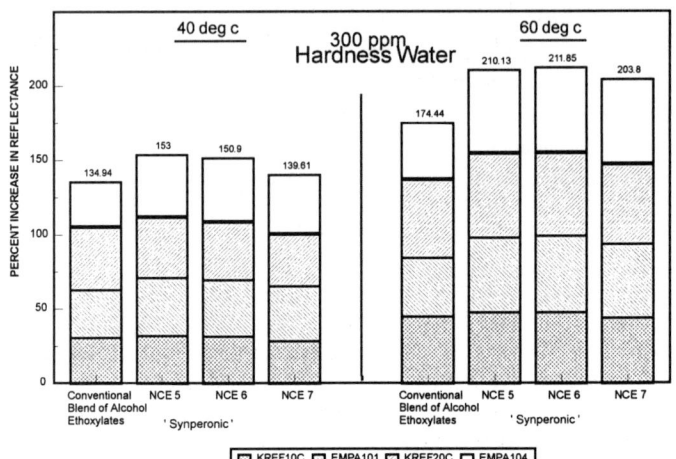

These results demonstrate that the newly developed product has comparable detergency performance compared to the conventional product.

15 POLOXAMER ETHYLENE OXIDE/PROPYLENE OXIDE CO POLYMERS

Another class of nonionic surfactants which have come under pressure because of environmental concerns are the ethylene oxide/propylene oxide block co polymers or poloxamers. The products of major commercial interest have the general formula:

$$H-(OCH_2CH_2)_a-(OCH(CH_3)-CH_2)_b-(OCH_2CH_2)_a-OH$$

These products are highly versatile with a wide range of properties. In particular they are low foaming which equips them for many industrial applications as shown in the following table.

16 SUMMARY OF GENERAL PROPERTIES/APPLICATIONS OF POLOXAMERS

Antifoamers	Emulsifiers
Defoamers	Emulsion Stabilisers
Wetters	ispersants
Rewetters	Demulsifiers
Rinse Aids	Antistatic
Degreasers	Viscosity modifiers
Detergents	Anti dust

During the 1970's and 1980's concerns were raised about the biodegradability of surfactants, particularly in detergent applications. This resulted in the EC Directive 82/242 which set standards for the primary biodegradability of nonionic surfactants when used in as detergents in cleaning applications.

Poloxamers do not meet the primary biodegradability standards set out in Directive 82/242 hence in Europe there use in cleaning formulations was phased out. This meant that replacement products had to be developed which satisfied those biodegradability requirements whilst still delivering the necessary level of performance. This resulted in the development of alcohol alkoxylates which have the general formula:

R - (EO)x(PO)y

By careful variation of the alkyl chain length, degree of ethoxylation and propoxylation it is possible to produce a range of products which satisfy both biodegradability and performance criteria.

This group of products is now widely used in a range of applications in which poloxamers were used, particularly in which low foaming and wetting are important.

However the nature of the products is such that some are toxic to aquatic organisms. Most of the products which provide the most desirable performance properties have LC50 of less than 1 mg/l which means that these products are classified as dangerous for the environment. This in turn causes problems for the formulator eg

16.1 Formulations containing such products will not gain approval under Eco Label schemes.

16.2 Preparations directive will include the category Dangerous For the Environment. Hence formulations containing such products may in turn be classified.

Once again there is a requirement for surfactants with improved environmental properties without any compromise in other performance properties, in this case particularly low foaming and wetting.

With the increasing understanding of the effect of structure of nonionic surfactants on aquatic toxicity and biodegradability[4] it has been possible to develop products which have the required balance between performance and environmental properties. For example the following table summarises the environmental performance of conventional alcohol alkoxylates and alcohol alkoxylates developed to have improved environmental properties. Comparisons are made for products with equivalent physical properties (eg cloud point, wetting and foaming).

17 ENVIRONMENTAL PERFORMANCE

PRODUCT	EC 50 Daphnia mg/l	Primary Biodegradability (Wickbold) %	Ultimate Biodegradability % DOC	Ultimate Biodegradability % CO2/ThCO2
Conventional Product 1	2.7	99	78	84
Development Product 1	0.17-0.25	99	64-87	33-42
Conventional Product 2	0.66-0.69	99	75	92
Development Product 2	6.8	92	62	29

18 PERFORMANCE/PROPERTIES

These products were evaluated in a range of tests relevant to the Household, Industrial and Institutional Cleaning industries and the results are summarised in the following table :

DYNAMIC FOAM HEIGHT
@ 30 C (mm)

PRODUCT	Surface Tension (0.1%w/w aqueous) at 20 C (mN/m)	Cloud Point C	Cotton Wettings Draves	1 min	2 min	8 min
Conventional Product 1	30.8	37		35	35	0
Development Product 1	34	39	28	40	40	0
Conventional Product 2	30	33	10	650		500
Development Product 2	35	36	5	600	660	130

19 CONCLUSIONS

The examples reviewed in this paper have demonstrated some of the issues and problems involved in developing nonionic surfactants with improved environmental properties. However with the improved understanding of the relationship between structure and effect it is possible to achieve the balance between environmental properties and performance.

With this improving understanding further developments are possible particularly with surfactant suppliers and end users working together to develop more effective and environmentally acceptable surfactants.

20 REFERENCES

[1]   Shinoda & Friberg, Emulsion Solubilisation, Wileg, New York, 1986, Chapter 2

[2]   CESIO Guidelines

[3]   Schick - Nonionic Surfactants P796

[4]   Shell Chemical Company 1983, Mayer and Ellersleck 1986

HLB: Is It a Valuable Concept or a Curiosity?

T. G. Balson

DOW EUROPE SA, BACHTOBELSTRASSE 3, 8810 HORGEN, SWITZERLAND

1 INTRODUCTION

Selecting an effective replacement surfactant for a particular application has always been an enigma for both the surfactant supplier and the end user. Occasionally it is essential to replace a particular surfactant whilst at least maintaining the same level of performance in the end application. This change of surfactant may be necessitated by environmental concerns, performance requirements, availability of raw materials or labelling concerns. This can create a dilemma for the end user and/or supplier of the surfactant.

The problem then becomes how to choose an effective replacement surfactant? Is it by trial and error in the laboratory, assuming the method used truly imitates the industrial conditions, or is there another technique which could be used to compare surfactants? Possible comparison methods could be via physical properties of the surfactant such as surface tension, spreading coefficient, cloud point, partition coefficient, interfacial tension, critical micelle concentration, etc., but is it possible to use the hydrophilic lipophilic balance (HLB) value of the surfactant?

2 HYDROPHILIC LIPOPHILIC BALANCE

The semi-empirical concept of HLB was introduced as a pragmatic method of characterising surfactants according to their water or oil solubility, and to their end application such as emulsifiers for oils and water in cosmetics.[1] The HLB value was based upon the weight percentage of the hydrophilic portion of the surfactant compared to the total surfactant. It was arbitrarily derived by dividing the weight percentage of the hydrophile by five, resulting in a dimensionless HLB scale of zero to 20. For nonionic surfactants the ethoxylate block and the polyhydric alcohol initiator (e.g. glycerol, sorbitol) were taken as the hydrophile, although in many articles this has become just the ethoxylate block as being the hydrophile, and many reported HLB values are simply weight percentage polymeric EO divided by 5 (HLB_G). Griffin's perceived usefulness of the HLB concept is that it is a guide for selecting a surfactant for a particular application, especially emulsification. This report will concentrate upon nonionic surfactants because of the apparent ease of calculating their HLB value.

Numerous attempts have been made to relate the HLB value of different nonionic surfactants to a physical property, such as those quoted in the introduction, but all have been unsuccessful.[2,3] Frequently a linear relationship exists for a physical property within a homologous series of a surfactant type, such as nonylphenol ethoxylates (NPE); however a similar but different relationship is shown for other nonionic surfactant series, such as fatty alcohol ethoxylates (FAE), sorbitol esters, polyethylene glycol esters of fatty acids, etc.[3,4,5] Consequently there is no standard graph, which is appropriate for all nonionic surfactants, which relates a physical property to HLB.

The major problem in various recent studies[3,10] seems to be that the HLB values quoted for nonionic surfactants are calculated (nearly always as HLB_G), rather than measured. These recent studies recognise that the nature of the hydrophobe and perhaps even the nature of the hydrophile, as well as other factors such as molecular weight and structure play a role in the HLB value, but are typically disregarded. The following table, where all the products listed have approximately the same 1% aqueous cloud point, illustrates this problem. The products selected are an NPE, two alkoxylate-capped FAE and both a PO and BO block copolymer.

Table 1 *HLB_G of products with the same aqueous cloud points*

Product	Chemistry	1% Aqueous cloud point (°C)	Weight % EO	HLB_G (%EO/5)
1	PO/EO block, 2,500 M. Wt.	30	30	6.0
2	BO/EO block, 1,900 M. Wt.	28-30	55	11.0
3	NPE, 7.5 moles EO	32	60	12.0
4	FAE with PO cap	30	38	7.6
5	FAE with BO cap.	32	56	11.2

Despite the fact that the products are all wetting agents (some low foam) with a similar order of performance, and that all have almost identical cloud points, the HLB_G values, calculated as %EO/5, range from 6 to 12. For nonionic surfactants which use a polypropylene glycol (PPG) as the hydrophobic block or as an end cap, it is much more difficult to determine the hydrophilicity because of the unique dual polarity character of the polymerised propylene oxide (PO) monomer.

To calculate a surfactant's HLB number only requires accurate knowledge of the weight percentage of the hydrophile. For nonionic surfactants such as FAE or NPE that seems to be fairly straightforward, and these products have the potential to be used as standards for determining the HLB number of other nonionic surfactants.

Nonionic surfactants are synonymous with polyglycols and this is the chemistry that will be studied in the remainder of this report, so a brief description of the pertinent details of polyglycol chemistry is included in this report.

3 CHEMISTRY AND POLARITY OF ALKYLENE OXIDES

The chemistry of polyglycols has been covered in many publications.[6,7] Polyglycols are formed by polymerising alkylene oxide(s) onto an initiator, yielding a normal distribution of oligomers around the desired molecular weight.

The initiator is a chemical with an active proton such as fatty alcohols, polyhydric alcohols, thiols, amines, etc. The main alkylene oxides used are ethylene oxide (EO) as the hydrophile and PO as the hydrophobe. Also gaining in importance is butylene oxide (BO) as a hydrophobic block or cap.

If two oxides are added to an initiator they can be added as a:
- Regular block (PO or BO first followed by EO).
- Reverse block (EO first then PO or BO).
- Mixed or Random feed (two oxides together, typically EO and PO)

The polyglycol-based surfactant contains a hydrophilic block, normally from EO but it could be from hydroxyl groups or other water-soluble moieties, and a hydrophobe either from the initiator, such as a fatty alcohol, and/or from an alkylene oxide such as PO or BO.

The character and properties of the polymerised alkylene oxide are derived from the balance between the number of carbon atoms and the ether oxygen. An alkylene oxide in a polymer has the structure $(-C_n-O-)$ where for EO n is 2; for PO n is 3 and for BO n is 4. With PO and BO the carbon chain is not linear, branching occurs at the carbon adjacent to the oxygen.

The hydrophobicity of the group of carbon atoms is balanced against the hydrophilicity of the ether oxygen, which originates from its hydrogen bonding properties. The polymerised alkylene oxide chain is normally characterised as the average of this polar/non-polar balance. Thus polymerised EO is represented as being water soluble, polymerised PO is typically represented as being water dispersible or water-insoluble and polymerised BO as very water insoluble. But practically, the character of polymerised alkylene oxides should be seen as many small alternating units of polar (ether oxygen) and non-polar (two to four carbons) entities combined with the hydroxyl group(s) present in the polyglycol.

Thus an alkylene oxide in a polymer will make its contribution to the total hydrophilicity or hydrophobicity regardless as to where it is in the polymer, i.e. present as a block, random or as a cap. The effect of increasing molecular weight, with the resultant decreasing contribution of the hydroxyl group(s), upon the total hydrophilicity of a polyglycol is shown by the aqueous cloud points of the PPG series in Table 13.

4 PREVIOUS MEASUREMENTS OF HLB

Traditionally the approximate HLB value or range was determined by adding the surfactant to water and observing the appearance of the mixture or solution. This technique bears some relationship to the value of the oil/water partition coefficient.[3] The potential application of surfactants was also based upon these solubility or dispersability observations.[8]

The assessment of the appearance of the surfactant in water should be done, particularly with polyglycol-based nonionic surfactants, at the temperature of the application. As an example, a polyglycol used as a foam control agent for treating an aqueous-based foam at a high temperature, such as in desalination, would be insoluble in water at that high temperature, but at room temperature would be water soluble.

Table 2 *HLB range and appearance in water and potential application*

HLB range	Appearance of mixture or solution	Application of surfactant
0 to 3	Insoluble	None quoted
4 to 6	Unstable dispersion	Water-in-oil emulsifier
7 to 9	Stable dispersion	Wetting agent
10 to 13	Hazy solution	8-13 Oil-in-water emulsifier
14 to 20	Clear solution	13-15 Detergent 15-18 Solubiliser

Although as mentioned earlier there is no standard graph, applicable for all nonionic surfactants, which relates a physical/chemical property to HLB, it was decided to review polyglycol block copolymers and fatty alcohol alkoxylates because these particular nonionic surfactants had been only partially reviewed.[9,10] In 1992, other methods were tried in our laboratory to determine if the HLB value of PO/EO-based surfactants and FAE capped with PO or BO could be measured. All of the methods used were comparative except for the Davies equation, which requires accurate knowledge of the surfactant's chemistry, but the complex polymeric distribution of the alkylene oxides is never considered and only the average molecular structure is used for the calculation.

The difficulty with the comparative method was to have a series of surfactants, with a known HLB value, which could be used as a standard. The obvious choice was a range of FAE or NPE, and eventually the NPE range was chosen as it offered a wider range of ethoxylation levels as well as liquid products. The NPE range was run in the quoted methods and a graph drawn of measured value against HLB_G number. The values for other products were related to this graph to give the comparative HLB number for the particular method used.

4.1 HLB Methods Used

Five methods were used, based upon ease of application and the chances of success, as a first attempt to measure HLB. The methods, with a brief description, were:

4.1.1 Davies Equation. Davies[11] derived his equation from the thermodynamics of micellisation and calculates HLB_D by assigning each functional group a value and then using the equation:

$$HLB_D = 7 + \Sigma hydrophiles + \Sigma hydrophobes. \tag{1}$$

The hydrophilic moieties have positive values and the hydrophobic moieties have negative values. As examples, EO is assigned +0.33 and a hydroxyl group is +1.9 but PO is -0.15, BO is -0.625 and a C-H, C-H$_2$ or C-H$_3$ group is -0.475.
The HLB_G and HLB_D values are shown in the following table for a range of NPE where the number after the letters NP is the moles of EO.

Table 3 *HLB values for NPE via Griffin and Davies*

Product	HLB_G (%EO/5)	HLB_D
NP2	5.7	2.4
NP4	8.9	3.1
NP6	10.9	3.8
NP8	12.3	4.4
NP9	12.9	4.8
NP10	13.3	5.1
NP12	14.1	5.7
NP20	16.0	8.4
NP30	17.1	11.7

The results in Table 3 show that the calculated HLB_D number does not correspond to the standard HLB_G value. The Davies scale is unsuitable for most nonionic surfactants.[12]

4.1.2 Water Number (WN). This is the surfactant's water-solubility factor determined by titrating deionised water into a clear solution of the surfactant (1.0g) in a dioxane/toluene blend (30 ml, 96/4 wt./wt.). The water number is the volume of water required to make the surfactant solution permanently turbid,[13] and is plotted against HLB_G.

4.1.3 Critical Micelle Concentration (CMC). The CMC values for a series of surfactants were derived from surface tension measurements by the incremental addition of surfactant solution, and were plotted against HLB_G.[14]

4.1.4 Thin Layer Chromatography (TLC). The polarity of a surfactant can be determined by chromatography.[10,15] For convenience TLC was chosen with methylethylketone as the mobile phase and silica gel as the stationary phase. The polyglycol surfactants were developed using Dragendorff's reagent. Because the surfactant's oligomers eluted at different times it was decided that the first oligomer would be taken for the retention time ratio (Rf value). The Rf values obtained were plotted against HLB$_G$.

4.1.5 Dielectric Constant (DC). DC is the amount of electrical charge that a substance can withstand at a given electrical field strength. A linear relationship has been shown to exist between HLB$_G$ and log DC for sorbitol esters and polyethylene glycol esters.[16]

4.2 Results of these HLB Determinations

The results are shown in Table 4 and highlight the differences between the various methods.

Table 4 *HLB determinations for the various methods*

Product	HLB$_G$ (%EO/5)	HLB$_D$	HLB$_{WN}$	HLB$_{CMC}$	HLB$_{TLC}$	HLB$_{DC}$
6	2.0	7.7	3.7	n/a	9.2	7.6
7	6.0	11.9	6.3	16.6	13.5	10.8
8	8.0	14.9	9.5	n/a	13.9	13.4
9	3.3	2.6	3.2	n/a	9.2	3.0
10	7.5	3.9	4.5	12.7	13.2	8.3
11	4.5	2.7	1.6	n/a	9.0	n/a
12	11.1	4.8	5.7	11.5	9.8	12.2

Samples 6 to 8 are PO/EO block copolymers with increasing EO content.
Samples 9 to 11 are fatty alcohol alkoxylates with varying amounts of EO, PO.
Sample 12 is an FAE capped with BO.

The true hydrophilicity value of a nonionic surfactant is probably at least equal to the weight percentage of EO in the product. Thus any measured HLB value in the above table which is less than the HLB$_G$ value must be incorrect, which eliminates the Davies and WN methods.
The HLB values from the TLC method fall in a very narrow band (9 to 14) and because the method used the elution of the first oligomer, it appears to be flawed.
The CMC method was limited to water soluble surfactants. The determination of the CMC value was difficult, because of the oligomers present in the polyglycol surfactant, resulting in the graph of surface tension vs. concentration being more of a curve than a distinct step.
The determination of HLB by DC was judged to be the most appropriate value. Unfortunately two different products (samples 10 and 12), although used in the same

low foaming application, gave very different HLB_{DC} values of 8.3 and 12.2 respectively.

The inaccuracy of the above methods for measuring HLB led to an hiatus for a few years. Then, based upon recent work, came a re-evaluation of the available techniques for determining the polarity of a surfactant. The use of chromatography seemed to be the best answer, particularly reverse chromatography.

5 REVERSE CHROMATOGRAPHY

This method uses the nonionic surfactant under study as the stationary phase and a blend of ethanol and alkanes is used as the mobile phase.[10,17] The retention times of ethanol (Rf_e) and the alkane (Rf_a) are deemed to be a measure of the hydrophilicity and hydrophobicity respectively of the surfactant. The ratio of the two retention times (Rf_e/Rf_a) is called the Polarity Index (PI) and is a measure of the total polarity of the surfactant. The use of the PI ratio, rather than individual retention times, helps to reduce any possible errors due to either the preparation of the individual chromatography columns, such as the packing density of the surfactant-coated support, or in the operating conditions of the gas chromatograph.

The PI can then be related to the HLB_G values of the surfactants. This technique has an advantage in that it averages the value of all the oligomers in a surfactant, particularly important for polyglycols, whereas in regular chromatography the oligomers elute at different intervals as per the TLC method detailed earlier.

The chromatography column was prepared by first dissolving the surfactant (0.66 g) in dichloromethane (10 g) in a 100 mL round bottom flask, adding Chromasorb PAW-DMCS 60/80 mesh (3.75 g) to form a slurry, then removing the solvent by rotary evaporation at room temperature. The surfactant-coated support was carefully packed into the column (glass, 100 cm length, 2 mm i.d., 6.35 mm o.d.) which was then assembled in the gas chromatograph, a Hewlett-Packard 5890 II model. The column was conditioned at 80°C, with nitrogen gas flowing at 20 mL/min, for 4 hours. The injection volume was 0.2 µL of a probe mixture containing equal parts by weight of ethanol, *n*-pentane, *n*-hexane, *n*-heptane and *n*-octane. The temperatures used were: injector 150°C, oven 60°C and the flame ionization detector 200°C. Each surfactant was tested in triplicate. The retention times, in minutes, were recorded automatically to three decimal places.

6 RESULTS OF REVERSE CHROMATOGRAPHY EXPERIMENTS

The PI values obtained for the NPE series and for several FAE series were compared to determine which series is the best one for use as a standard to derive the HLB numbers (HLB_{PI}) of other nonionic surfactants.

Many PI values were obtained for a range of polyglycols, and for convenience the results will be presented in various sub-sections, coupled with a discussion of the derived HLB_{PI} numbers.

These sub-sections will compare:
- Products with the same cloud point, but different chemistry.
- Products with the same chemistry (glycerol alkoxylates), but different cloud points due to the position of the oxides in the alkoxylate.
- Nonionic surfactants which have been successfully replaced in industrial applications by other nonionic surfactants.
- The PPG series.

6.1 NPE and FAE Results

The PI values measured for a range of NPE and several ranges of FAE were plotted against their HLB as calculated from the following equation, which is slightly different from HLB_G but closer to the original concept of the total hydrophilicity of a surfactant.

$$HLB_{EO/OH} = (\text{weight \% EO} + \text{weight \% OH}) \div 5 \qquad (2)$$

The PI values measured for the various alkanes and ethanol, used as the mobile phase in reverse chromatography, gave similar curves when plotted against $HLB_{EO/OH}$ for both the FAE and NPE.
The correlation coefficients for the FAE curves were:
n-pentane: 0.939; *n*-hexane: 0.973; *n*-heptane: 0.981; *n*-octane: 0.975.
The results for *n*-heptane were chosen as being representative of the alkane series and having the best correlation coefficient.

Table 5 *Reverse Chromatography Results for NPE using n-heptane*

NPE	$HLB_{EO/OH}$	Rf_e (minutes)	Rf_a (minutes)	PI (Rf_e/Rf_a)
NP2	6.8	2.313	2.071	1.12
NP4	9.7	2.646	1.874	1.41
NP6	11.6	2.492	1.492	1.67
NP7	12.3	2.652	1.478	1.79
NP8	12.9	2.553	1.340	1.91
NP9.5	13.6	2.332	1.224	1.91
NP12	14.6	2.410	1.121	2.15
NP14.5	15.3	2.557	1.164	2.20

All the NPE samples, except one, were from one European manufacturer and form a regular series. The NP9.5 sample was from a USA manufacturer and so may not fit exactly the same nomenclature or specification pattern as the other European NPE.

Table 6 *Reverse Chromatography Results for FAE using n-heptane*

FAE	HLB$_{EO/OH}$	Rf$_e$ (minutes)	Rf$_a$ (minutes)	PI (Rf$_e$/Rf$_a$)
C$_{12-15}$OH 4EO	10.0	2.804	2.566	1.09
C$_{12-15}$OH 6EO	11.9	2.740	2.147	1.28
C$_{12-15}$OH 10EO	14.1	2.816	1.812	1.55
C$_{12-15}$OH 12EO	14.8	2.593	1.508	1.72
C$_{12-15}$OH 14EO	15.3	2.789	1.572	1.77
*C$_{12-15}$OH 6EO	12.1	2.740	2.031	1.35
*C$_{12-15}$OH 6.75EO	13.1	2.968	2.072	1.43
*C$_{12-15}$OH 8.4EO	13.6	2.771	1.818	1.52
*C$_{12-15}$OH 8.5EO	13.7	2.816	1.768	1.59
*C$_{12-15}$OH 10.5EO	14.5	2.883	1.672	1.72
C$_{13/15}$OH 6EO	11.9	2.984	2.343	1.27
C$_{13/15}$OH 7.3EO	12.8	2.718	1.916	1.42
C$_{13/15}$OH 11EO	14.5	2.477	1.473	1.68
C$_{13/15}$OH 13.9EO	15.3	2.663	1.427	1.87
C$_{13/15}$OH 18EO	16.2	2.797	1.340	2.09

* Indicates a C$_{12-15}$ alcohol with a different carbon chain length distribution from the other C$_{12-15}$ alcohol.

The results from Tables 5 and 6 were then prepared as a graph for ease of comparison.

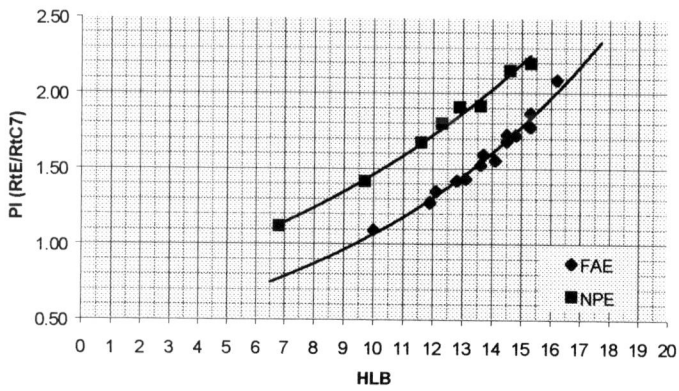

Figure 1 *Graph of HLB$_{EO/OH}$ values vs. PI (Rt$_{ethanol}$/Rt$_{heptane}$) for FAE and NPE*

All the NPE and FAE samples in Figure 1 are polymers and are made to a specification range. It is inevitable that there will be slight variations from the mid point of the specification range.

Figure 1 clearly shows that for the FAE, regardless of the fatty alcohol used (from C_{12} to C_{15}), the results lie on a reasonably well-defined line. The results for the NPE also lie on a similar, but different, line. The pertinent question is, which is the best group, FAE or NPE, to use as an HLB standard?

It is reasonable to expect two products with similar PI values, i.e. the same ratio of total hydrophilicity to total hydrophobicity, to have the same HLB number. But in Figure 1 the FAE and NPE products with the same PI do not have the same HLB, as could be anticipated from previous work.[3,10] To take specific examples, NP4 and $C_{13/15}OH$ with 7.3EO, have the same PI values (1.41 and 1.42 respectively), but have very different $HLB_{EO/OH}$ values (9.7 and 12.8 respectively). The rationalisation of these $HLB_{EO/OH}$ differences is in sections *6.1.1* and *6.1.2*.

6.1.1 Water Solubility of Hydrocarbons. To help explain the difference in PI vs. HLB curves for the NPE and FAE series, a study was made of the solubility of various hydrocarbons in water.[18] The aliphatic hydrocarbon range shown in the graph (Figure 2) is for normal paraffins from methane to decane. It should be noted that the water solubility values for branched paraffins are of a similar order to the normal paraffins. The solubility data quoted for the aromatic hydrocarbon range is for linear alkylbenzenes from benzene to butylbenzene, but again as per the paraffins the various isomers, e.g. xylene or ethylbenzene, have a solubility in water which are of the same order.

This information shows that the aromatic ring is very hydrophilic and makes some contribution to the hydrophilicity of the surfactant. This indicates that the NPE series is not the best one to use as a standard for measuring other HLB values. This is examined in Section *6.1.2* which discusses the PI results of NP9.5 after hydrogenating the aromatic ring to a cyclohexane derivative.

On the other hand, the very limited water solubility of the C_{12} to C_{15} aliphatic carbon chains makes any FAE with this carbon chain length an ideal HLB standard. Thus at least for the C_{12} to C_{15} FAE, the $HLB_{EO/OH}$ value can be regarded as being closest to the true HLB value, assuming that polymeric EO is totally hydrophilic.

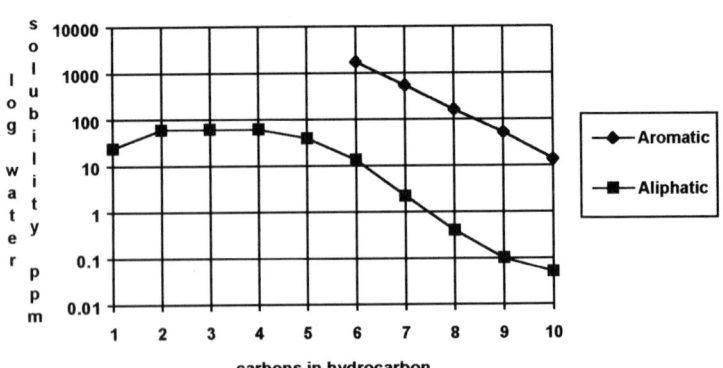

Figure 2 *Water solubility of aliphatic and aromatic hydrocarbons*

6.1.2 PI and HLB$_{PI}$ values of a hydrogenated NPE using FAE as a standard. In order to confirm if the hydrophilicity of the phenyl group was a factor in the determination of the HLB value of an NPE, an NPE sample (NP9.5) was reduced by catalytic hydrogenation to a nonylcyclohexanol ethoxylate (NCH9.5) and the structure was confirmed by NMR.[19] The PI value for NP9.5 was measured as 1.91 which places it on the NPE curve, but for NCH9.5 the PI is 1.61, which places it almost on the FAE curve. This confirms that the hydrophilicity of the aromatic ring is the reason for the differences shown in Figure 1 between the NPE series and the FAE series.

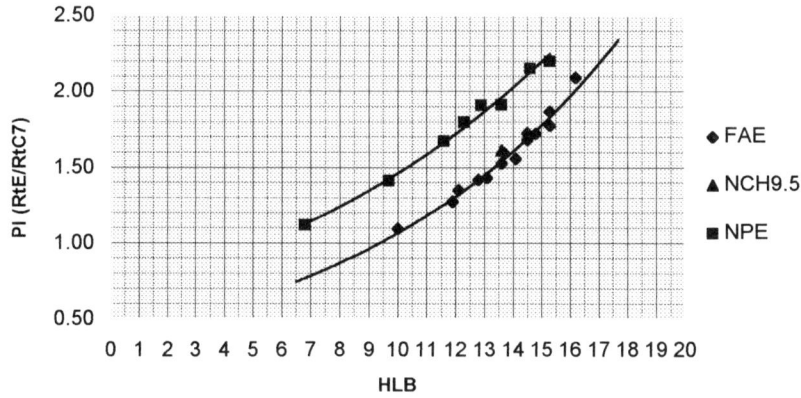

Figure 3 *Figure 1 re-produced to include the value for NCH9.5*

In the following table the concept of "Apparent EO" is introduced which refers to the hydrophilicity of the surfactant's hydrophobe. The difference between HLB$_{EO/OH}$ (hydrophilicity due to EO and OH) and HLB$_{PI}$ (total hydrophilicity) is related to the hydrophilicity of the hydrophobe. This is converted to a %EO equivalent figure by using Griffin's equation (HLB = %EO/5).

Table 7 *Comparison of NP9.5 and its hydrogenated homologue*

Product	Wt % EO	HLB$_{EO/OH}$	HLB$_{PI}$	"Apparent %EO" [5 x (HLB$_{PI}$ - HLB$_{EO/OH}$)]
NP9.5	65.5	13.64	15.7	10.3
NCH9.5	65.1	13.55	14.2	3.3

This "Apparent EO" result is in line with expectations because the water solubility of benzene (1,760 mg/L) is much higher than cyclohexane (56 mg/L) which in turn is much higher than nonane or higher paraffins (<0.2 mg/L). This water solubility data probably explains why the data point for NCH9.5 (PI: 1.61; HLB: 13.55) lies slightly above the FAE line in Figure 3.

The "Apparent EO" result of NCH9.5 also suggests that the nonyl group plays a separate role from the ring structure with respect to hydrophobicity, because nonylcyclohexane must have a very low water solubility (<0.1 mg/L) based upon the above water solubility of nonane and the information that 1,4-dimethylcyclohexane (closest available data to nonylcyclohexane) has a solubility of 3.8 mg/L.
In other words, perhaps the hydrophobicity of an NPE, and maybe its performance as a surfactant, may be due to the nonyl group and the phenyl group acting separately, rather than acting as nonylphenyl, a single moiety.

  6.1.3 Universal HLB equation from FAE data. The data in Table 6 (Figure 1) of measured PI values against the calculated $HLB_{EO/OH}$ number for all the FAE series was used to establish the following universal equation for nonionic surfactants relating PI to HLB (termed HLB_{PI}).

$$HLB_{PI} = [\ln(PI_{heptane}) - \ln(0.3835)]/0.1022 \qquad (3)$$

The HLB_{PI} value can potentially be used in two different ways:
1. To calculate the hydrophilicity of the nonionic surfactant's hydrophobe.
2. To replace one surfactant in an application by another which has the same HLB_{PI} value.

6.1.4 HLB_{PI} values of NPE using the Universal HLB equation. The PI values for the NPE series were substituted in Equation 3 to give their HLB_{PI} number as detailed in the following Table. The HLB_{PI} numbers were used to determine the apparent hydrophilicity of the hydrophobe.
Taking into account the increase in weight percentage of ethoxylation and the decrease in the hydroxyl contribution, there is a consistency in the "Apparent %EO" value in the contribution of the nonylphenyl hydrophobe to the overall hydrophilicity of the NPE.

Table 8 *HLB values of NPE and the calculated hydrophilicity of the hydrophobe*

Product	Wt % EO	$HLB_{EO/OH}$	HLB_{PI}	"Apparent %EO" $[5 \times (HLB_{PI} - HLB_{EO/OH})]$
NP2	29	6.8	10.5	18.5
NP4	44	9.7	12.8	15.5
NP6	55	11.6	14.4	14
NP7	58	12.3	15.1	14
NP8	62	12.9	15.7	14
NP12	71	14.6	16.9	11.5
NP14.5	74	15.3	17.1	9

6.2 Products With The Same Cloud Point, But Different Chemistry

The products are those detailed in Table 1, with the exception of the BO, EO copolymer. Table 1 is now partially reproduced with the HLB_{PI} values obtained from Equation 3, including the calculation of the hydrophilicity of the surfactant contributed by the hydrophobe.

Table 9 *Reassessment of HLB values from Table 1 using HLB_{PI}*

Product	Wt % EO	HLB_G (%EO/5)	HLB_{PI}	"Apparent %EO" [5 x (HLB_{PI} - HLB_G)]
1	30	6.0	13.8	39
3	60	12.0	15.4	17
4	38	7.6	9.8	11
5	56	11.2	11.4	1

Product 1 is a PO/EO block copolymer.
Product 3 is NP7.5.
Products 4 and 5 are both FAE with a PO and BO cap respectively.

This indicates that the PO block, whether used as the major hydrophobe or as a cap to reduce foaming, makes a major contribution to the hydrophilicity of the surfactant. Conversely BO makes almost no contribution (1% "Apparent %EO" vs. 39% for PO).
It also shows that despite the products having similar cloud points, and the same order of performance in similar applications, the HLB_{PI} values are quite different. These results suggest that the hydrophilicity of the hydrophobe is not a factor in relation to the surfactant's performance.

6.3 Glycerol Alkoxylates With Different Cloud Points

The previous group of products all had the common feature of similar cloud points. This group of three products employs exactly the same chemistry, glycerol reacted with a uniform amount of EO and PO, the only difference being that they are structurally different, having been prepared as a regular block, reverse block and random feed.
The products are used in different applications, the blocks being surfactants (e.g. wetting agents, foam control agents) and the mixed feed is a lubricant.
The following Table 10 shows the products with their respective 1% aqueous cloud points, Rf values, PI ratio, HLB_{PI} numbers and Ross-Miles foaming test data (foam height immediately after 200 mL of a 0.1% solution of the chemical has fallen 100 cm into 50 mL of the same solution, and then the foam height after 5 minutes).

Table 10 *HLB$_{PI}$ results from glycerol alkoxylates having the same composition*

Product	1% Aqueous cloud pt. (°C)	Rf$_e$	Rf$_a$	PI	HLB$_{PI}$	Ross-Miles foam test
Regular	42	2.409	1.361	1.77	15.0	23mm, 0mm
Reverse	47	2.654	1.476	1.80	15.1	30mm, 2mm
Mixed feed	55	2.480	1.368	1.81	15.2	1mm, 0mm

The regular and reverse block copolymers have typical surfactant structures, as indicated by the Ross-Miles foaming test. The mixed feed copolymer, with its alternating EO/PO polymeric oxide structure, does not have a surfactant structure, as shown by the lack of initial foam in the Ross-Miles foaming test. The HLB$_{PI}$ results indicate that the total polarity of all the three products is the same. Clearly the HLB$_{PI}$ number by itself does not indicate if a product is a surfactant, it needs to be judged together with the structure of the polyglycol. The results in Table 10 and those in Table 9 confirm the conclusion that all the alkylene oxide units within a polyglycol contribute to the polarity of the total molecule, albeit very slightly for the hydrophobic BO, regardless of their position within the molecule.

6.4 Nonionic Surfactants Replaced In Industrial Applications

1. An NPE was used as an emulsifier for an oil-in-water system, but due to legislation the NP5.5 had to be replaced. Extensive work, based upon trial and error, led to a replacement product related to FAE chemistry. The following Table 11 details the HLB results.

Table 11 *Comparison of HLB values of NPE and its replacement surfactant*

Product	HLB$_G$	HLB$_{PI}$
NP5.5	10.5	14.0
FAE derivative	8.7	11.2

2. In another example an alkylphenol alkoxylate (APA), used as an emulsifier for an oil-in-water system, was replaced due to potential toxicological concerns. Extensive work led to a polyglycol block copolymer being successfully used as a replacement emulsifier

Table 12 *Comparison of HLB values of an APA and its replacement surfactant*

Product	HLB$_G$	HLB$_{PI}$
APA	6.7	11.4
Block copolymer	8.0	15.2

In both cases neither of the HLB numbers (HLB_G or HLB_{PI}) of the two new surfactants show any similarity to those of the replaced surfactants. Thus in this instance an effective replacement surfactant could not be obtained by the HLB technique.

6.5 Polypropylene glycols

PPG are normally thought of as products used in non-surfactant applications such as lubricants, reactants, co-solvents, heat transfer fluids, etc. However PPG are used in some applications, e.g. foam control agents, where they may be perceived as having surfactant characteristics. This would suggest that the PPG structure contains a hydrophobe and a hydrophile. The 1% aqueous cloud points of the PPG series are shown to indicate the hydrophilic character of the products. For the PPG range the HLB_G number is irrelevant (all being zero) and this is replaced in Table 13 by $HLB_{EO/OH}$ (which is %OH/5). The "Apparent %EO" figure is also calculated for the PPG range. The molecular weight of the PPG is indicated by its name.

Table 13 *HLB values and hydrophilicity of PPG series*

Product	1% Aqueous cloud pt. (°C)	$HLB_{EO/OH}$	HLB_{PI}	"Apparent %EO" [5 x (HLB_{PI} - $HLB_{EO/OH}$)]
P-400	soluble	1.7	16.9	76
P-1000	25	0.7	11.5	54
P-2000	15	0.3	9.4	46
P-3000	12	0.22	9.1	44
P-4000	9	0.17	8.2	40

The hydrophilicity of both PO and the PPG series is shown by the cloud point data as well as the "Apparent %EO" figure. The hydrophilicity of the two OH groups plays a major role in determining the nature of the lower molecular weight PPG. The higher molecular weight members show consistency in their "Apparent %EO" value, this is probably due to the decreasing weight percentage, and hence influence of the hydroxyl groups on the polarity. The "Apparent %EO" range of 40 to 46% is comparable with the 39% of the PO/EO block copolymer in Table 9.

6.6 Review of "Apparent %EO" calculations

From the various values obtained by calculating "Apparent %EO", such as for PO, BO, nonylphenyl group, etc., it becomes clear that the hydrophilicity of a nonionic surfactant cannot be expressed by just reviewing the EO block. The total hydrophilicity of a nonionic surfactant is due to many factors and is best expressed by Equation 4, where *f* represents the factorial influence (from zero to 100%) that the particular chemical species plays in creating hydrophilicity.

Hydrophilicity = fOH + fEO + fPO + fBO + $f$hydrocarbon/initiator +
$\qquad\qquad\qquad$ $f$miscellaneous (e.g. hetero atoms, unsaturation).   (4)

But even this equation may be an oversimplification because the hydrocarbon (e.g. nonylphenyl) may act as two moieties rather than one. In reviewing the contribution to hydrophilicity of the alkylene oxides, it is clear that BO has a minimum influence, whilst PO has a major contribution, but does EO have a 100% contribution? What is the hydrophobicity of the two methylene groups in polymeric EO?
The hydroxyl group plays a much larger role than EO in creating hydrophilicity as shown by the PPG series in Table 13.
Ironically this concept of hydrophilicity as defined in Equation 4 brings us partially back to Davies and the contribution of the different species. The difference being that reverse chromatography is a technique which measures polarity, via standards, rather than just calculating it as per Davies.

7 CONCLUSIONS

Previous studies of HLB have generally concentrated upon the hydrophilicity of the surfactant's obvious hydrophile, such as an ethoxylate block, for calculating the HLB number. The conclusion was reached in these previous studies that various series of surfactants could be related to HLB_G but only on an intra series basis. There was no inter series relationship between HLB_G and a physical or chemical property of the different surfactants. The hydrophilicity within the hydrophobe was not considered, although it was recognised by some studies.[10]
$\qquad$ The very low water solubility of the carbon chain (at least a C_{12} unit) within an FAE makes it the best standard surfactant for determining HLB values. The NPE is less suitable due the solubility of the hydrophobe.
$\qquad$ The use of reverse chromatography, which measures the total hydrophilicity of the surfactant, has allowed a means of indirectly measuring HLB via the PI value. Using the values obtained from different FAE, a universal surfactant equation relating PI to HLB was derived. This equation was used to relate PI values for other surfactants to their HLB_{PI}.
$\qquad$ The HLB_{PI} is derived from the total hydrophilicity of the surfactant and can be used to determine the hydrophilicity of the hydrophobe. This shows that a PPG block contributes about 40% hydrophilicity, and that this inherent combination of hydrophobicity/hydrophilicity contributes to the overall performance of the PO/EO block copolymer surfactants and to the PPG products. A polybutylene glycol block only contributes about 1% to the hydrophilicity.
$\qquad$ The nonylphenyl group typically contributes about 15% hydrophilicity. This result coupled with data from the hydrogenated NPE, may explain that the hydrophobicity of an NPE could come from both a nonyl moiety and a phenyl moiety, rather than from the nonylphenyl group, which may help explain the good performance of the NPE series.
$\qquad$ Nonionic surfactants with the same 1% aqueous cloud point did not have the same HLB_{PI} or HLB_G number.

The HLB$_{Pl}$ numbers for the three chemically-identical, but structurally-different, glycerol alkoxylates are the same, so HLB$_{Pl}$ does not distinguish between surfactant structures and non-surfactant structures. Thus to be useful the chemical structure has to be known as well as the HLB$_{Pl}$ number.

Products that were successfully introduced into applications had different HLB$_{Pl}$ numbers from the products they replaced. Thus HLB$_{Pl}$ cannot be used to determine replacement surfactants.

The total hydrophilicity of a nonionic surfactant is due to many different moieties in the molecule, including such unexpected species as the hydrocarbon backbone and BO. In general the hydrophilicity of the hydrophobe does not seem to be related to the surfactant's overall performance.

The alkylene oxides provide their particular level of hydrophilicity regardless of where they are in the molecule, such as block, cap or random.

Unless there is an intrinsic error in the determination of HLB by reverse chromatography with the use of FAE as standards, then the answer is that HLB remains a scientific curiosity rather than a useful concept.

Acknowledgements

The author wishes to thank Dow Europe S.A. for permission to publish this paper and his colleague, Dr. Mark Nace, for many useful discussions.

References

1. W. C. Griffin, *J. Soc. Cosmetic Chem.* 1949, **1**, 311-326.
2. P. Becher, *J. Dispersion Sci. Technol.* 1984, **5**, 81-96.
3. H. Schott, *J. Pharm. Sci.* 1995, **84**, 1215-1222.
4. H. Schott, *J. Pharm. Sci.* 1984, **73**, 790-792.
5. J. Broniarz; M. Wisniewski and J. Szymanovski, *Tenside Detergents* 1974, **11**, 27-30.
6. R. H. Whitmarsh, 'Nonionic Surfactants', Ed. V. M. Nace, Marcel Dekker: New York, 1996; **60**, Chapter 1.
7. T. G. Balson, 'Chemicals in the Oil Industry', Ed. L. Cookson, P. H. Ogden, The Royal Society of Chemistry: Cambridge, 1998, pp 71-79.
8. P. Becher, M. J. Schick, 'Nonionic Surfactants', Ed. J. S. Schick, Marcel Dekker, New York, 1987, **23**, Chapter 8.
9. J. Szymanovski, *Fette Seifen Anstrichmittel*, 1982, **84**, 245-248.
10. V. M. Nace, 'Nonionic Surfactants', Ed. V. M. Nace, Marcel Dekker, New York, 1996, **60**, Chapter 4.
11. J. T. Davies, *Proc. Int. Congr. Surface Activity, 2nd*, Butterworths, London, 1957.
12. H. Schott, *J. Colloid Interface Sci.* 1989, **133**, 527-529.
13. H. L. Greenwald, G. L. Brown and M. N. Fineman, *Analy. Chem.*, 1956, **28**, 1695-1697.
14. H. Schott, *J. Pharm. Sci.* 1969, **58**, 1131-1133.
15. V. Huebner, *Analy. Chem.*, 1962, **34**, 488-491.

16. W. G. Gorman and G. D. Hall, *J. Pharm. Sci.* 1963, **52**, 442-446.

17. P. Becher and R. Birkmeier, *JAOCS,* 1964, **41**, 169-172.

18. C. L. Yaws, H-C. Yang, J. R. Hopper and K. C. Hansen, *Chemical Engineering,* April 1990, Hydrocarbons: water solubility data, 177-180.

19. K. A. Burdett, The Dow Chemical Company, Private communication, January 1998.

Surfactants Derived from Secondary Alcohols

Kenji Rakutani*, Yoshiyuki Onda and Toru Inaoka

OSAKA OFFICE, NIPPON SHOKUBAI CO., LTD., KOGIN BLDG., 4-1-1 KORAIBASHI, CHUO-KU
OSAKA, 541-0043, JAPAN

1 Introduction

Since 1972, Nippon Shokubai has been producing higher secondary alcohols and their ethoxylates, or SAEs, by liquid phase air-oxidation of normal paraffins. These products have been widely applied in the manufacture household detergents and for industrial uses as highly safe and environmentally-friendly surfactants with unique characteristics.

In this report, the properties and applications of SAEs are described and compared with those of primary alcohol ethoxylates (PAEs) and nonylphenol ethoxylates (NPEs), generally known as nonionic surfactants. A new SAE manufacturing process, recently developed by Nippon Shokubai, is also described. SAEs produced by Nippon Shokubai were given the brand name SOFTANOL. The production capacity is 27,000 Mt/y as 3-mole ethoxylate (SOFTANOL 30). The production site is in Kawasaki, Japan.

1.1 Manufacturing Technology

The manufacturing process for secondary alcohols has been described in several reports (1-6).

A mixture of secondary alcohols is obtained through several processes, such as, liquid phase air-oxidation of normal paraffins in the presence of a boric acid catalyst, the recovery of normal paraffins, and the refining of alcohols (Figure 1).

Addition of ethylene oxide to secondary alcohols is difficult in the presence of an ordinary alkali catalyst. Thus, ethylene oxide is added to secondary alcohols in the presence of an acid catalyst, and then the unreacted alcohols are stripped and recovered by distillation to obtain the 3-mole ethoxylate (SAE (3EO)). These processes are continuous up to the final product of SAE (3EO).

Ethoxylates with the mole number of addition higher than 3 are obtained by addition of ethylene oxide in the presence of a base catalyst. This process is used as a continuous process or as a batch-wise process depending on the end products desired.

1.2 Chemical Charactetristics

Figure 2 shows the general structural formula of SAE.
Table 1 lists the characteristics of SAEs from a structural point of view.

SAEs consist of a linear alkyl chain to which polyoxyethylene chain can be added at random positions.   The carbon number of the alkyl chain is in the range of 12-14. SAEs contain almost no unreacted alcohols and few ethoxylates with a low mole number of addition. Thus, they show a narrow EO adduct distribution.

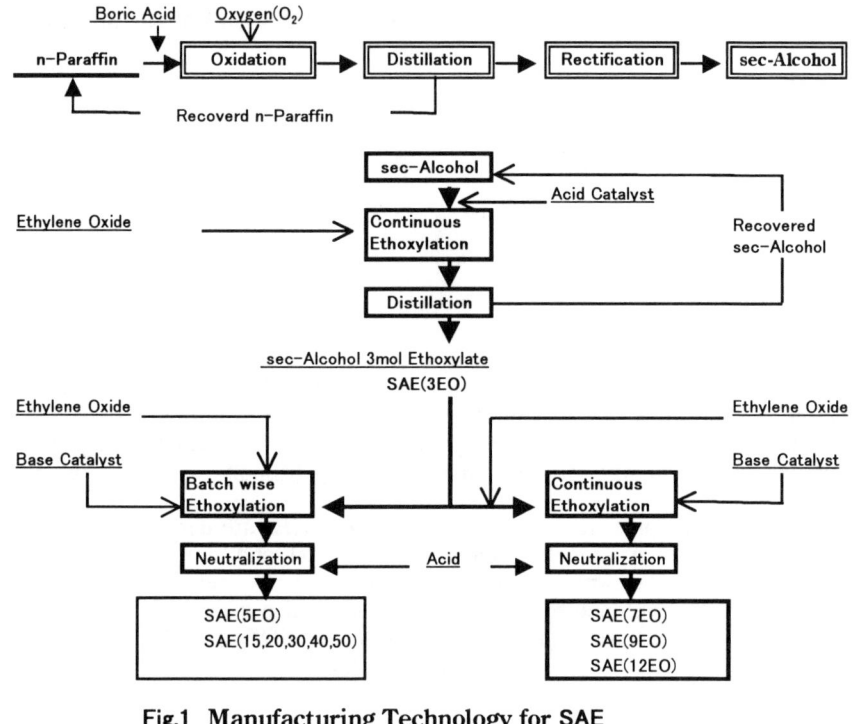

Fig.1 Manufacturing Technology for SAE

$$CH_3-(CH_2)m-CH-(CH_2)n-CH_3$$
$$|$$
$$O(CH_2CH_2O)xH$$

$$m+n=9\sim11$$
$$x=3,5,7,9,12...$$

Fig. 2 Chemical Structure of SAE

Table 1 Chemical Characteristics of SAE

- Linear alkyl chain
- 12-14 alkyl carbon range
- Non free Alcohol
     Odor less
     Low toxicity and low irritation
- Narrow   EO adduct distribution

Figure 3 shows the distribution of EO units in SAE (3EO) and PAE (3EO). Concentrations (% by weight) are plotted as a function of the EO mole number. SAE (3EO) contains no free alcohols and exhibits a narrower distribution than PAE (3EO).

Figure 4 shows the distribution of EO units in SAE (9EO) and PAE (9EO). Similarlly to SAE (3EO) and PAE (3EO) described above, SAE (9EO) contains no free alcohols and exhibits a narrower distribution of EO units compared to PAE (9EO).

2 Physical Properties

Table 2 lists the physical properties of SAEs characteristically associated with their structure. Their physical properties of note are high fluidity, low viscosity, low pour point and narrow (or no) gel range. SAEs contain no free alcohol, and thus have no characteristic alcoholic odor; they also have a low irritation index for eyes or skin.

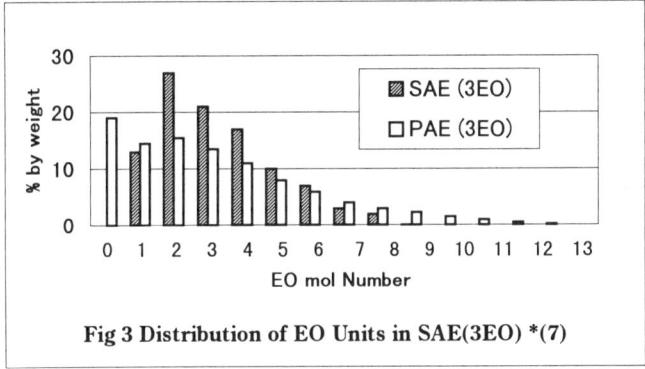

Fig 3 Distribution of EO Units in SAE(3EO) *(7)

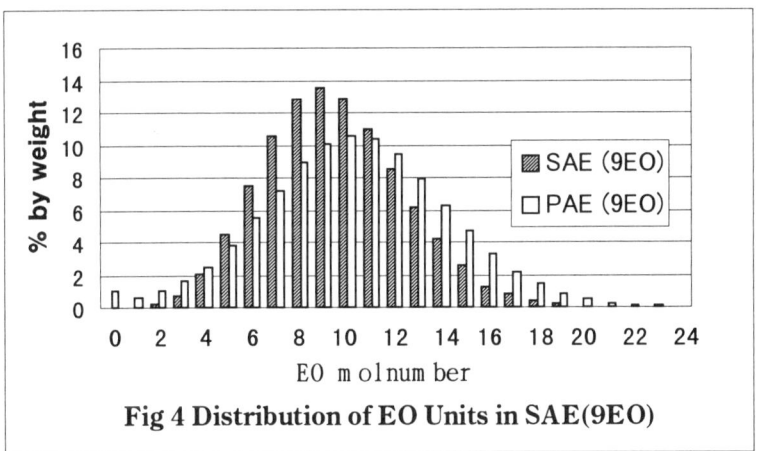

Fig 4 Distribution of EO Units in SAE(9EO)

Table 2   PHYSICAL PROPERTIES
- High Liquidity
- Low viscosity
- Low pour   point
- Narrow (or no) gel range
- Odorless
  No characteristic alcoholic odor
  because   of containing no free alcohol

2.1 Viscosity and Pour Point

Figure 5 shows the relationship between temperature and surfactant viscosity. Viscosity is plotted as a function of temperature. The viscosity of PAE increases significantly at temperatures below 25°C, resulting in a gel or solid state. Compared with PAE, SAE is still a liquid even at 10°C and is therefore much easier to handle. NPE has a higher viscosity than SAE at any temperature.

Figure 6 shows the pour points for various nonionic surfactants. Pour points are plotted as a function of Griffin's HLB value. At any HLB value SAE has a lower pour point than PAE. A similar behavior is observed between SAE and NPE, although the tendency is reversed at HLB values below 10, resulting in a lower pour point for SAE.

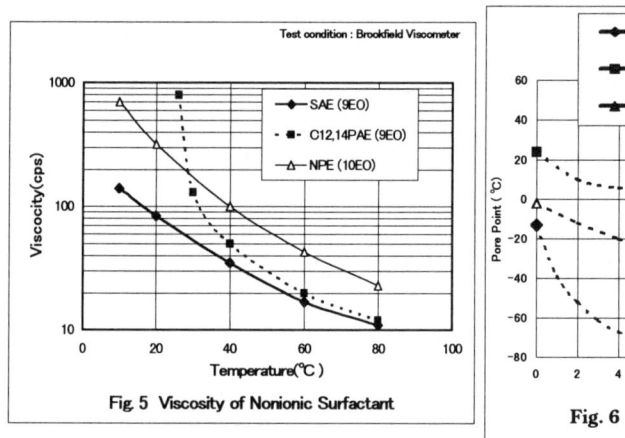

Fig. 5 Viscosity of Nonionic Surfactant

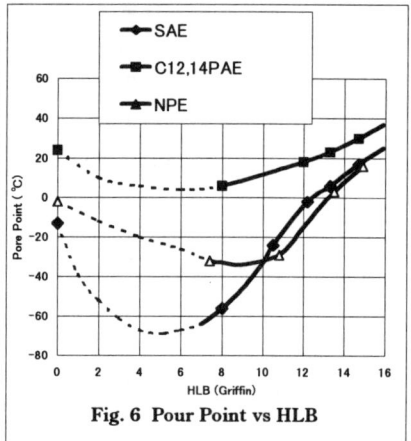

Fig. 6 Pour Point vs HLB

2.2 Gel Range

Figure 7 shows viscosity of Aqueous Solutions. The viscosities of some aqueous surfactants are plotted as a function of surfactant concentration.

PAE and NPE have viscosities as high as 10,000 cps or more in the concentration range of 40-70%, resulting in gel formation. On the contrary, in the same concentration range SAE has lower viscosity without gel formation, and thus is easier to handle. This result means that with SAE there is a very high degree of freedom in formulating liquid detergents and the solubilizing agents or viscosity control agents are not necessitated.

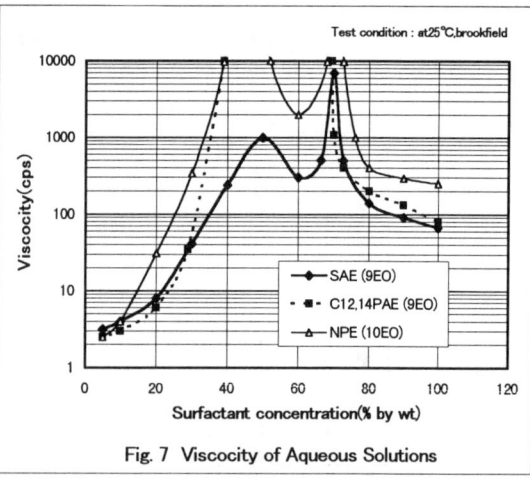

Fig. 7 Viscocity of Aqueous Solutions

Figure 8 shows the range of gel formation when Dodecyl benzene sulfonate (LAS) and water are mixed with nonionic surfactants, SAE (9EO), PAE (9EO) or NPE (10EO). As this figure shows, SAE has almost no gel range in a variety of compositions, while both PAE and NPE have wide gel ranges.

Due to such properties, SAE, when compared with both PAE and NPE, allows a higher degree of freedom in formulating liquid detergents and is suitable for liquid detergent formulations containing a high concentration of surfactant.

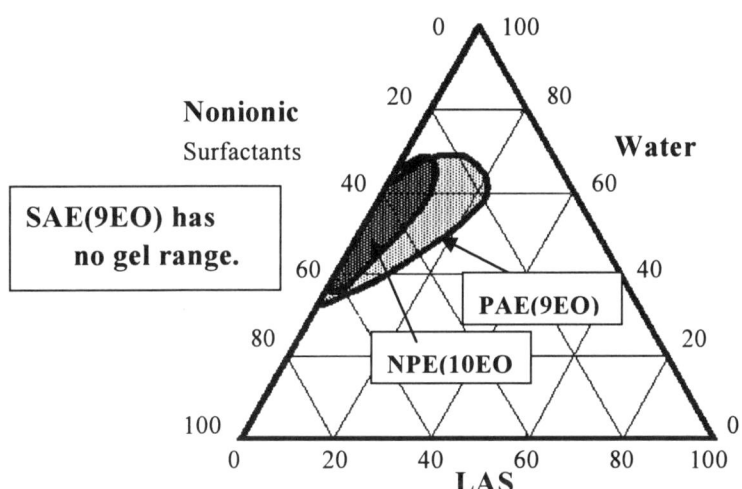

Fig. 8   Viscosity of Detergent based on Nonionic Surfactants/LAS/Water system

3.   Surface Phenomena

SAEs have excellent surface tension reducibility, high wetting power, unique foaming properties and good foam breakability.

3.1 Surface Tension Reducibility

Figure 9 shows the surface tension .   The surface tensions are plotted as a function of the EO mole number. SAEs exhibit a higher surface tension reducibility than both PAEs and NPEs at any EO mole number.

3.2 Wetting Power

Figure 10 shows the wetting power on wool.   The wetting times are plotted as a function of Griffin's HLB value for several nonionics. SAE demonstrated better wetting power when compared with PAE as described in previous papers by MacFarland (8) and Zika (9,10). The test method is in accordance with JIS K3362 of Japanese industrial standards.   In practice, a strip of woolen cloth was set afloat into the surfactant solution at 25°C, and then the time was measured until the solution permeated the cloth and made it sink.

SAE demonstrated much higher wetting power than both PAE and NPE. The same tendency was observed when the test was performed using either polyester cloth or cotton cloth.

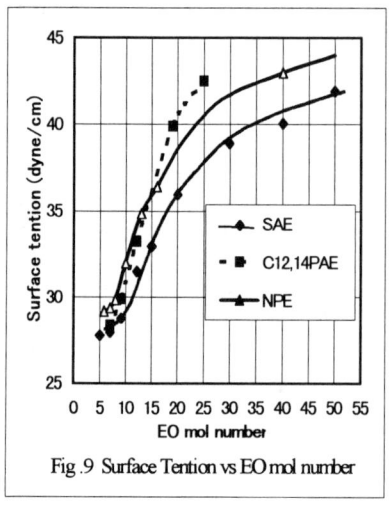

Fig.9 Surface Tention vs EO mol number

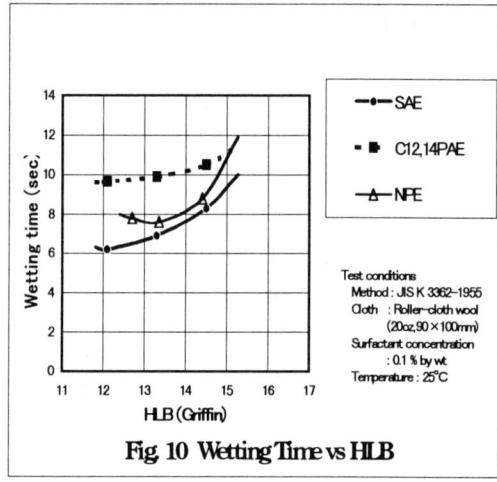

Fig. 10 Wetting Time vs HLB

Figure11 shows the wetting power on cotton. The wetting times are plotted as a function of surfactant concentration. SAE exhibited much higher wetting power at lower concentrations than PAE (9EO). On the other hand, SAE (9EO) showed almost the same wetting power for cotton fabric as that of NPE (10EO).

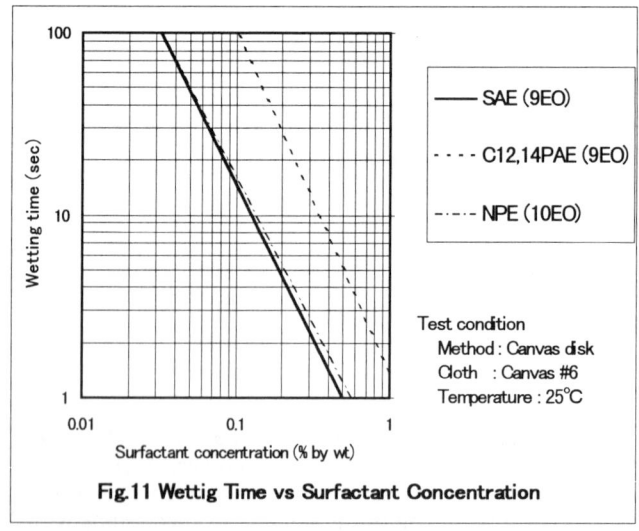

Fig.11 Wettig Time vs Surfactant Concentration

3.3 Foaming Property

SAE has very unique foaming properties. A previous paper (1) reported that SAEs show better foam breakability than both PAE or NPE.

Figure 12 shows the behavior of foam, such as time course changes in foam volume and foam breakability measured by the Ross-Mailes method. The foam heights are plotted as a function of time.

As seen in the figure, SAE foams in the same manner as both PAE and NPE in the initial stage, but after that, displays excellent foam breakability. Accordingly, it is considered that SAE rinses easily and that the waste water after washing has very little impact on the environment.

3.4 Low Foam Derivatives of SAE

Figure 13 shows the general structural formula of low foam derivatives of SAE.

The derivatives of this type are secondary alcohol ethoxylate propoxylates (SAEPs) (Nippon Shokubai's brand name: SOFTANOL EP series), obtained by adding propylene oxide to SAE.

The foaming property of SAEP is shown in figure 12. SAEP displays very poor foam formation and excellent foam breakability, compared with SAE, PAE and NPE. This means that SAEPs are much more suitable in the fields where foam formation is undesirable, particularly for its applications in industrial detergents and in other detergents used in automatic dish washers.

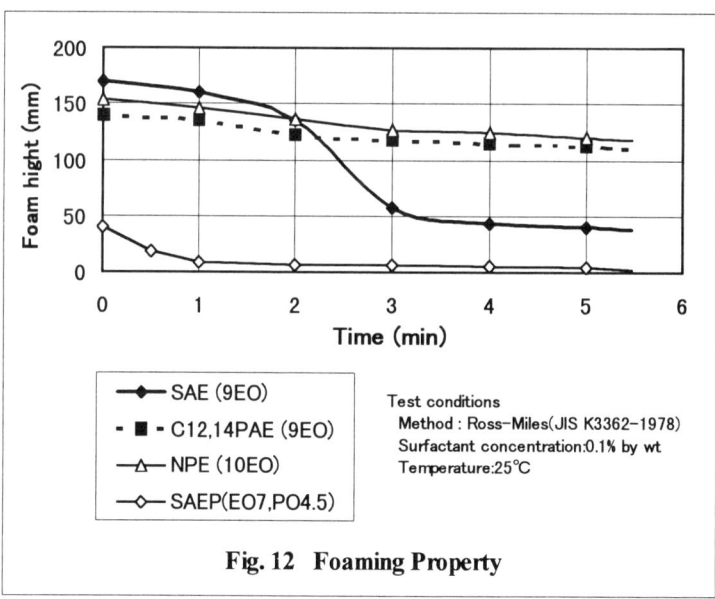

Fig. 12   Foaming Property

$$CH_3-(CH_2)m-CH-(CH_2)n-CH_3$$
$$\mid$$
$$O(CH_2CH_2O)x(CH_2CHO)yH$$
$$CH_3$$

m+n=9~11
x=5,7,9...
y=2.5~5...

Fig. 13   Chemical Structure of SAEP ~low foam surfactants~

4   Environmental Effects and Safety of SAEs

SAEs exhibit excellent safety and low toxicity, and have a low irritation index for eyes and skin.

NPEs have been utilized in various fields due to their excellent performance and low prices.   In recent years, however, their effects on the environment have been the subject of several reports.   NPEs have aromatic rings, and thus exhibit a very poor biodegradability, resulting in the accumulation of residue in rivers and others (11). In addition, the intermediate products induce estrogenic effects (12,13).

SAEs are highly biodegradable, while their performance is equivalent to that of NPEs.   Thus, SAEs can be suitably substituted for NPEs, even in the field where PAEs cannot be substituted in view of performance. In addition, SAEs demonstrated a much higher safety than PAEs.

4.1 Biodegradability

The biodegradability of SAEs has already been described in several reports.(14-17)

Figure 14 shows the rate of biodegradation in a batch system.   The degradations are plotted as a function of time.   The sludge used was collected from waste at the municipal sewage-treatment plant, Kawasaki-city, Japan.   The degradation was obtained by the BOD method [performed at 25°C for 8 days in compliance with OECD test guide lines (OECD 302C, amended NITI test method)].

SAE and PAE showed a high degradation at almost the same rate.   On the other hand, NPE was degraded by about 30%, indicating its poor biodegradability.

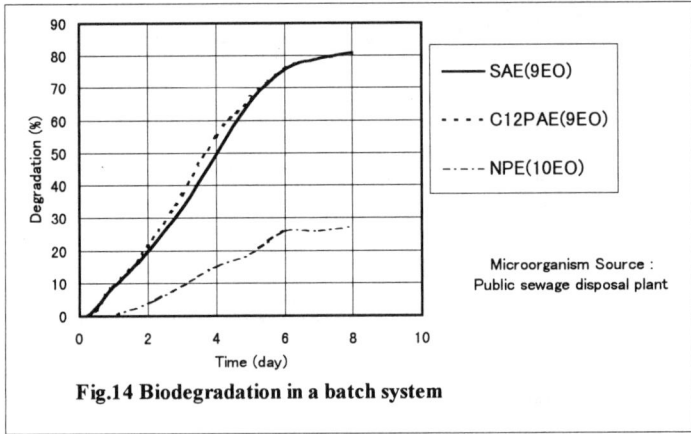

Fig.14 Biodegradation in a batch system

Table 3 lists the biodegradability tests condition and method in continuous-flow activated sludge tests. (16)

This test was conducted by Nippon Shokubai at a high concentration of surfactant (almost 200 mg/L) using sludge collected at the Kawasaki Ukishima Plant.

The biodegradation was followed by methods using cobalt thiocyanate active reagents (CTASs) (JISK-3364) , chemical oxygen demand [COD (Cr) and COD (Mn)] (JISK-0102), and total organic carbon (TOC).

4.2 Safety of SAE

Table 4 shows the safety data for SAEs and PAEs.(18)
SAEs have slightly lower toxicity than PAEs. The point to note here is that the irritation index for the skin is much lower for SAEs than that for PAEs. Such a low irritation index is considered to be due to the low content of free alcohols as well as ethoxylates with small EO mol numbers.
In fish SAE toxicity is lower than that of PAEs. This is considered to be a correlation between surface tension of water and fish toxicity. Since fish cannot breathe through their gills in water with a surface tension below 48 dynes/cm. Accordingly, in the table, the comparison of data for PAEs and SAEs indicates that both have similar surface tensions. From these data, SAEs are judged to exhibit a lower toxicity than PAEs.

Table 4   The Comparison of Safety Data in Alcohol Ethoxylates

	SOFTANOL 90 (sec.-$C12\sim14$-9EO)	Coconut Alc.-9EO (Natural Fatty Alcohol	prim.-$C12\sim15$-9EO (Synthetic Alcohols)
Acute Oral Toxicity, Rats, $LD50$ (mg/kg)	1,800	1,600	1,600
Primary Skin Irritation, Rabbits (Draize)	None to Mild	Moderate	Severe
Human (Patch Test)	None	--	Very slight
	(20% aq. soln.)		(15% aq. soln.)
Acute Toxicity to Fish, $LC50$, 48hr (mg/l)			
Gold fish	5.1	1.9	1.4
Bulegill fish	9	--	8
Mutagenic Effects Ames test	Negative	--	--

5. The Application of SAEs

SAEs are used in various fields because they exhibit excellent detergency due to their outstanding penetrating power.

5.1 Application in Household Detergents

In the household field, SAEs are suitable for the applications listed in table 5.
Recently, clothing detergents have been required to contribute to energy conservation in a compact size, by having high detergency as well as multipurpose functions in a compacted form.   Under these circumstances, the amount of nonionic surfactant used is on the increase.   Thus, both the liquid and powdered form of laundry detergents contain a very high percentage of SAEs.(19.20)
SAEs exhibit excellent detergency without any builder even at low temperatures or in hard water.   SAEs are excellent also in detergency against oil stains, and thus suitable for dish washing detergents.
In addition to these applications, SAEs are used for the following special cleaners: prewash spotter applied to spots or stains on collars and sleeves before washing; hard surface cleaner for bathtubs, toilets and ventilation fans; softening agents; and shoe cleaners.

Table 5   Apprication of "SAE" in Household Detergents

Main use
Liquid and Powder Detergents for Laundry
Light Duty Liquid Detergents
Dish Washing and Hard Surface Cleaner
Special Detergent
Silk and Wool Detergents (Softening Agents)

5.1.1   Laundry Detergents

Table 6   shows the detergency test conditions employed practically.   As seen from Table 6, the test was conducted using polyester/cotton cloth and soil of the fats/clay type.

Table 6   Aprication of "SAE" for Laundery Detergents

[Test condition]
- Cloth   Polyester/Cotton
- Soil   Fats/Clay type

[Terg-O-tometer Test Condition]		[Soil Component]	
• Surfactant Conc.	0.03 wt%	Myristic Acid	8.3 wt%
Temperature	25°C	Oleic Acid	8.3
• Agitation Speed	100rpm	Tristearin	8.3
Wash Time	5min	Triolein	8.3
• Rinse Time	5min	Cholesterol	4.4
Surfactant Solution	600ml/Pot	Cholesterol	1.1
• Water	Tap Water	Stearate	
•	(CaCO₃ as 50ppm)	Paraffine Wax	5.5
• Cloth Load	3/Pot	Squalene	5.5
	(10cmX10cm)	Clay	49.8

Figure   16   shows   the   comparison   of   detergency   between   several   nonionic surfactants.   The relative detergency with that of SAE (9EO) as 100 is plotted as a function of the HLB value.   The detergency rate was determined by measurement of reflectance. Both SAE and PAE tended to have a peak detergency around an HLB value of 13, but SAE showed a much higher detergency than PAE.

Figure 17 shows the comparison of detergency between several surfactants including anionic ones when they were used alone.

SAE (9EO) and NPE (10EO) both exhibited the highest detergency at an equivalent degree.   On the other hand, PAE had a little lower detergency, and the anionic surfactants had the lowest detergency.

Figure 18 shows synergistic effects when SAE and LAS were combined.

The relative detergency with that of SAE (9EO) as 100 is plotted as a function of the quantity ratio of SAE to LAS. Figure 18 also shows the effects of the concentration of sodium tripolyphosphate (STTP) used as the builder.

As seen from this figure, the combined use of SAE and anionic surfactant results in a higher detergency, due to synergistic effects of SAE and LAS.   The peak detergency was observed when SAE and LAS were in the ratio of 1:1 without STTP and 1:3 with STTP.

Table 3 The biodegradability tests condition and method			
Sample		**SAE (9EO)**	**NPE (10EO)**
Influent Composition			
Surfactant	**(mg/l)**	**200**	**200**
Urea	**(mg/l)**	**25**	**25**
Phosphoric Aci (mg/l)		**10**	**10**
Ave. Anal. Value in Influent			
COD(Cr)	**(mg/l)**	**406**	**339**
COD(Mn)	**(mg/l)**	**156**	**210**
CTAS	**(mg/l)**	**200**	**200**
TOC	**(mg/l)**	**123**	**120**
Residence Time	**(hrs)**	**48～24**	**48～24**
MLSS	**(mg/l)**	**2000±500**	**2000±500**

The test results are shown in figure 15.

The residual concentrations of both SAE (9EO) and NPE (10EO) are plotted as a function of time.   The residence time in each interval is shown along the vertical axis.   This test was performed on a continuous basis for about 250 days.

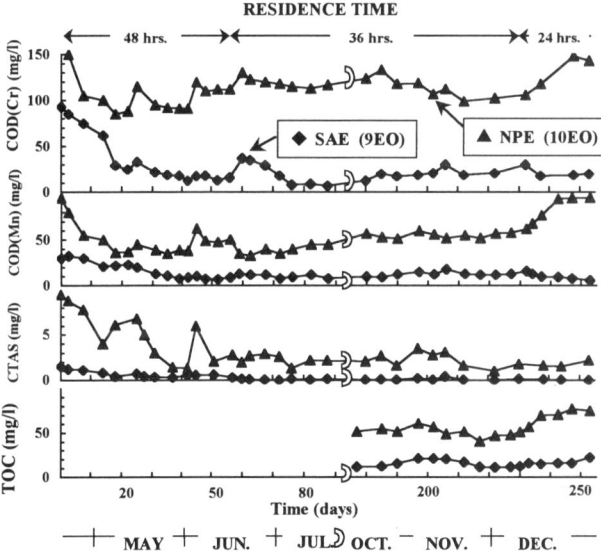

Fig.15 Biodegradability

The residual concentrations differed considerably according to the respective evaluation methods.   By the CTAS method the residual concentrations are the lowest, because they are measured by coloring of the remaining ethoxylates.   In the COD method (JISK-0102), Mn with higher oxidation power leads to lower residual concentrations, and thus even NPEs seem to demonstrate excellent biodegradability, resulting in a high degradation.   TOC is believed to produce the most accurate values, based on which the biodegradability of SAE is found to be higher than that of NPE.

When STTP as the builder was not mixed in, the detergency of LAS decreased significantly, although that of SAE changed only a little, suggesting a large difference between nonionic and ionic detergents.

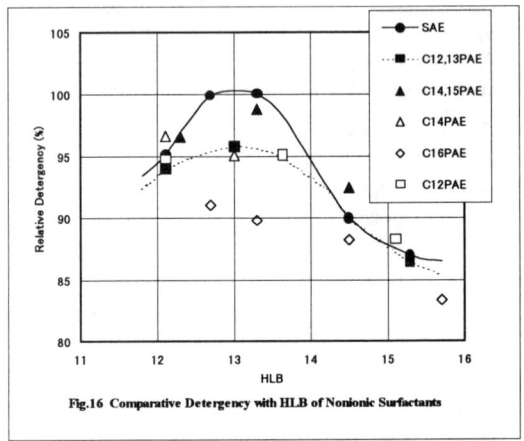

Fig.16 Comparative Detergency with HLB of Nonionic Surfactants

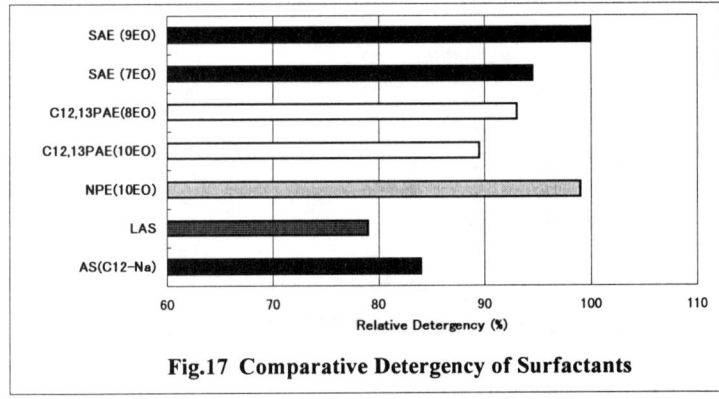

Fig.17 Comparative Detergency of Surfactants

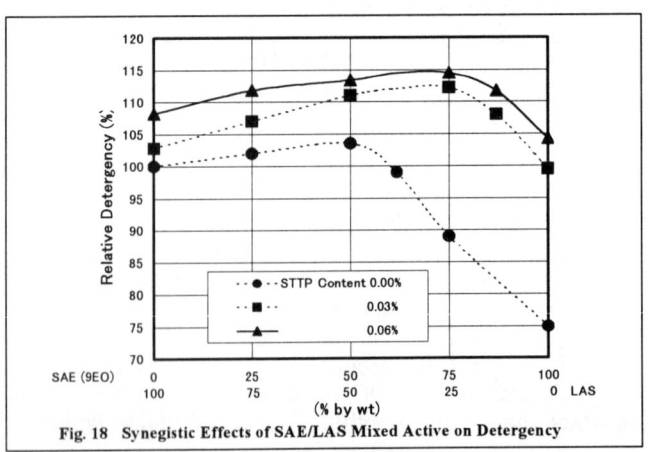

Fig. 18 Synegistic Effects of SAE/LAS Mixed Active on Detergency

Figure 19 shows the relationship of water-hardness to detergency.

The detergencies are plotted as a function of $CaCO_3$ concentration.

SAEs, nonionic surfactants, are not so affected by water-hardness, while anionic surfactants such as LAS exhibit a significant decline in detergency with a rise in water-hardness.

Interestingly enough, SAE and LAS used in a ratio of 1:1 demonstrated very little decline in detergency with a rise in water-hardness, compared with LAS alone. The synergistic property of the SAE and LAS combination was not an arithmetic mean of their respective properties.

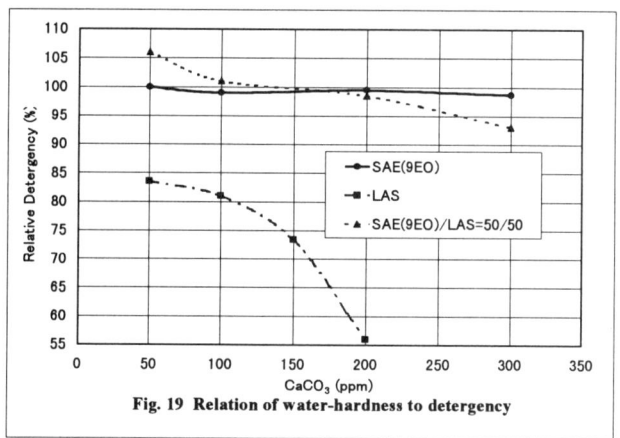

Fig. 19 Relation of water-hardness to detergency

SAEs are used for not only a liquid but also in powder types of detergent.

SAE have a high liquidity, therefore the dry mixing method, as shown in Figure20, is favored over the spray drying method to manufacture detergent powder containing SAE.   Figure 20 shows an example of the composition using SAE at a concentration of 15%.

Features of this method are as follows: Crystallized water is formed by spraying water at the initial stage to improve the affinity of inorganic powder with SAE; SAE is then blended in; and water is sprayed again to improve powder flow.   This method

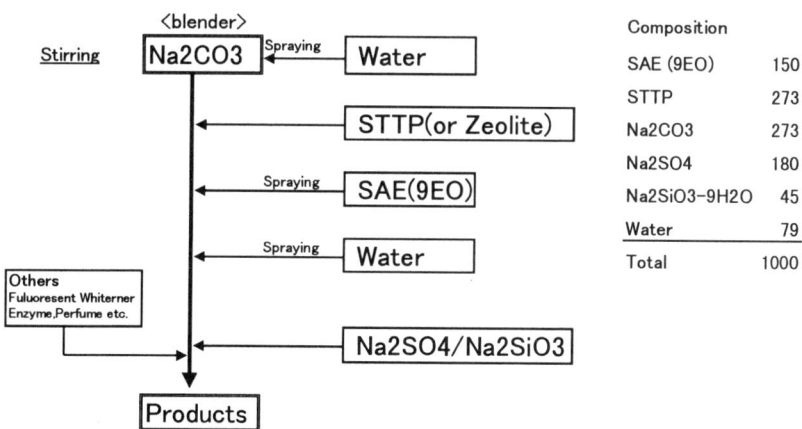

Fig. 20   Producing process of powder type detergent using "SAE"
~by Dry-Mixing Method~

facilitates SAE blending of up to 17%.   The blending up to about 25% is possible if some type of absorbent is used.

Table 7   shows examples of the composition of general types of powdered and liquid detergents.   When SAE is used, it is possible to formulate high-concentration liquid detergents even without any solubilizer such as ethanol.   The recent trend for detergent compositions are as follows: Concentrations of surfactant are as high as 40% or more in many of the compositions; the amount of nonionic surfactant used is large; and acrylic maleic acid polymer is used in many types of powdered detergent.

Table 7 Typical example of Heavy-Duty Detergents using "SAE"

Type	Powder Detergents	Liquid Detergents
[Compponents,% by wt.]		
SAE(9EO)	15	
SAE(7EO)		40
LAS(Na)	15	5*
SOAP	5	1
Zeolite	20	–
Na2CO3 and/or K2CO3	5	–
Na2SiO3·9H2O	20	–
Na2SO4	5	
SiO2 (White Carbon)	5	–
Triethanolamine	–	3
Etylene Glycol	–	1
Sodium Citrate	–	2
Fluoresent Whitener	q.s	–
Perfume,Antiseptic agent. Dye etc.	q.s	q.s
Acrylic–Mareic	3	–
Water	7	Balance

5.1.2   Light Duty Detergent

Figure 21 shows the state of foaming and the detergency during dish washing.   The numbers of dishes washed are shown on the vertical axis, while the types of the surfactants are listed on the horizontal axis.   In the test, dishes were washed with a brush until the foam vanished.   Then, the number of dishes washed and the number of dishes with oil removed were counted by the naked eye.

As seen in the figure, the foam diminished rapidly with SAEs as well as PAEs, and only very slowly with LAS.   In addition, the possible number of dishes washed was the largest with SAE (12EO), indicating that SAE had a higher degree in excellence of detergency (12EO) compared to PAE (9EO).

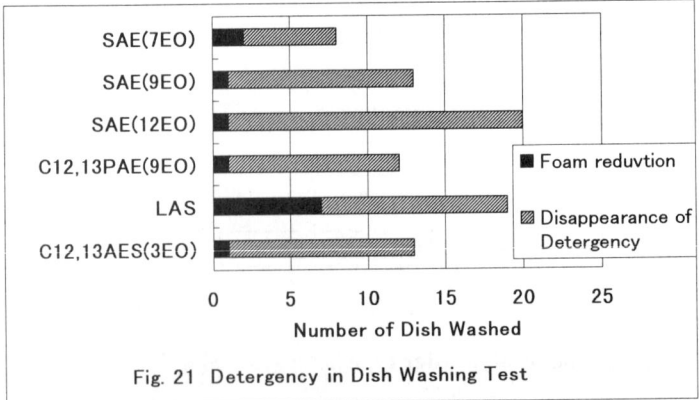

Fig. 21 Detergency in Dish Washing Test

5.2 The Industrial Applications

The representative fields are textile industry (9,21,22), leather industry, pulp and paper industry(23,24), hard surface cleaner (or metal industry) ,wool scouring industry and Others. (25~28)

5.2.1 Textile industry

SAEs are utilized in the field of the textile industry due to the ease of handling, excellent biodegradability, and excellent detergency. In recent years, SAEs have drawn attention as surfactants which can be substituted for NPEs in view of the performance, and have already been substituted in practical applications.(20,21)

Figure 22 shows the types of SAEs used in various processes in the textile industry.

SAEs are suitable for high speed washing due to their excellent detergency even at low temperature, which contributes to the labor/energy conservation and decreased processing time.   They have frequently been used in the scouring process in particular. SAE which has both a high emulsifying and dispersing power as well as excellent detergency is also used in the scouring process for synthetic fibers to remove finishing oil and or sizing agents such as polyvinyl alcohol . SAE is effectively used in solvent cleaning as well as in conventional alkali scouring .

Table 8 shows examples of the composition of scouring agents for synthetic fibers.

Composition-A is the standard type in which SAE is included as the main component , and composition-B is the low-foam type in which SAEP is included.

Table 9 shows the scouring test for polyester fibers . This test is a scouring test in which caustic soda at 80℃ is used for alkali washing. The effective rate (%) of cleaning after scouring with SAEs was equivalent to that for NPE.

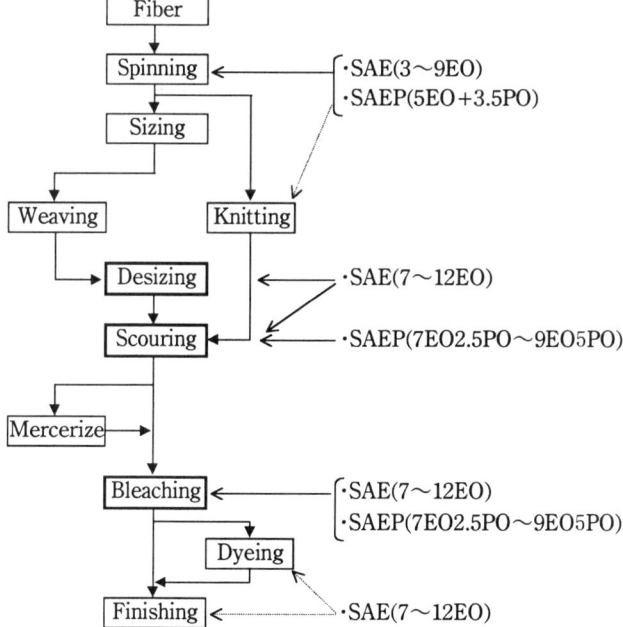

Fig.22 Aplication of SAE in Main Textile Process

Table 8 Typical Example of Scouring Agent for Synyhetic Fiber using "SAE"

Type	Composition–A Standard	Composition–B Low Form
Components		
SAE(3EO)	5	6
SAE(7EO)	10	6
SAE(12EO)	40	10
SAE(40EO)		10
SAEP(7EO+4.5PO)		20
LAS(Na)	5	4.5
Coco–dietthanolamide	5	4
IPA	5	
Etylene Glycol		9
NTA(3Na)	0.5	0.5
Water	34	30
Active Contents(%)	66	70

Table 9 The scouring test for polyester fibers

[Scouring conditions]

Scouring bath:	Caustic soda	0.2 %
	Surfactant	0.2 %
	Temperature	80 ℃
Test cloth :	Sudan Ⅱ dyed ,mineral oil immersed polyester cloth	
bath ratio:	50	
Immersing time :	15 minutes	

[Results]

Surfactant	cleaning effect 1) (%)
SAE(7EO)	7 5
SAE(9EO)	8 5
SAE(12EO)	8 5
NPE(10EO)	8 5

1)Cleaning effect is indicated by reflectance .

5.2.2 Wool Scouring

Figure 23 shows the scouring test of greasy Wool fiber . This test is performed with and without Soda ash. Test condition and results are shown figure 23. The residual grease (%) are plotted as a function of the number of 5g greasy wool.

Figure 24.25 shows the number of grams (g) of wool which can be washed by 1g of surfactant.   Without soda ash, the performances of individual surfactants improve most in the vicinity of HLB13   and the performance is the best for SAE(9EO). With soda ash ,SAE is equivalent to NPE, while PAE is a little inferior.

5.2.3 Metal industry

Figure 26 shows examples of the application in the metals industry.

This test is a dewaxing test by the modified Leenerts test. Figure 26 is an illustration of the device   The test oils used were slushing oil and rolling oil , and the test piece is a plate of Stainless Steel (SUS304).

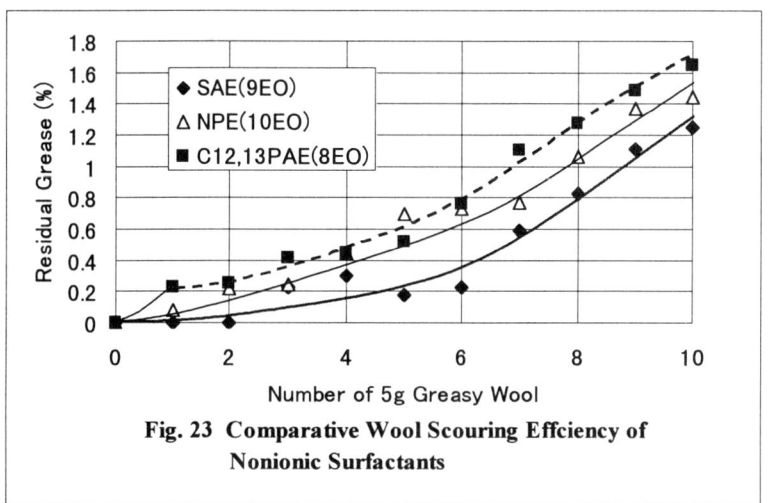

Fig. 23 Comparative Wool Scouring Effciency of Nonionic Surfactants

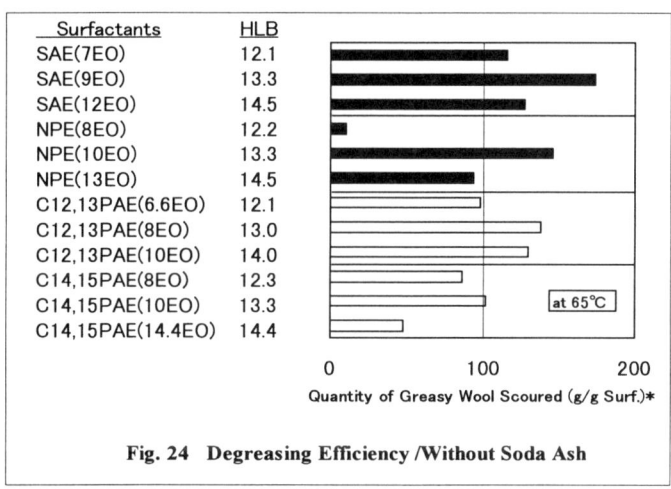

Fig. 24 Degreasing Efficiency /Without Soda Ash

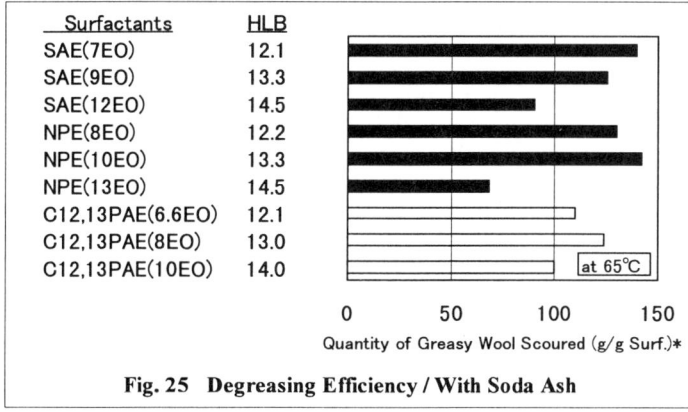

Fig. 25 Degreasing Efficiency / With Soda Ash

The dewaxing rate (%) after washing with SAEs was equivalent or higher than that of NPE.(Figure 27)

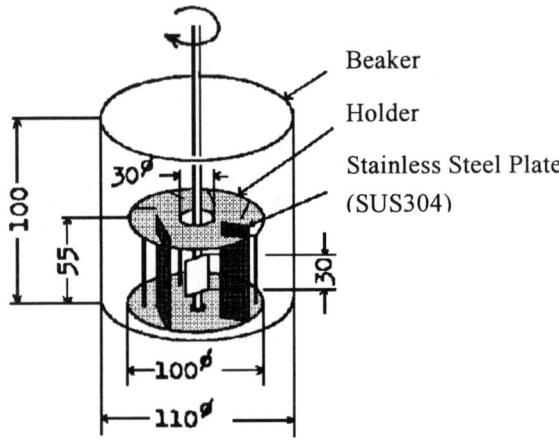

Fig. 26   Device of Modified "Leenerts"

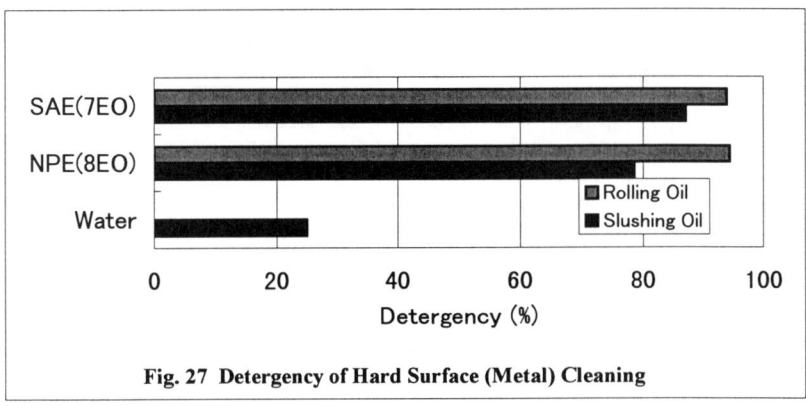

Fig. 27 Detergency of Hard Surface (Metal) Cleaning

6. New production process of SAE

Figure 28 shows the reaction formula.   In this process, SAE (1EO) is obtained by adding ethylene glycol under a catalyst directly to olefin as the raw material.

Then, ethylene oxide is added to SAE (1EO) as the raw material under an alkali catalyst to obtain the products with high EO mol number.

Characteristics of this new process is listed in Table 10.

SAEs conventionally provided by Nippon Shokubai have a mixture of alkyl chains with carbon numbers ranging from 12 to14.   By this new process, however, variations of carbon numbers are obtained, leading to the production of a single article with C12, C14 or C16.

In this method, ethoxylates are obtained by processes which do not require alcohol, and thus the resulting product has a high purity and is alcohol-free.   In addition, the distribution of EO adducts in the product is narrower than that when alcohol is used

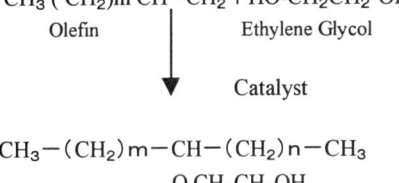

$$CH_3(CH_2)m\ CH= CH_2 + HO\text{-}CH_2CH_2\text{-}OH$$

Olefin   Ethylene Glycol

Catalyst

$$CH_3-(CH_2)m-CH-(CH_2)n-CH_3$$
$$O\ CH_2CH_2OH$$

SAE(1EO)

Fig. 28   New Process of SAE

Table 10   Characteristic of New SAE Process

- Variation of Alkyl Chain
  Single Articles of C12 or C14 or C16
- High Purity , Narrow EO Adduct Distribution
  Alcohol Free→Low Irritation
- Gentle to Enviroment
  Little Abandonments due to High Selectivity
- Conversion of Raw Materials
  n-Paraffin → Olefin
- Derivative
  Alkyl Ether Sulfates (AES)

as the raw material.

This process is an addition reaction using a catalyst under mild conditions, and results in almost no waste due to its high selectivity, compared to the oxidative reaction process. This means that this process is much gentler to the environment.

The products derived from SAEs include high purity alkyl ether sulfates (AESs) with 1 mole of ethylene oxide added.

As described above, the new SAE process produces SAEs with a very high commodity value. This process is now under examination on a pilot scale, and the new SAEs will be introduced in the very near future.

7 Summary

Properties of SAEs previously described are summarized in Table 11.

Among them, the most outstanding properties are high liquidity, excellent wetting power and good biodegradability. It is hoped that by promoting the advantages of their characteristic properties , SAEs will be utilized in many fields as a substitute for NPEs, as a matter of course.

Table 11   Characteristics of SAE

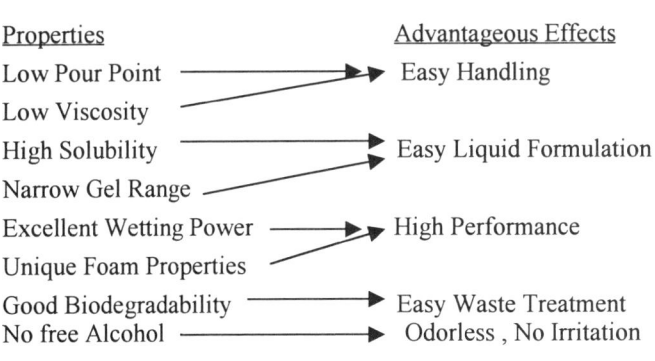

References

1, N. Kurata , K. Koshida, Hydrocarbon Processing,57,1,145(1978)

2, N. Kodo, S. Kaneko , N. Kurata , Yukagaku,24,427(1975)

3, I. Maeda, H.Yokoyama , N. Kurata , Y. Okuda , Japanese Patent
S48-37242,(1973)

4, N. Kurata , K. Koshida, , H.Yokoyama , T. Goto , ACS Symposium series 159
(1981)

5, N. Kurata , K. Koshida, ,M. Tuchino, Yukagaku 41,12,1203(1992)

6, N. Kurata , T. Goto , K. Rakutani, Yukagaku 42,5,388(1993)

7, H.Pushmann,Tenside,5,207(1968)

8, J. H. MacFarland , P. R. Kinkel , J. Am. Oil Chem. Soc. ,41,742(1964)

9, H. T. Zika, Textile Chemist & Colorist.,1,15,26,(1969)

10, H. T. Zika, J. Am. Oil Chem. Soc.,48,273,(1971)

11, E. S. Lashen , K. A. Booman , Water and Swage Works,18,119 (1967)

12, Kikuchi ,Yukagaku kyokaisi,45,10,1189 (1996)

13, T. Colborn , D. Dumanoski , J. P. Myers , 'Our Stolen Future' ,A Dutton Book,
128 (1997)

14, Z. Inoue, J. Fukuyama, J. Honda, Mizushori Gjyutsu,18,119(1977)

15, N. Kurata , K. Koshida, Yukagaku,24,879(1975)

16, N. Kurata , K. Koshida, Yukagaku,25,499(1976)

17, R. A. Conway, G. T. Waggy, American Dyestuff Reporter,Aug.1,33(1966)

18, N. Kurata , K. Koshida,T. Fujii, Yukagaku,26,115 (1977)

19, T.P. Matson, "Detergents in the Changing Scene", A Short Course Sponsored
by The Am. Oil Chem. Soc.s' 'Education Committee, Pensylvania',28-32(1975)

20, K. W. Dillan, E. D. Goodard, D. A. Mackenzie, J. Am. Oil Chem.
Soc. ,56,59(1979)

21, Union Carbide Corp., Textile and Paper Chemicals , Tergitol 15-S
Nonionic surfactants for Use in Textile Lubricants, F-44456(1973)

22, H. T. Rain, American Dyestuff reporter, Nov. 20,55(1967)

23, K. Sakuma, S. Miyanaga , S. Tatehara , Japanese pat. S50-22606 (1975)

24, K. Sakuma, S. Miyanaga , S. Koyasu , Japanese pat. S53-7522 (1978)

25, K.Kishi , M. Yamada , R. Ohshima , Japanese pat. S54-2999 (1979)

26, K. Yamazaki , Y. Ogata , Y.Ishikawa , K. Kawaguchi , S. Takeuchi ,
Japanese pat. S54-36951 (1979)

27, K.Kishi , M. Yamada , R. Ohshima , Japanese open pat. S49-30539 (1974)

28, T. Honda , K. Sugiyama , T. Tanaka , Japanese open pat. S50-98409 (1975)

Sulphobetaines and Ethersulphonates: Unique Surfactants via Sulphopropylation Reactions

Peter Köberle

RASCHIG GMBH, MUNDENHEIMER STRAßE 100, D-67061 LUDWIGSHAFEN, GERMANY

1.) Introduction

In aqueous systems, sulphonates provide good stability, improved water solubility, lower sensitivity against the addition of inorganic salts and the resulting material has a better dyeability. Micro-heterogenous systems such as certain solids in water can be effectively stabilised by the help of sulphonate carrying molecules. Although the advantageous properties of the sulphonate group are known for a long time, the defined incorporation of a sulphonate group into an organic molecule often poses a synthetic challenge[1,2]. Raschig has a long standing experience in sulphopropylation chemistry and offers a variety of functional sulphonates to industrial customers. The versatility of the reaction with 1,3-propanesultone allows reactions with highly hydrophilic carbohydrate based surfactants as well as with functional hydrophobic silicones. Moreover, the unique characteristics of the sulphopropyl group is often favourable in biological and medical applications.

2.) The sulphopropylation reaction

Nowadays sulphonation reactions are mainly based on the oxidation of mercaptans, sulphoxidations or direct sulphonations with SO_3 or aqueous sulphonate salts[1,2]. These reactions often have suffer from two major drawbacks. First, the harsh reaction conditions can be incompatible with sensitive functional groups and lead to only poorly defined products. Second, miscibility can be a problem when a hydrophilic sulphonating agent is reacted with a hydrophobic organic material. Within Raschig, we use a different approach allowing to incorporate propanesulphonate groups into a variety of amines, alcohols and carboxylic acids. This is achieved by a ring opening reaction of a cyclic sultone[3,4]. The reagent 1,3-propanesultone is hydrophobic allowing reactions in homogeneous hydrophobic phase. But also reactions in water are feasible, because hydrolysis of the sultone in water is slow[5] e.g. compared to its reactivity with amines. On the other hand, the reaction requires only relatively mild conditions and there is a distinct sequence in reactivity between non equivalent functional groups in a given educt[6]. The products are in most cases highly defined because the components react in equimolar quantities with high yields. Besides, the sultone alone has no tendency to polymerise. Homo- and copolymers of propanesultone[7] can only be obtained in non hydrolysing environment, and the products decompose readily in water. The reactions of 1,3-propanesultone are schematically illustrated in figure 1.

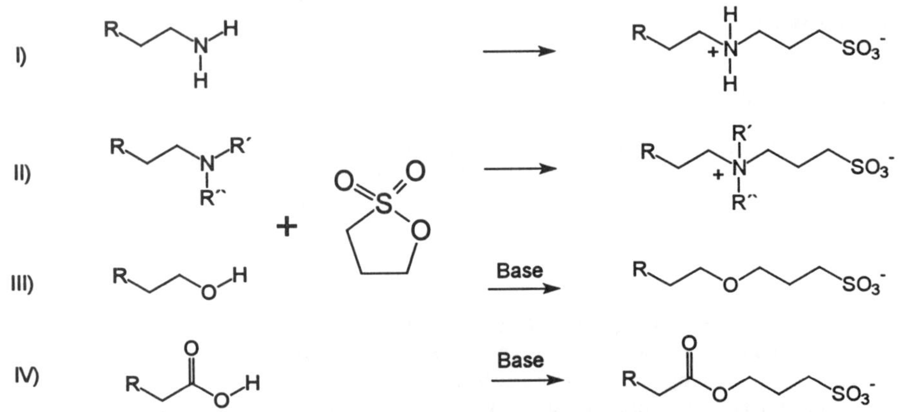

fig.1) Reaction of 1,3-Ppropanesultone with I) a primary amine to monoalkylated ammoniumpropanesulphonates, II) a tertiary amine to ammoniumsulphopropylbetaines, III) an alcohol to sulphopropylethers and IV) a carboxylic acid salt to sulphopropylesters.

An other advantage of the reactions I and II in Fig.1 is the complete absence of any kind of inorganic salts during the synthesis. Furthermore, because of the strong difference in hydrophilicity between educts and product the latter often precipitates during the reaction and can easily be isolated. Propanesultone is a toxic substance and therefore at Raschig it is directly converted into the non-toxic sulphopropylated products.

The reaction of sultones with less nucleophilic compounds affords stronger conditions. 1,3-Propanesultone in combination with alcohols produce sulphopropylethers (fig. 1, III) and in the case of carboxylic acids the analogous esters (fig. 1, IV) are obtained. For the latter reactions basic conditions have to be used. Therefore these products are no longer completely salt-free.

The different reaction products of the nucleophilic educts in figure 1 with propanesultone exhibit a number of characteristic features (fig. 1). Whereas the reaction with a primary or secondary amine yields an ampholytic ammoniopropanesulphonate (fig. 1, I), a tertiary amine is converted into the true sulphobetaine (fig. 1, II). Some ampholytes of this type are well established e.g. as biological buffers such as MOPS (Fig.2) and HEPPS[8]. In contrast, a sulphobetaine-molecule is not subject to protonation or deprotonation equilibria. The quaternary ammonium nitrogen carries four alkyl substituents which can not be replaced by hydrogen and the low basicity of the sulphonate group is responsible for the prevailing of the true zwitterionic form over the whole pH-range. Together with the low complexing properties of the sulphonate group, these are the main differences between the properties of sulpho- and carbobetaines. From their preparation, these systems are completely free of inorganic counterions. Furthermore, compared to the similar sulphate group the sulphonate provides a much better hydrolytic stability and improved solubility. Because of the fusion of the 2 educt molecules to one product molecule in the reaction also the amount of waste is minimised. Pyridines can easily be quaternised to the analogous pyridinium-sulphopropylbetaines in the same way and phenols can be converted to the sulphopropyl-phenylethers (Fig. 2).

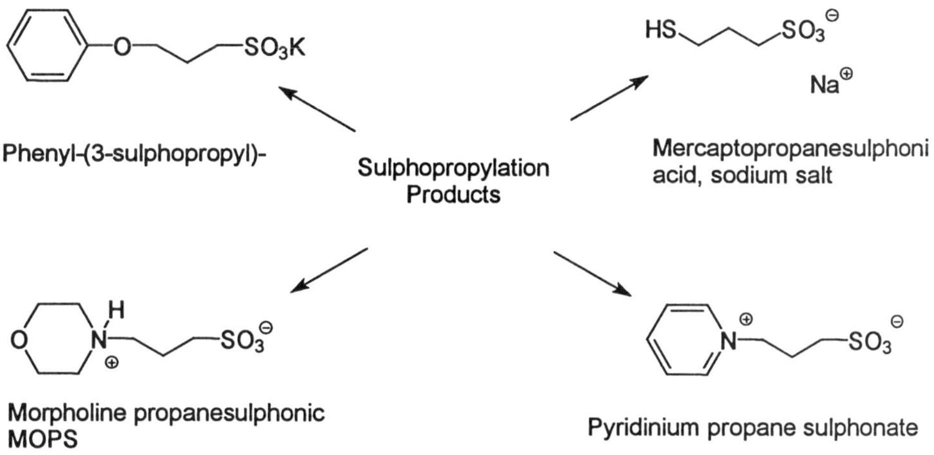

fig.2) Selected sulphopropylation products

It has also been shown that 1,3-propanesultone is suitable for the introduction of sulphopropyl group into a variety of polymers via polymer-analogous reactions[9]. Examples are the quaternisation of synthetic polyvinylpyridines[10] and the etherification of natural polymers such as starch and cellulose[11]. The versatility of the reaction with propanesultone can convert highly hydrophilic carbohydrate based surfactants as well as functional hydrophobic silicones into the corresponding sulphopropyl products[12]. The unique characteristics of the sulphopropyl group is often favourable in analytical, biological and medical applications.

The use of sulphopropyl group is particularly advantageous because of it's often better stability and water solubility in comparison to the analogous ethyl or butyl derivatives. Other reagents able to introduce sulphoalkyl-groups are shown on in fig.3 :

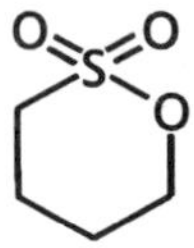

1,4–Butane sultone:
- less reactive
- less available

Isethianic systems:
- limited solubility / C2–spacer
- less pure

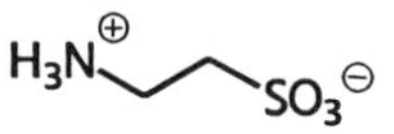

Taurine:
- limited solubility / C2–spacer
- less pure

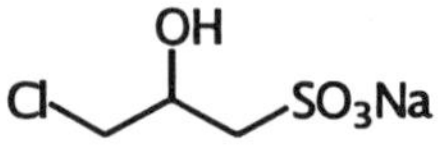

1–Chloro–2–hydroxy–propanesulfonic acid
- limited solubility / C3–spacer
- less pure

Fig.3) Other sulphoalkylating agents with their solubilites in hydrophobic solvents, purity and reactivity

Most of these sulphoalkylating molecules are hydrophilic and only slightly soluble in hydrophobic solvents. The only exception is the butane sultone which on the negative side has only limited reactivity compared to propane sultone. Nevertheless, some surfactants based on taurine, isethianic acid and chlorohydroxypropansulphonic acid are established in the market[13].

3.) Structure and Properties of the Sulphobetaine Group

As mentioned above the reaction of 1,3-propanesultone yields sulphobetaines. Their structure for a sulphopropylbetaine is shown in fig. 4.

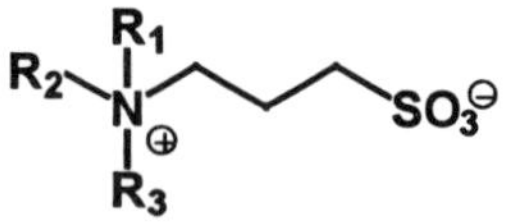

fig.4) Sulphopropylbetaine structure ($R_x \neq$ Hydrogen)

Sulphobetaines consist of an anionic sulphonate group and a cationic ammonium group. Generally sulphobetaines are zwitterionic[14], independent of pH (Tab.1).

Are Sulphobetaines Amphoteric ?	
No !	**Yes**
permanent cationic charge at fully alkylated ammonium	sulphonate group can be protonated in strong aqueous acids
anionic sulfonate group difficult to protonate, weak base	
no IEP behaviour / soluble in water over whole pH-range	
→ Sulphobetaines have zwitterionic character	

Tab. 1)   Arguments for and against the amphoteric character of sulphobetaines

Sulphobetaines derived from sultones are pure, free of inorganic salt and highly stable. Therefore, they are perfect candidates for defined chemical applications with the requirement of a zwitterionic hydrophilic group.

Sulphobetaines exhibt good water solubility and have high melting points because of their inner salt character. In some cases, the molecules decompose before the melting point is reached. Occasionally liquid crystalline phases are observed. Different alkyl spacers between the opposite charges are possible but in practice the C_3-spacer is dominating. Apart from the pure akyl spacer also other linkers between the charges can occasionally be found which are not always true sulphobetaines. Examples are given in fig.5:

fig. 5)   a) salt-containing OH-sulphobetaine[15)]
b) sulphobetaine with hydrophilic spacer[16)]
c) hydrolysable ammonioethylsulphate

Sulphobetaines have peculiar properties useful for numerous applications. Their structure is compatible with cationic and anionic materials. In addition, the sulphobetaine group is biocompatible but at the same time has a certain biocidal potential. The latter being at least partially attributed to the cationic moiety. Generally, there are remarkable similarities in structure and properties between the synthetic sulphobetaine- and the naturally occurring phosphatidylcholin-group[17]. Both structures can be used on surfaces in contact with biological material and are able to prevent the adsorption of proteins. In addition, molecules carrying the sulphobetaine headgroup exhibit an interesting adhesion behaviour at inorganic surfaces. Sulphobetaines themselves are insulators and do not migrate in an electrical field. The high stability of sulphobetaines derived from sultones in water against redox, pH, hydrolysis and salt addition are the basis for their use under challenging conditions. Some sulphobetaines show salt effects often following the Hofmeister lyotropic series[18]. Their solubility in water can to some extent be influenced by the addition of salt. In spite of their cationic ammonium group, sulphobetaines have not been used as phase transfer catalysts yet.

4) Sulphobetaine Surfactants

If the ammonium group of the sulphobetaine carries one or two longer alkyl substituents, one deals with sulphobetaine surfactants. The majority of the commercially available products have a single alkyl chain with lengths between 8 and 18 methylene units. Most common is their combination with the hydrophilic dimethylammonioalkylsulphonic headgroup. An example is shown in fig. 6:

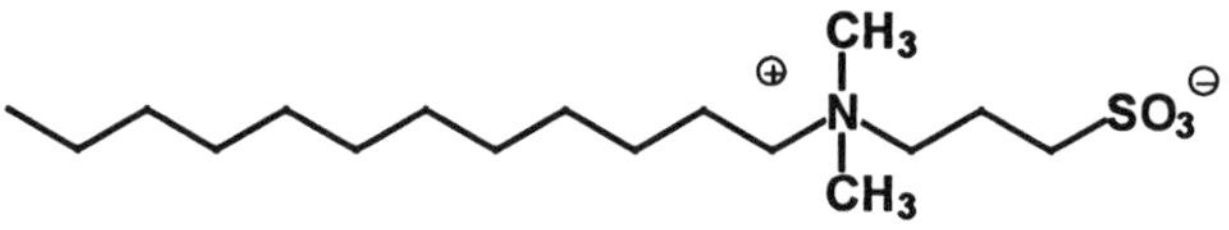

fig. 6) N,N-Dimethyl-N-dodecyl-N-(3-sulphopropyl) ammonium betaine Ralufon® DL[19]

Typically, the pure products are white crystalline solids with high melting points. Sulphobetaine surfactants prepared from 1,3-propanesultone are free of any kind of inorganic salts and have no counterions. The Krafft – temperatures of dimethylammoniumsulphobetaine surfactants with chain lengths up to 14 methylene units usually are below room temperature. For the C_{18}-analogon, it rises to about 75°C. Although the materials are hydrophilic they are only slightly hygroscopic. Typical surface tensions of the dimethylammoniopropanesulphonic headgroup are in the range of 30-40 mN/m. With fluorocarbon or silicon substitution, lower values can be obtained. Salt-in effects are observed for some surfactants. Some characteristic data of dimethylammoniopropansulphonate surfactants is given in tab.2.

Hydrophobic Chain Length	CMC [mmol]	Aggregation Number	Product Name
8	-	-	
10	25-40	41	Ralufon® DD
12	2-4	55	Ralufon® DL
14	0.01-0.04	83	Ralufon® DM
18	-	-	Ralufon® DS
CHAPS	6-10	4-14	

Tab. 2) Properties of Dimethylammoniopropanesulphonate surfactants[20]

An example for the use of sulphobetaines in life science is the solubilisation of proteins[21]. Here this type of surfactants is able to solubilise certain proteins, such as membrane proteins, efficiently without denaturation. Therefore products of the Ralufon® series are well established in diagnostic and analytic applications. A simplified model for this solubilisation is given in fig. 7.

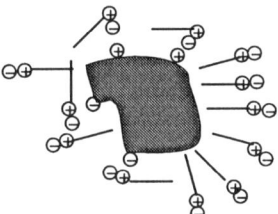

fig. 7) Sulphobetaine surfactants for the solubilisation of proteins

In contrast to many common ionic surfactants, the sulphobetaines do not form strong complexes with the ionic charges of the protein. This facilitates the removal of the surfactant in a following step. Practically this can be done e.g. by dialysis. In that way solubilised proteins can be further purified by an additional electrochemical step. This is possible because of the absence of counterions or salt in the sulphobetaine and the internal charge compensation within each individual molecule. The products are also used in chromatographic separations and purification of membrane proteins. Sulphobetaine surfactants, such as the cholic acic derivative CHAPS[22], are to some extent also used in electrophoresis. The sulphopropyl based biological buffers (see e.g. MOPS in fig. 2) are used in biological applications, too. They keep the pH constant without taking part in biochemical reactions or being involved in pH shifting caused by CO_2 equilibria.

Recently, the use of non-detergent sulphobetaines in a new biological application has been described. These low molecular weight molecules, such as the pyridinium propane sulphonate in fig. 2, are used to re-naturate proteins into their original structure[23]. This is believed to be partially caused by a reduction of the amount of aggregation. Some sulphobetaine surfactants are active against bacteria and fungi[24].

Exemplary for the use of sulphobetaines in material science is their use as dispersing agents in certain technical applications. Since a long time it is known that sulphobetaine surfactants are very efficient lime soap dispersing agents[25]. However, recent publications indicate that this effect is not limited to lime soap systems. Reetz[26] and co-workers showed that with the help of sulphobetaine surfactants one is able to obtain high concentrations of metallic nanoparticles in

water. The metal cluster under investigation had been electrochemically produced preferably
from precious metals such as Platinum. It is assumed that the stabilisation has it's origin in the
interaction of the cationic ammonium charge of the surfactant with the anionic surface charge
of the metal particle (fig. 8). The sulphobetaine surfactant surrounds the particle like a
protecting shell.

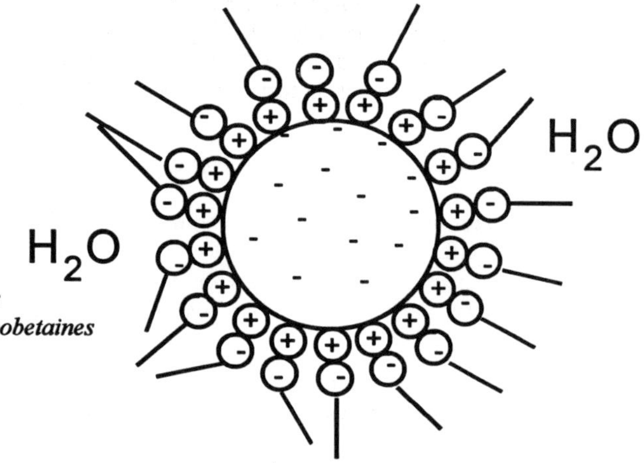

*fig 8)   Stabilisation of metallic
         nanoparticles with sulphobetaines
         in water*

Here the negatively charged sulphonate groups point into the water phase and are responsible
for the stability of the dispersion. By this method a concentration of 1 mol/l Platinum in Water
were obtained. Also metallic pigments can be dispersed with sulphobetaine surfactants.

A similar effect could play a role in the use of certain sulphobetaines in the electroplating
industry e.g. Nickel baths. As most natural systems possess negatively charged surfaces,
suphobetaines might be helpful to get otherwise insoluble material into water.

Sulphobetaine surfactants have also found entry in household and personal care applications.
Here usually the less pure OH-sulphobetaines are used in formulations. The hydrophobic unit is
often being an alkylamidopropyl moiety, similar to the carbobetaine surfactants.

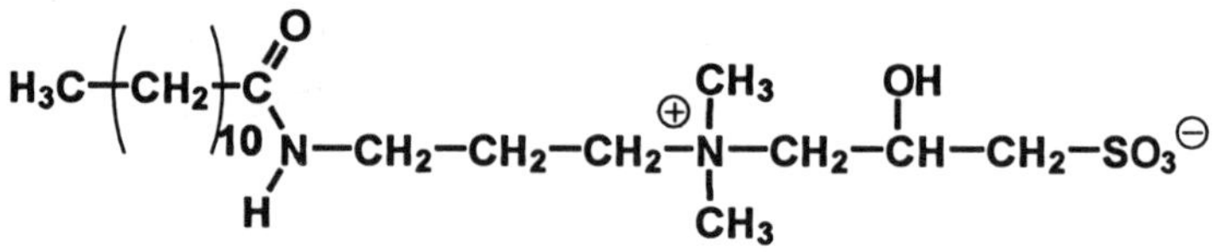

fig. 9) OH-sulphobetaine, Ralufon® CAS-OH

In baby products and shampoos, these structures are well known for their mildness. Other
applications are acidic and alkaline cleaners.

Although sulphobetaines belong to the class of rather less well known chemicals their use is
established or has been patented in a variety of fields. Sulphobetaine surfactants are used as
cleansers, as dispersing auxiliaries, in biochemistry and medicine, in electroplating, in the textile
industry, in personal care, in inks and in the photo industry.

5) Structure and Properties of Ethersulphonate Surfactants

The sulphopropylation of alcohols yields sulphopropylethers. If the alcohol component carries a long enough hydrophobic moiety then sulphopropylether surfactants can be obtained. Is the starting material already a nonionic surfactant with a reactive OH-group then the products are sulphopropylated nonionic surfactants. In analogy to the now upcoming ethersulphate surfactants these are named ethersulphonates[13]. Despite the similarity in their names both types of surfactants differ strongly in their structure (fig. 10) and properties.

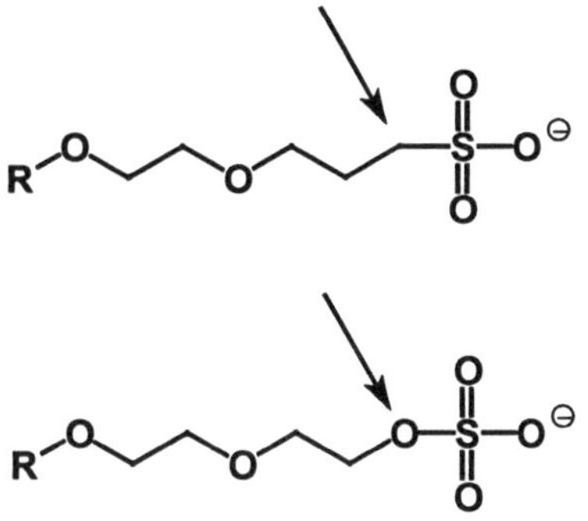

fig. 10)Ethersulphonate (top) and Ethersulphate (below)

Whereas ethersulphonates have a covalent CH_2-S bond between the hydrophobic part and the anionic headgroup the ethersulphate are linked via the CH_2-O-S sequence. As a consequence ethersulphates can be easily hydrolysed whereas Ethersulponates possess a much higher stability. The ethersulphates degrades into a nonionic surfactants and a sulfate salt.

The number of possible ethersulphonate surfactants is only limited by the number of nonionic starting materials. The surfactants do not exhibit a cloud point and are generally obtained in high purity. They are almost free of inorganic salt and completely free of halogens. An example is given in fig. 11:

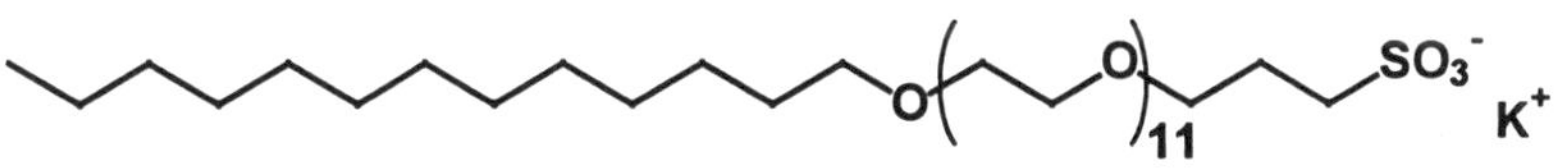

fig. 11)   *Ralufon® F11-13*
                    (fatty alcohol-EO-sulphopropylsurfactant K-salt)

Two main Ralufon® ethersulphonate product lines are established. One is based on ethoxylated branched or linear fatty alcohol and the other on ethoxylated nonylphenols.

a)

b)

Ethersuphonate surfactant	hydrophobic unit	EO(av.)	Form	Remarks
RALUFON® F 5-13	C_{13}-C_{15}-lin.-Oxo-Alcohol	5	Paste	
RALUFON® F 7-1	C_{13}-C_{15}-lin.-Oxo-Alcohol	7	Paste	
RALUFON® F 11-13	C_{13}-C_{15}-lin.-Oxo-Alcohol	11	Paste	
RALUFON® F 4-I	Iso-Tridecyl (branched.)	4	Paste	
RALUFON® N 9	p-Nonylphenol	9	Paste	
RALUFON® N 20-90	p-Nonylphenol	20	Solution	10% water
RALUFON® NAPE 14-90	Naphtol propoxylated.	14	Solution	10% water
RALUFON® EA 16-90	Ethylhexanol	15	Solution	10% water

fig. 12)   *a) nonylphenole-EO-sulphopropylsurfactant K-salt (Ralufon® N types)*
          b) fatty alcohol-EO-sulphopropylsurfactant K-salt (Ralufon® F types)
          Commercially available Ethersulfonate surfactants

Parameters such as foaming and biodegradability of ethersulphonate surfactants is mostly determined by the nature of the hydrophobic unit and can be varied by choosing the appropriate starting material. The surface tensions of ethersulphonate surfactants above their CMC lie between 45 and 30 mN/m as shown in Tab. 3.

Product	CMC (Critical Micellation Concentration) in [%]	Surface Tension above CMC [mN/m]
RALUFON® N9	0,1	44,4
RALUFON® NAPE 14-90	0,3	40,3
RALUFON® EA 15-90	3	38,2
RALUFON® F 11-13	0,006	30,4

Tab. 3) CMC and surface tension of Ethersulphonate surfactants

The outstanding feature of ethersulphonates is their remarkable stability. This was also the reason why they were recommended for enhanced oil recovery (EOR) in the 1970s[27]. Their particular application were oil fields with high salt concentration and also at elevated temperatures. Nonylphenol-based products turned out to even more stable than the linear ethersulphonates. Also in metal treatment, this class of surfactants has been tested successfully.

A special dispersing agent is the naphtol-based surfactant Ralufon® Nape 14-90 (fig.12). It is used to effectively solubilise aromatic chemicals but also inorganic substances.

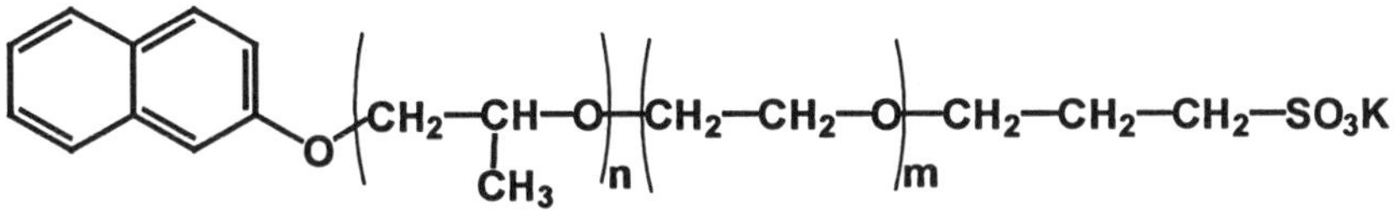

fig. 12)   *Ralufo®n Nape 14-90*
              (sulphopropylated polyalkylated beta-naphtol, K-salt)

Ethersulfonates have found application in cleansers, emulsion polymerisation, paper, textile, electroplating, photography and ink formulations.

1,3-Propanesultone also allows the production of estersulphate surfactants[28]. An example are the sulphopropylalkyl maleate surfactants. These are hydrolytically less stable.

6) Unusual Polymeric Sulphonates

Ionic polymers are used for a variety of applications, such as waste water treatment, detergents or in any kind of aqueous dispersions[29]. In particular, acrylic monomers are known to be very flexible in the synthesis of homo- and copolymers. Up to now, most anionic vinylic polymers are based on acrylic or methacrylic acid. These carboxylate systems often have a number of disadvantages. They exhibit a sometimes complex solubility in water and are sensitive towards the addition of salts[30]. To overcome these drawbacks the carboxylic unit has to be at least partially replaced by a less complexing and better water soluble group. Here sulphonate groups can be the solution but the number of acrylic monomers containing these groups is rather limited. Some of these few existing systems have stability problems or do not copolymerise well. Raschig has developed a variety of monomers based on the sulphopropylation reaction (fig.13)

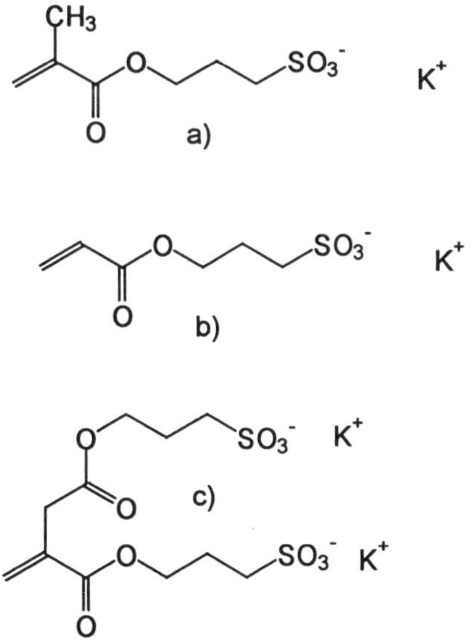

fig.13)
  a) Methacrylic acid-(3-sulphopropyl)ester,
     potassium salt (SPM)
  b) Acrylic acid-(3-sulpho-
     propyl)ester, potassium salt (SPA),
  c) Itaconic acid-bis-(1-propylsulphonic
     acid-3-)ester, di-potassium salt (SPI)

These monomers combine the strong hydrophilicity of the sulphonate group with the organic acrylate moieties. The propyl linkage acts as a small „spacer" underlining the bifunctional character of the molecules. This makes the molecules interesting for the synthesis of amphiphilic copolymers. The compounds can be used to lower the necessary amount of low molecular weight surfactants in emulsion polymerisation and should yield a better water resistance of the final polymer film. As the monomer becomes part of the resulting polymer, this is an important step towards the „Emulsifier free emulsion polymerisation". Possible fields of applications for all above mentioned monomers are e.g. aqueous dispersions, thickeners, hydrogels, hydrophilized foils, paper coatings and ion-exchange resins.

The quaternization of polymerisable tertiary amines with 1,3-propanesultone leads to ammoniosulphobetaine monomers. In contrast to the above discussed anionic sulphonate monomers, all charges are covalently bound in one repeat unit. As a consequence of the lack of mobile counterions the properties of polymers derived from sulphobetaine monomers differ strongly from normal polyelectrolytes. Neither inorganic halogen nor metal ions are present in these products. In homopolymers each individual polymer chain has exactly the charge zero. The solutions of the polysulphobetaines in water are low viscous. In aqueous solutions, there is no increase in the reduced viscosity at low polymer concentrations and the solubility of the polymer in water is enhanced upon the addition of inorganic salts. Because of these unusual effects the polysulphobetaines are often classified as „antipolyelectrolytes"[31]. Moreover, the observed salt effects follow certain selectivities. They depend on the character of the salt-anion and salt-cation as well as on the salt concentration. Raschig produces at present three different sulphobetaine monomers based on methacrylic acid-, methacrylamide- and 2-vinylpyridin-structures. The formulas are shown in figure 14.

s

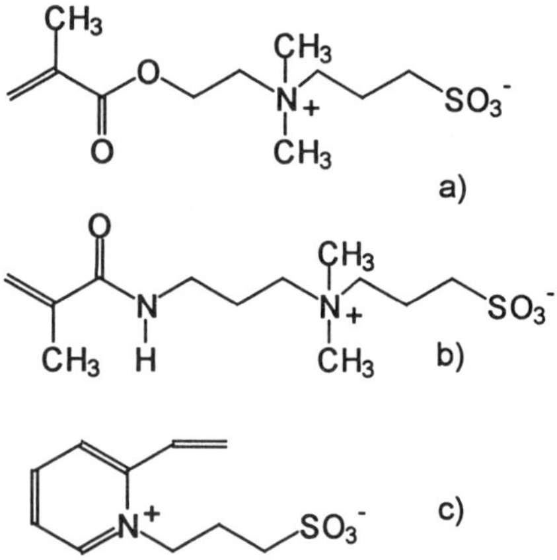

fig.14)

a)N,N-Dimethyl-N-methacryloxyethyl-N-(3 -sulphopropyl)-ammoniumbetain (SPE)

b) N,N-Dimethyl-N-methacrylamido-propyl-N-(3-sulphopropyl)-ammoniumbetain (SPP)

c) 1-(3-Sulphopropyl)-2-vinylpyri-diniumbetain (SPV)

Sulphobetaine homopolymers have a high structural ionic content and are normally hygroscopic materials with high glass transition temperatures. In copolymers they introduce good dyeing properties. The pure polymers can form homogeneous blends with stochiometric amounts of several inorganic salts[32], and should in appropriate combinations be suitable for new conductive or high refractive index materials. Because of this unique ability to mix with inorganic salts, they might even be suitable in formulations for solid state batteries. The dry polysulphobetaines without water or salt are strong insulators[33]. All charges are bound to the polymer and can not be separated in an electrical field. Furthermore, the swelling behaviour of these polymers in hydrogels[34] is sensitive towards the addition of salt which may be of use for several membrane applications. The presence of e.g. NaCl is destroying the betaine group associations and a large increase in swelling can be observed. Temperature and salt-concentration have an influence on these salting-in effects. The question whether the sulfobetaine groups aggregate preferably inter- or intramolecularly is not yet answered satisfactory[35].

6.) Conclusion

Sulphonate groups offer a number of advantages in aqueous systems. Sulphopropylation reactions are a mild, defined and convenient way to introduce the anionic sulphonate group into a variety of organic molecules. The sulphopropylation chemistry is described taking into account the latest industrial developments. Focus is put on surfactants, but also monomers bearing sulphopropyl or ammoniumpropanesulphonate are described. The reaction is already widely used for the preparation of additives for the electroplating and photographic industry and will surely enter new fields in the future.

7.) References

1) Th. F. Tadros, editor „Surfactants", Academic Press, London (1984)
2) S.J. Gutcho, „Surfactants and Sequesterants", Noyes Data Corporation, Park Ridge (1977)
3) D.W. Roberts et al., Tetrahedron (1987) **Vol. 43 No. 6**, pp 1027-1062
4) P. Metz, J.prakt.Chem. (1998) **340**, 1-10
5) T. Nilson, Ph.D. Dissertation, University of Lund, Sweden (1946)
6) P. Köberle et al, Macromolecules (1994) **27**, 2165-2173
7) T. Saegusa et al., Macromolecules (1975) **Vol.8 No. 3**, 259-261
8) N. Good et al., Biochemistry (1966) **5**, 467
9) Klingenberg et al. Polym. Mater. Sci. Eng. (1993) **69** , 353-4
   M.L. Gieselman, J.R. Reynolds, Macromolecules (1993) **26**, 5633
   S.A. Chen, G.W. Hwang, J.Am.Chem.Soc. (1994) **116**, 7939
   X.-L. Wie et al., J Am.Chem.Soc. (1996) **118** , 2545-2555
10) J.C. Galin et al., Polymer (1984) **Vol 25**, 121
11) Toshiyuki et al., Sen'i Gakkaishi (1995) **51(12)**, 571-9
12) R. Wagner, Ph. D. Dissertation, University of Dresden, Germany (1993)
13) H. Stache editor, „Tensid Taschenbuch", CarlHanser Verlag, Munich (1979)
14) R.G. Laughlin, Langmuir (1991) 7, 842-847
15) D.W. Osborne, Langmuir (1989) **5**, 924-926
16) P. Le Perchec et al., Tetrahedron (1989) **Vol 45 No 11**, 3370
17) T. Kunitake et al., J. Coll. & Interface Sc. (1981) **Vol 82 No 2**, 401
18) F.Hofmeister, Arch. Exptl. Pathol. Pharmakol. (1888) **24**, 247

19) B. Sesta et al., Langmuir (1990) **6**, 728-731

20) B. Sesta et al., Colloid Polym Sci (1989) **267**, 748-752

21) D. Soler, Journal of Protein Chemistry (1995) **Vol 14 No 7**, 511
    S.E. Rapper et al, Hepatology (1992) **16**, 433

22) Y. Chen et al., Biochemistry (1992) **31**, 2415

23) L. Vuillard et al., J. Cryst. Growth (1996) **168(1-4)**, 1-4

24) Raschig GmbH, unpublished results

25) N. Parris et al., J. Am. Oil Chemists Soc. (1973) **Vol 50**, 509

26) M.T. Reetz et al., Angew. Chem. (1995) **107 No 20**, 2461

27) K. Kosswig, Chemie in unserer Zeit (1984) **Vol 18 No. 3**, 87-95

28) H.v. Berlepsch, Langmuir (1995) **11**, 3676-3684

29) R.D. Lundberg, "Ionic Polymers" in "Encyclopedia of Polymer Science and Technology",
    2nd edition, Vol 8, p 393, Wiley-Interscience, New York (1985)

30) E.A. Bekturov, Z.Kh. Bakauova, "Synthetic Water-Soluble Polymers in Solution", Hüthig
    and Wepf, Basel, Heidelberg, New York (1986)

31) J.C. Galin et al., Polymer (1984) **25**, 254

32) A. Laschewsky et al., Macromol Symp. (1994) **88**, 165-175

33) S.A. Rozanski et al., Macromol. Chem. Phys. (1995) **196**, 877-890

34) M.B. Huglin, Macromolecules (1993) **26**, 3118-3126

35) J.L. Brédas et al., Macromolecules (1988) **21**, 1633-1639

Alkyl Phosphates as Cosmetic Emulsifiers: A Contribution to Non-ethoxylated Emulsifiers

R. Aigner, M. Löffler, A. Overweg and A. Turowski

CLARIANT GMBH, FRANKFURT, GERMANY

1 INTRODUCTION

For over four decades, *o*-phosphoric esters of fatty alcohols have been used as very effective emulsifiers in cosmetics. Ongoing work in this field had the objective of developing an ideal emulsifier or an ideal emulsifier mixture with the following properties (Figure 1):

- Defined substances, colourless/odourless, no ethoxylation
- High interfacial activity, low interfacial tension against oils of different polarity
- Fast spreading capacity at the interface
- Very high stability of emulsions at low concentrations
- Easy product handling (liquid or pourable)

This paper describes the various methods of synthesis, the physicochemical parameters and the stability tests of emulsions (Figure 2).

Figure 3 shows the structure of alkyl phosphoric esters. Varying the C chain length of the original alcohol, the degree of esterification and the type of cation and degree of neutralization produces different characteristics, with specific areas of suitability.

1.1 Synthesis

Different processes are available for producing alkyl phosphoric esters (Figure 4):

1. Reacting phosphoric oxytrichloride with fatty alcohols or fatty alcohol ethoxylates ("POCl$_3$ process")
2. Direct esterification of the phosphoric acid
3. Reacting phosphorus pentoxide with fatty alcohols or fatty alcohol ethoxylates ("P$_4$O$_{10}$ process") and
4. Direct esterification of polyphosphoric acid

There is a considerable variation in the quantitative composition of the reaction products obtained in each case. Basically, mono- and diesters have an acid character. Depending on the application, they can then be adjusted to the required pH value using an appropriate alkaline solution. With the POCl$_3$ process, a composite product is created consisting of mono-, di- and triesters with a relatively high proportion of neutral components. Due to the presence of chlorine, this composite product also contains

Objectives

To develop an optimized emulsifier system which provides the following benefits:

> Well defined chemicals, free of EO and chlorine, colourless and odourless

> High interfacial activity, reduction of inter-facial tension vs. oils of different polarity

> Fast spreadability

> Highest stability at low concentration

> Easy to handle

Figure 1

Tok 003 E

1. > New Products by traditional Processes

⊐ Manufacturing

⊐ Characterization by physical/chemical and optical methods
– Surface tension
– Interfacial activity against oils of different polarity

2. > Emulsifiability

⊐ Screening stability test

⊐ Guide formulations

3. > Summary & Perspectives

Figure 2

Tok 004 E

Structure of Alkyl Phosphoric Acid Ester

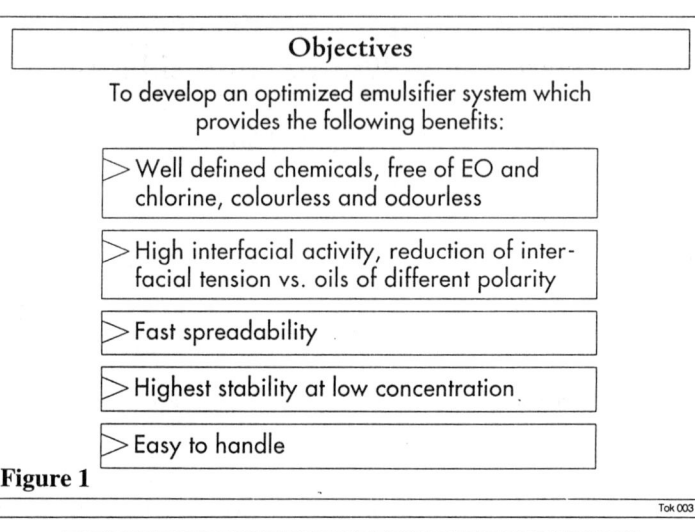

| Monoester | Diester | Triester | [Fatty alcohol, phosphoric acid] |

Figure 3

Tok 005-1 E

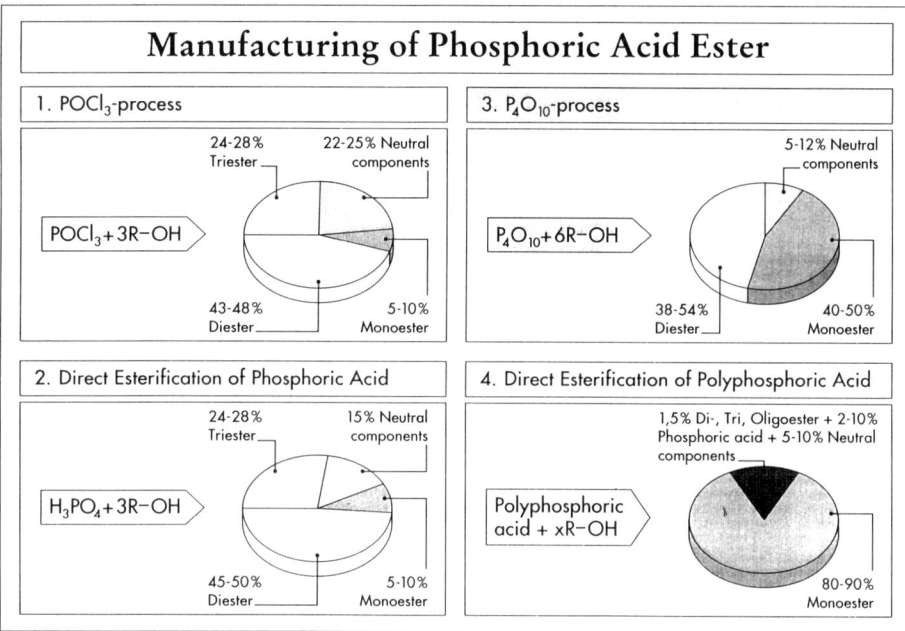

Figure 4

Tok 007-1 E

Physicochemical Characteristics (P_4O_{10}-Process)

Phosphate Ester	C–Chain	Consistency	m.p. [°C]	Paraffin RT/80°C		Soja oil RT/80°C		Isopropyl-palmitate RT/80°C		Water RT/80°C	
Lauryl phosphate	C_{12}	solid	47	□	○	□	○	□	○	□	□
Oleyl phosphate	$C_{16/18}$	solid waxy	45	□	○	□	○	□	○	□	□
Stearyl phosphate	$C_{16/18}$	solid	62	□	○	□	○	□	○	□	□
Isostearyl phosphate	C_{18}-iso	high viscous liquid	3	○	○	○	○	○	○	□	□
Octyldecyl phosphate	C_{18}-Guerbet	liquid	<0	○	○	○	○	○	○	□	□
Behenyl phosphate	C_{22}	solid	82	□	○	□	○	□	○	□	□
Oleth-3 phosphate	C_{18}	viscous liquid	−2	△	○	△	○	△	○	△	△

Solubility 3% Al: ○ Clear   □ Not Soluble   △ Turbid

Tok 006-1 E

Figure 5

chlororganic compounds. Direct esterification of phosphoric acid, on the other hand, is a far more elegant process. It permits a similar ester distribution – and thus creates comparable emulsifying properties – but produces no chlororganic compounds.

Due to the special structure of the P_4O_{10} molecule, a trigonal pyramidal configuration around every tetrahedrally bound phosphorus atom, converting phosphorus pentoxide with alcohols produces equal proportions of mono- and diesters with a low proportion of triesters and non-converted fatty alcohol.

Neutralizing this product mix requires a rather more alkaline solution than the process described previously, and this has to be taken into account when thickening the aqueous phase with salt-sensitive polymers. The polyphosphoric acid process produces the highest amount of monoesters with a possible 90%.

Depending on the reactants, however, sometimes a relatively high proportion (up to approx. 30%) of phosphoric acid (or phosphate salts after neutralization) is unavoidable.

Emulsifiers with minimal by-products, which are well-suited for use in cosmetics, can be produced with the P_4O_{10} method of synthesis. By varying the fatty alcohol reaction-components, products with different physicochemical properties can be obtained.

1.2 P_4O_{10} method – varying the alkyl substituents

The phosphoric esters discussed below were produced directly by reacting P_4O_{10} with the corresponding fatty alcohols. They are characterized in Figure 5 by a number of physicochemical data. An oleth-3 phosphate emulsifier is used as a benchmarker. Linear alkyl esters are all solid at room temperature. However, esters with alkyl branching are liquid down to temperatures of approximately 3°C, greatly simplifying the use of these products in continuous handling. Ethoxylated esters such as oleth-3 phosphates are also liquid at room temperature. With melting points over 50°C, products have been found to flake easily or form pellets, whereas melting points of 47°C or 45°C are not suitable for forming pellets due to caking. All phosphoric esters are insoluble in water and, even though oleth-3 phosphate dissolves in warm water, it gels on cooling. At a concentration of 3%, the branched products isostearyl and octyl decyl phosphate are soluble in paraffin oil, soya oil and isopropyl palmitate, both at 80°C and at room temperature. All other products are only soluble at elevated temperatures.

1.3 Interfacial Activity

An emulsifier's application performance is made up of very different properties, a few of which can be described and quantified by physical and physicochemical effects. Surface tension and interfacial tension are used here as an example of parameters which are extremely important to the performance of an emulsifier.

The surface activity of alkyl phosphates at the water/air interface was studied using a ring tensiometer made by the company Lauda. The solutions were made individually, as problems occurred with automatic dilution due to insolubility. All tests were performed at 25°C and a pH value of 7 (adjusted with NaOH). In the case of stearyl and behenyl derivatives, no reasonable data could be measured due to their low solubility.

Figure 6 shows plots of static surface tension against the logarithm of surfactant concentration. This type of plot normally shows a break point at the critical micelle concentration. Noticeable differences can be seen between the different phosphoric esters. The curve of the ethoxylated oleic acid ester is very flat, but has a break point at the lowest concentrations.

In contrast, isostearyl ester shows a break point at the highest concentration, but the curve gives the greatest reduction in surface tension. This effect cannot be explained by

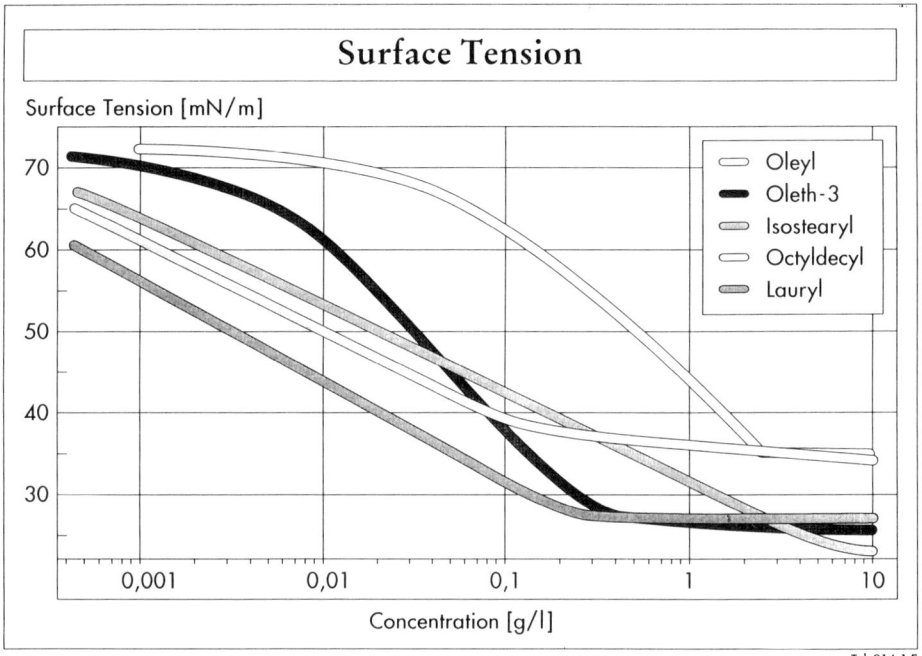

Figure 6

Table 1

Phosphate ester	MMW Na-salt	CMC		C$_{20}$		Surface Tension at CMC
		[g/l]	[mmol/l]	[g/l]	[mmol/l]	[mN/m]
▷ Lauryl	363	0.23	0.63	0.002	0.006	27.5
▷ Oleyl	469	2.43	5.18	0.405	0.83	35.2
▷ Isostearyl	485	4.06	8.37	0.013	0.027	23.4
▷ Octyldecyl	488	0.22	0.46	0.028	0.057	27.1
▷ Oleth-3	630	0.10	0.16	0.007	0.01	37.3

CMC -Value from Surface Tension

Tok 020-1 E

the only slight difference in length in the alkyl chain. A possible explanation is the different water solubility; oleth-3 ester dissolves far better than the isostearyl product. The curves of octyl decyl phosphate and lauryl phosphate approach each other at the break point and lead to the same reduction in surface tension.

Additional information on the curve is provided by the C_{20} value, which is listed in Table 1 together with CMC values and corresponding minimum surface tensions. This value denotes the concentration at which surface tension is reduced by 20 mN/m, *i.e.* in the range where there are still no micelles. Lauryl phosphate achieves this reduction even at concentrations as low as 0.0002% or 0.006 mmol/l. Interfacial tension of 1% aqueous emulsion solutions was measured with a Lauda drop volume tensiometer against three oils of different polarity (Figure 7). It is widely known that due to the greater electrostatic repulsion between head groups, ionic surfactants are not as active interfacially as non-ionic surfactants. This is also confirmed by the present results.

Phosphoric octyl decyl ester, however, is clearly the exception, showing the lowest interfacial tensions in relation to all three oils. Figure 8 shows interfacial tension against paraffin oil and soya oil as a function of logarithmic of surfactant concentration for stearyl and octyl decyl phosphate. As the method does not permit continuous dilution, individual measurements were carried out with specific concentrations.

Values of the CMC of the octyl decyl ester in the oil/water system do not agree with these obtained from measuring surface tension. This means that the emulsifier partitions between the aqueous and in the oil phases.

For paraffin, a noticeable shift in the CMC can be seen towards higher surfactant concentrations. The plateau value with paraffin, on the other hand, is lower than with soya oil. If the linear C_{18} is compared with the strongly branched phosphoric ester, the plateau values are far higher with both oils.

2   INVESTIGATIONS   (DIFFERENTIAL   INTERFERENCE   CONTRAST MICROSCOPIC INVESTIGATIONS)

Initial investigations into the phase behaviour of the phosphoric acid esters were conducted as measurements with a Zeiss standard polarizing microscope. The methods used to prepare the samples for the contact preparations and 5% aqueous solutions were the same for all emulsifiers. To produce the 5% emulsions, the esters were mixed at room temperature and pH 2.4 with a stirrer. The phase behaviour was observed for at least 48 hours at 25°C.

As a rise in temperature in ionic surfactants causes spherical micelles to re-form from anisotropic micelles, the optical investigations were also conducted at elevated temperature. The results are summarized in Table 2. At the high surfactant concentrations existing in the contact preparation, the behaviour was not seen to vary at the different temperatures, whichever ester was involved. In the case of lauryl phosphate, a liquid-crystalline lamellar phase with clearly identifiable oily streaks occurs in the high-concentration range. Adjacent to these there are crystalline regions. In the binary system comprising phosphoric acid ester and water, a liquid crystalline phase composed of anisotropic micelles forms at above 50°C. A texture typical of the anisotropic mesophase is displayed by oleth-3-ester after establishment of equilibrium in the high-concentration gradient of the contact preparation. In 5% surfactant solution it is possible to identify lamellar Maltese crosses alongside an isotropic phase.

The long-chain C_{18} and C_{22} esters display very similar phase behaviour. As the water flows slowly into the crystalline structure the resulting change is only minimal. Only after the temperature is raised to 70°C and the crystals have undergone homogenization do

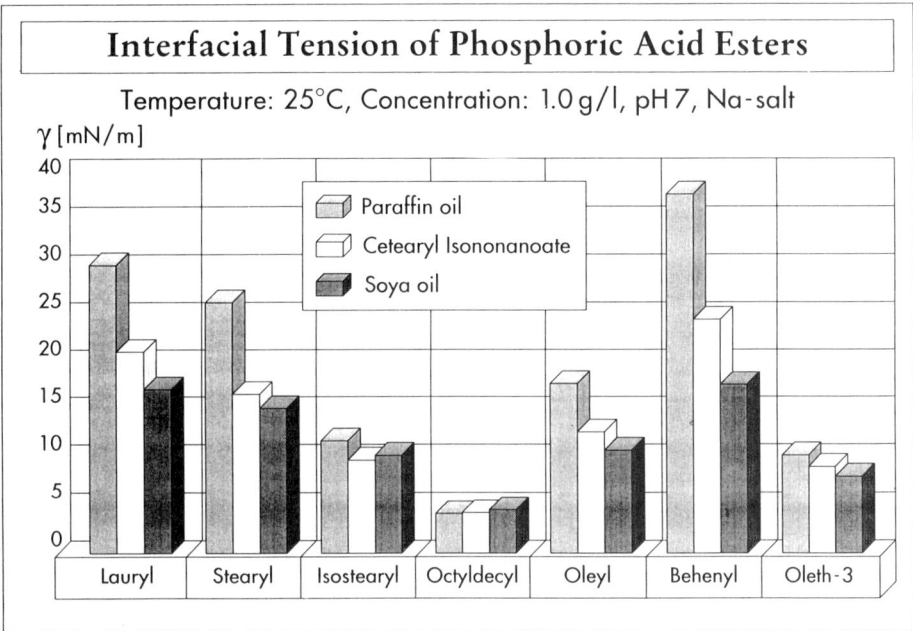

Figure 7

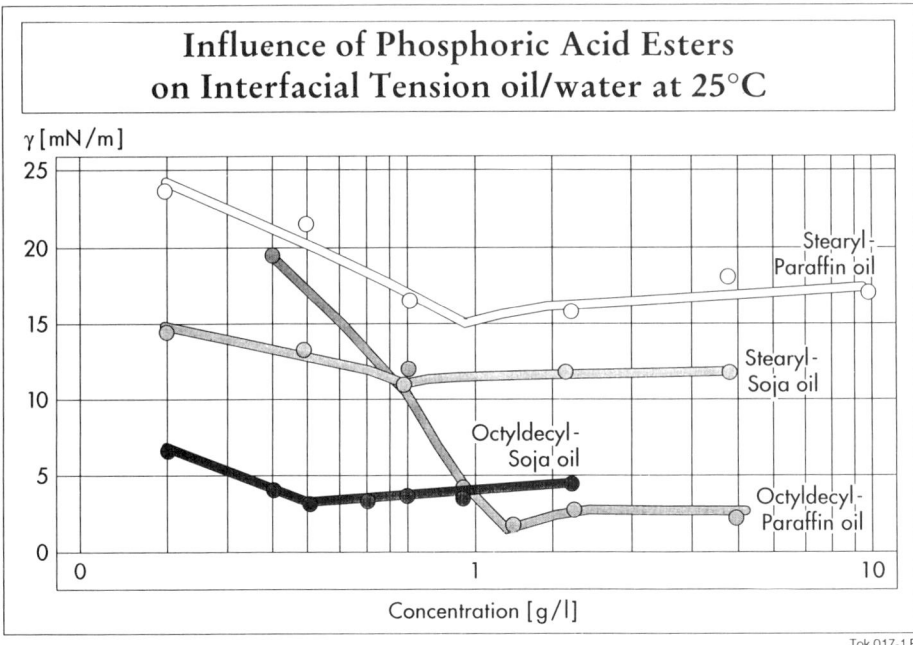

Figure 8

Table 2

Differential Interference Contrast Microscopy			
Phosphate ester	Contact Preparation with water	5% A.I. in water	
	RT / 70°C	RT	70°C
> Lauryl	lamellar Lα crystalline	smectic	liquid crystalline
> Oleyl	crystalline texture	two-phase isotropic + crystalline texture	two-phase
> Stearyl	dispersed crystals	dispersed crystals	
> Isostearyl	two-phase isotropic	two-phase isotropic	
> Octyldecyl	two-phase isotropic	two-phase isotropic	
> Behenyl	crystalline texture	dispersed crystals	
> Oleth-3	lamellar Lα crystalline	two-phase lamellar isotropic	

Tok 019 E

Micrograph with Polarized Light
C$_{12}$-Phosphoric Acid Ester, at 25°C, 100x

| after 1h self-seeding of cristallisation | lamellar drops + tubuli in two-phase region | Lα with crystals |

Tok 028-1 E

Figure 9

double refracting films form. In the binary system there are double refracting aggregates alongside an isotropic phase. Probably because of its better solubility and low melting range, the unsaturated ester has two isotropic phases at 70°C. When the temperature drops to approximately 62°C, the crystallization produces not only an isotropic texture but also a second, double refracting texture.

Throughout the entire temperature range from 25 to 70°C, in both the binary system and also at very high concentrations, the branched isostearyl and octyl decyl esters display two isotropic-liquid phases, which form an emulsion.

The next four illustrations show photomicrographs of typical textures as seen through a polarizing microscope. Figure 9 shows the crystallization behaviour (left-hand and right-hand photos) of lauryl phosphate and the liquid-crystalline lamellar droplets and tubuli in the two-phase region (centre photo). Figure 10 shows the L• texture with very clearly identifiable oily streaks of oleth-3-phosphate. The left-hand and centre pictures of Figure 11 show two isotropic phases, while the right-hand picture is an example of a homogeneous double refracting crystal film.

3 EMULSIFYING CAPACITY – THERMAL STABILITY OF THE PHOSPHORIC ACID ESTERS

Oil-in-water emulsions are affected by a wide variety of factors, such as emulsifiers, consistency modifiers, the oil phase and the manufacturing process. These factors determine the consistency at room temperature, the flow temperature in creams, and also both high- and low-temperature stability.

The object of the investigations presented here is to ascertain the emulsifying behaviour of the phosphoric acid esters on oils of varying polarity. The method of assessment chosen was to conduct the thermal stability test on various screening recipes at 40, 45 and 50°C. The liquid-crystalline gel structure in the external emulsion phase is built up by polyacrylates.

The results for the recipe comprising 18.8% lipids, 80% water, 1% emulsifiers and 0.2% consistency modifier are shown in Figure 12. In each instance the lipid phase is formed from one oil, the oils used being paraffin, cetearyl isononanoate, isopropyl palmitate, soya oil and squalene. The method of assessment for the liquid emulsions used a 10-point scale covering the descriptions which are normally used. These extend from complete coalescence, through distinct separation, sharp colour contrast and only slight coalescence, to completely homogeneous, unchanged emulsion.

All test results for the three temperatures have been summarized as a mean score for each oil and are depicted in a type of "spider's-web" graph for each ester. The best results are achieved by the oleyl, octyl decyl and behenyl phosphoric acid esters. After a storage time of 30 days the emulsions have not changed when assessed by any of the assessment criteria.

The stearyl and isostearyl esters score 80 in the emulsification of non-polar paraffin. In this comparison the C_{12} ester achieves an emulsification score of 100 but only with soya oil.

In another model recipe (Figure 13), which contains a higher lipid content of 26.7%, along with 70% water, 3% emulsifier and 0.3% thickener, the ratio of polar to non-polar components is changed within the oil mixture. The stability tests are carried out over a period of three months at raised temperatures of 40, 45 and 50°C. Figure 14 summarizes the results for oil mixture I, which consists of 50% paraffin, 30% IPP and only 20% of a vegetable triglyceride. The bars in the background give the results at 40°C, the front row

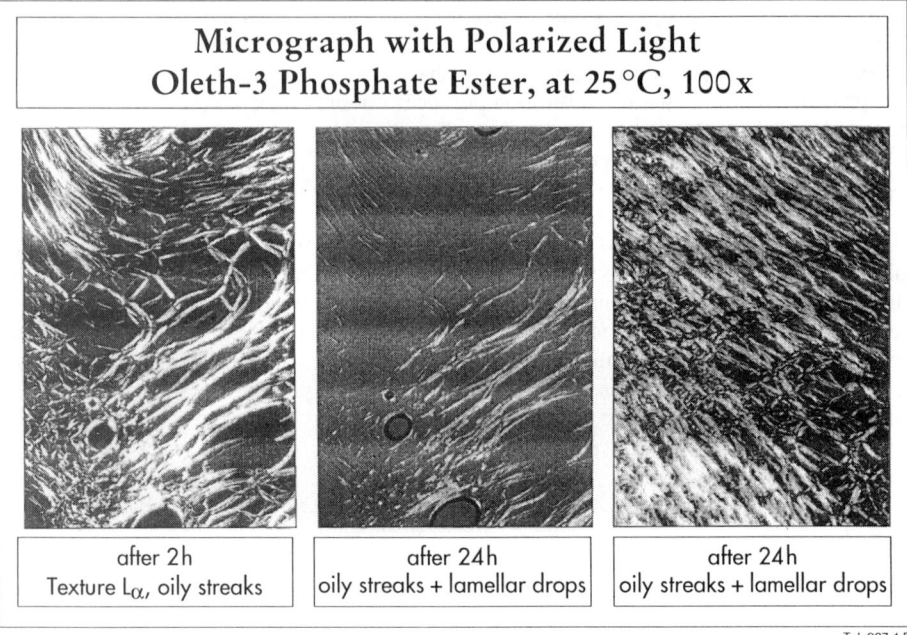

Figure 10

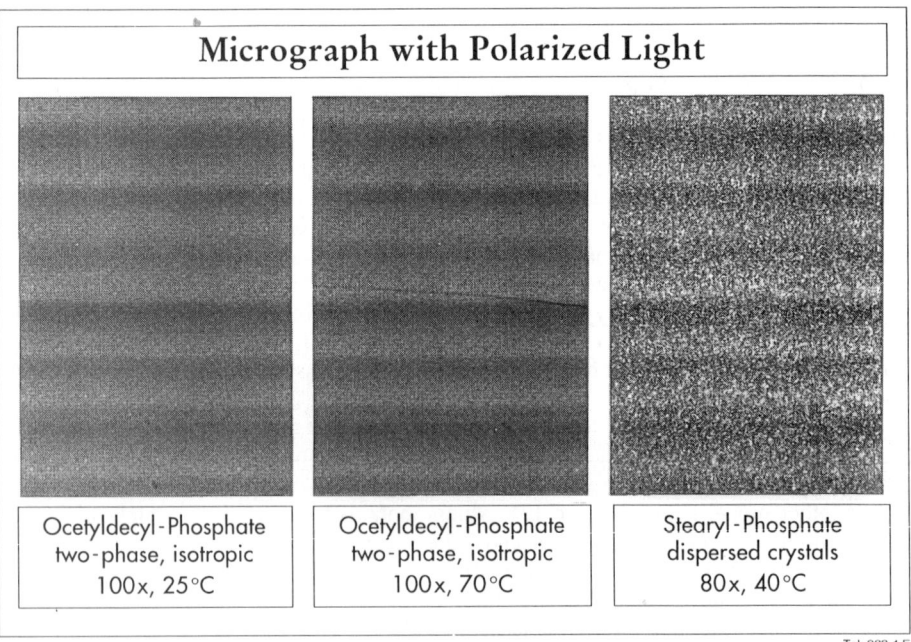

Figure 11

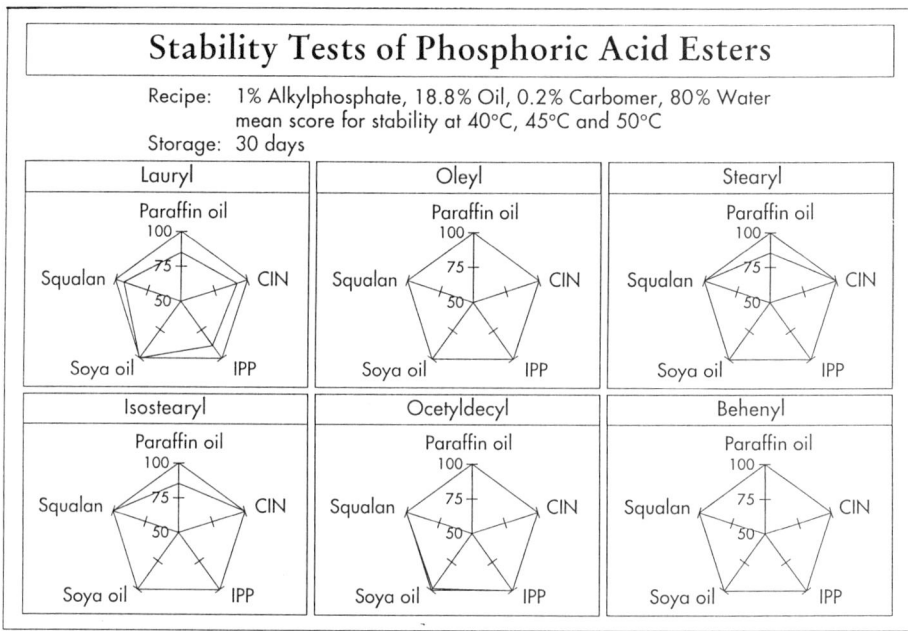

CIN: Ceteraryl Isononanoate, IPP:Isopropylpalmitate   Tok 026 E

Figure 12

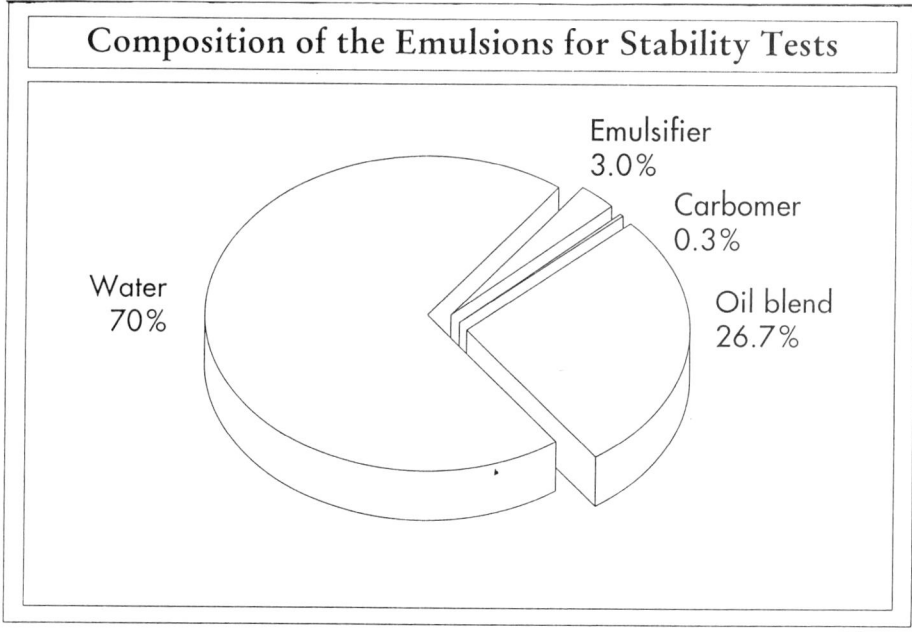

Figure 13

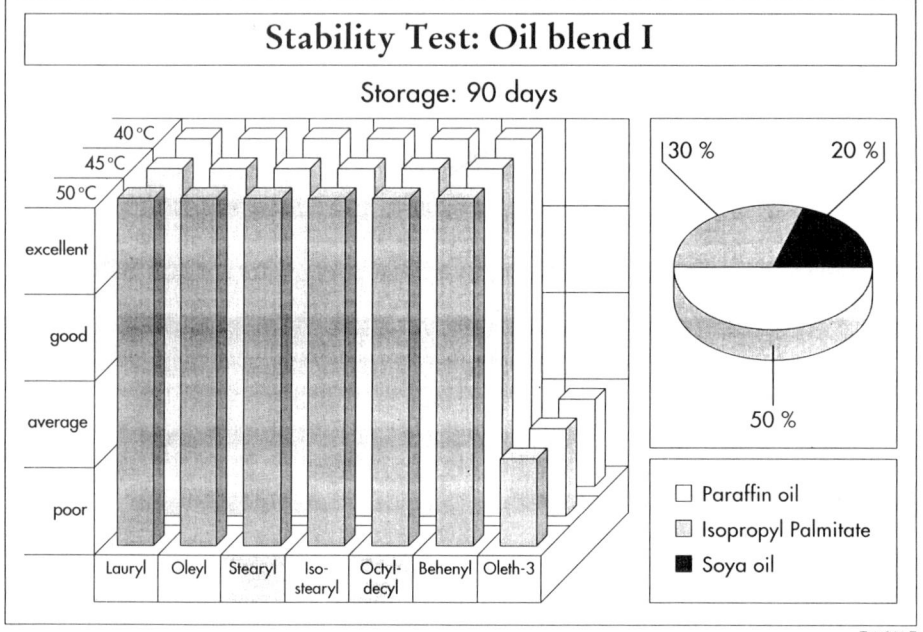

Figure 14

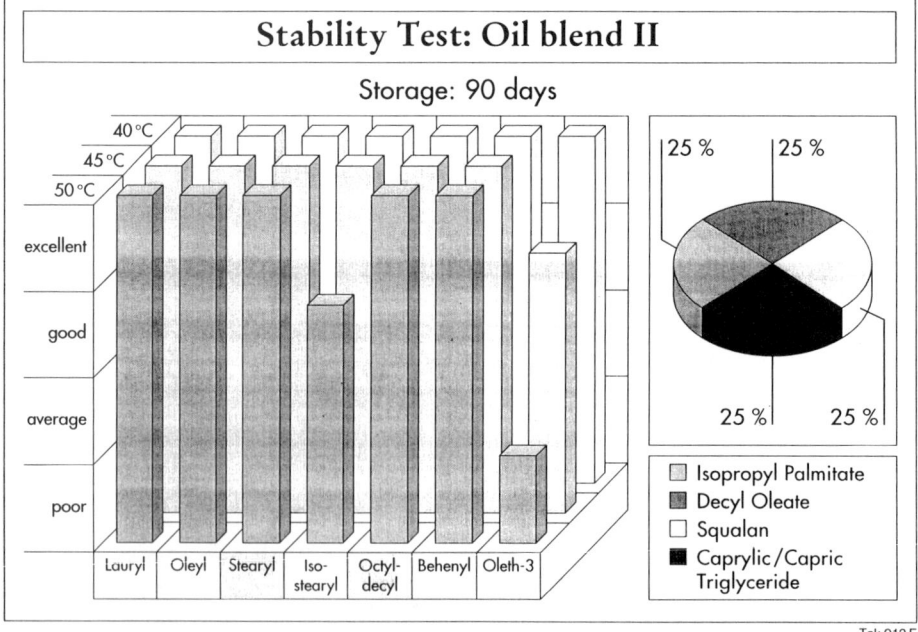

Figure 15

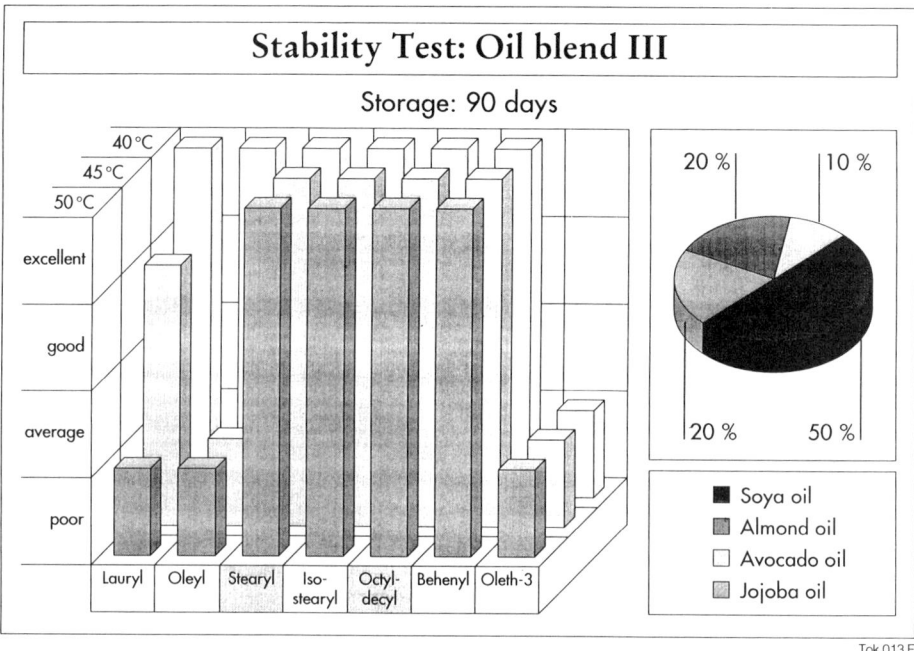

Tok 013 E

Figure 16

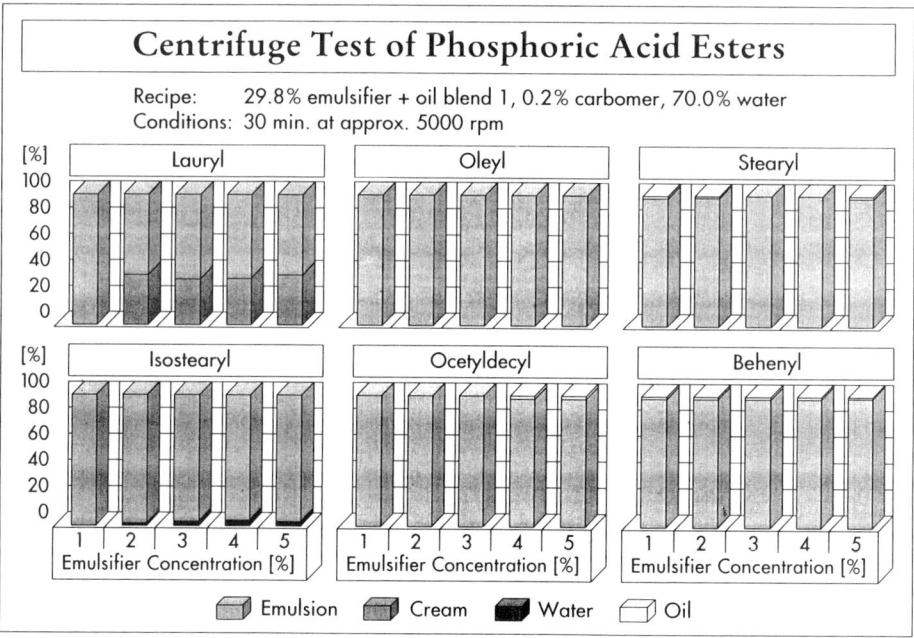

Tok 025 E

Figure 17

of bars the results at 50°C. All phosphoric acid esters based on the P_4O_{10} process show excellent results, whereas those for the ethoxylated oleic acid ester fall off sharply.

In the next test series (Figure 15) the oil mixture comprises equal parts of squalene, IPP, decyl oleate and caprylic/capric triglycerides. The results show a slight drop in values in the 50°C test with isostearyl ester and in both the 50 and 45°C tests with oleth-3 phosphate.

The oil mixture in the next test series (results in Figure 16) consists entirely of vegetable oils. These oils represent a challenge to any combination of emulsifier and consistency modifier. The recipes with the stearyl, isostearyl, octyl decyl and behenyl esters all show excellent stability (Figure 17).

3.1 Guide Formulations

The development of guide recipes was concentrated on stearyl and octyl decyl phosphates, both of which have so far displayed excellent emulsifying properties.

All emulsions were manufactured by the direct method, a technique that is most probably the favourite method currently used in the cosmetic industry. In this process, the emulsifiers and consistency modifiers are added to the oil phase. If heating is needed owing to the nature of the ingredients, or if it is desirable in any case, the oil and water phases are heated separately to 65-85°C and then mixed to form a low-viscosity oil-in-water emulsion. After being homogenized to a specific degree of dispersion (approx. 4 •m), the emulsion is cooled with moderate stirring. If polyacrylates are used as consistency modifiers, the emulsion is afterwards rehomogenized (at approx. 35°C). if other thickeners are used (alcohols/glycerin monostearate = GMS), final homogenization is not always essential.

Oil-in-water lotions can be produced by using 1.5% stearyl phosphate or 1.5% octyl decyl phosphate. The oil mixture contains only 25% paraffin; the remainder consists of vegetable oils/esters. These recipes are also ideal as basic recipes for sunscreen lotions. Besides being free from ethoxylates, they have a simple emulsifier system in a low concentration (single product).

All of our experiments have also shown that the hydrophilic nature of the stearyl and octyl decyl esters is in sufficient to produce stable oil-in-water creams. A strongly hydrophilic component (HLB>15) will be needed, though only in very small quantities.

This role may be performed by an acyl glutamate in a concentration of 0.6% as is, or 0.15% relative to 100%. The polyglyceryl-2-monostearate acts both as a co-emulsifier and also as a consistency modifier.

4 SUMMARY AND PROSPECTS

To summarize the most important features of alkylphosphates are described in Figures 18 and 19, *i.e.*

- Phosphoric acid esters are synthesized by the P_4O_{10} process based on the C_{12}, C_{18} and C_{22} chains including branched Guerbet ones.
- All phosphoric acid esters display good emulsifying power. The results obtained with the stearyl and octyl decyl esters in particular are excellent. Octyl decyl phosphate has proved to be an excellent emulsifier for oils of widely differing polarity.
- Under normal conditions (RT = approx. 25°C), phosphoric acid esters are extremely insensitive to pH over range extending from approximately pH 2 to pH 12. Even after one year in AI, which in 100% form has a pH of approximately 2.5, octyl decyl phosphate shows no sign of hydrolysis.

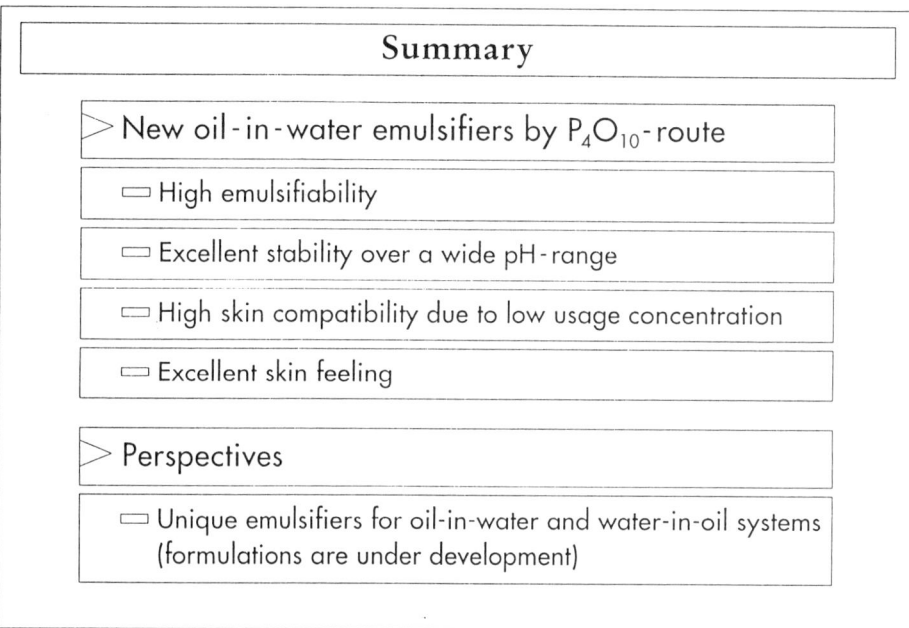

Figure 18

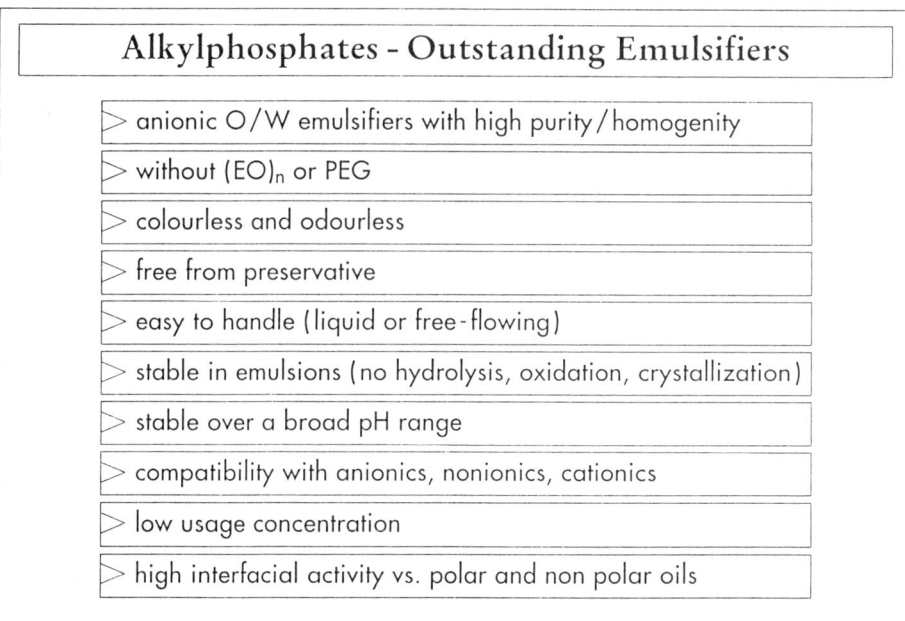

Figure 19

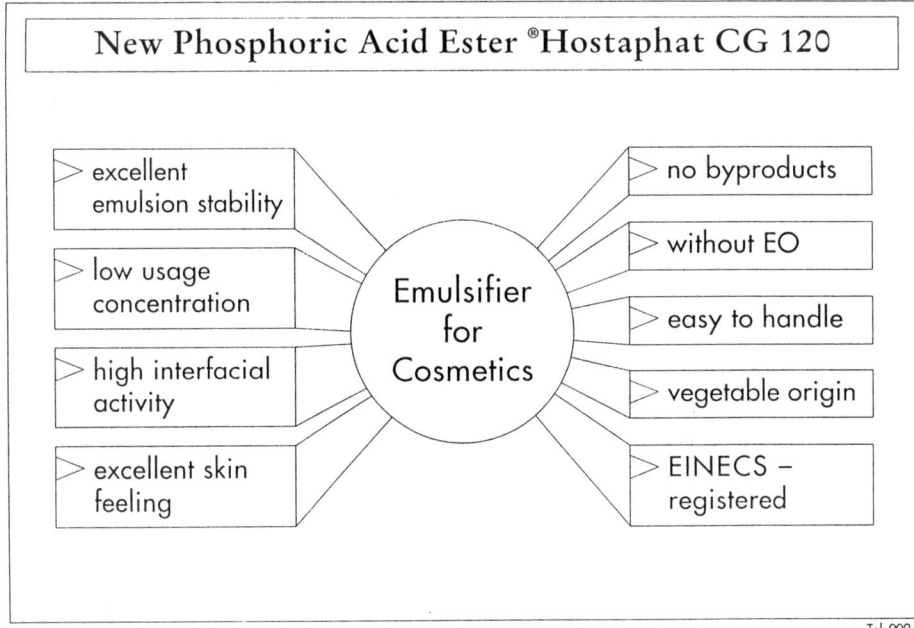

Figure 20

Product data ®Hostaphat CG 120

Emulsifier for the manufacture of oil-in-water emulsions
for the cosmetic industry

⊳ Chemical composition	Phosphoric acid octadecyl ester, branched and linear
⊳ INCI designation	Octyldecyl Phosphate (registration in progress)
⊳ Appearance (20°C)	clear, colourless liquid

⊳ Chemical and physical data

– Acid value	185-195 mg KOH/g	DIN 53402
– Iodine colour number	max. 1.0	DIN 6162
– pH-value (tel quel)	2.0-4.0	DIN 53996
– Viscosity (20°C)	300-400 mPas	DIN 53015

Tok 014-1 E

Figure 21

- The small quantities of emulsifier used and good compatibility with stabilizers/consistency modifiers helps ensure good skin tolerance. In epicutaneous testing on 50 subjects, both octyl decyl and stearyl phosphates (5% in paraffin oil) produced no positive or doubtful reactions. There was therefore no indication of any primary irritative effect on the skin.
- In an in-house test (blind study) model emulsions with octyl decyl phosphate or stearyl phosphate were estimated found to have a very pleasant feeling on the skin.

The main properties and performance advantages of the new, cosmetic grade phosphoric acid ester, ®Hostaphat CG120, are given in Figure 20 and typical product data shown in Figure 21. This product is based on a C_{18}-Guerbet alcohol and produced by the P_4O_{10}-process (Figure 22).

The development work has shown that the octyl decyl and stearyl phosphoric acid esters can be used to meet the requirements imposed by a large number of differently formulated oil-in-water emulsions. There were clear indications that the emulsifiers are also suitable for use in the production of water-in-oil emulsions. Development work along these lines is now proceeding.

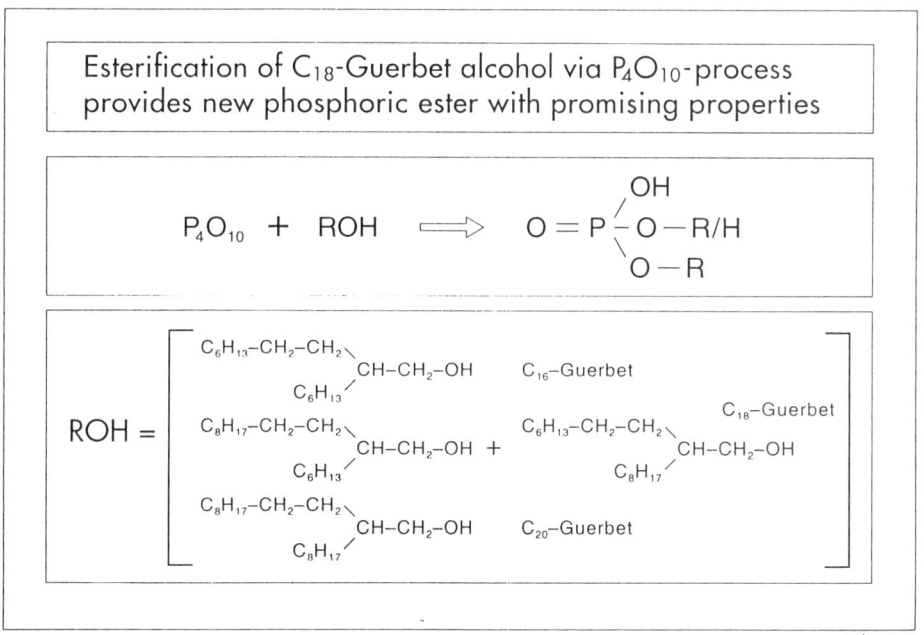

Tok 008-1 E

Figure 22

The Application of Surfactants in Preventing Gas Hydrate Formation

N. J. Phillips[1] and M. A. Kelland[2]

[1] TECHNICAL SERVICES GROUP, TR OIL SERVICES, DYCE, ABERDEEN AB21 0GP, UK
[2] RF-ROGALAND RESEARCH, 4004 STAVANGER, NORWAY

1 INTRODUCTION

In today's highly competitive and cost conscious oil and gas industry there is an increasing demand for improved efficiency in design and operation for both existing and new offshore field developments. To improve the economic recovery of some marginal reserves, and to enhance those already operating, complex gathering networks have been installed often transporting partially or unprocessed production streams. These systems are predominantly subsea, some at depths over 1000 meters and are designed to transport multiphase fluids comprising produced water, liquid and gaseous hydrocarbons. However, the combination of these more extreme conditions can result in hydrate formation, and, if not treated may restrict or block production of pipe and flowlines with serious consequences.

The formation of natural gas hydrates has long been recognised as a potential problem to the oil and gas industry ever since Hammerscmidt identified pipeline hydrates in the 1930's[1,2]. During recent years the general trend within the industry to make ever more efficient designs and to minimise cost wherever possible has lead to a considerable effort both to understand more about hydrate formation and also its prevention.

Gas hydrates are crystalline compounds formed by the hydrogen bonding of water molecules around non-polar gas molecules forming a lattice structure with interstitial cavities of varying size. When a minimum number of these cavities are occupied by the gas molecules then the crystalline structure becomes stable forming a solid gas hydrate[3].

The conventional chemical treatments for the prevention of hydrate formation are glycol (mono-ethylene glycol) and methanol for thermodynamic inhibition and tri-ethylene glycol for gas dehydration. However, it is increasingly recognised that the cost of operating the supply, deployment and regeneration systems associated with the use of glycol or methanol[4] can be prohibitively expensive for new developments. Additionally, the operating efficiency of existing mature assets could be significantly improved by replacing these conventional treatments with alternative low dose hydrate treatment. Health, Safety & Environmental (HS&E) concerns associated with the use of large volumes of chemical has also prompted a review in to current treatment regimes. This is particularly true of methanol which is both toxic and flammable. Although strict controls are currently applied to the use and disposal of chemicals in most producing regions there is potential for further legislation to restrict both the use and disposal of all chemical treatments. The oil and gas production industry has responded to both the economic and HS&E concerns by identifying alternative low dose chemical hydrate inhibitors[5-7]. These inhibitors can be divided into two groups, kinetic inhibitors and anti-agglomerator (AA) agents.

Kinetic inhibitors are usually polymers with surfactant properties which delay hydrate nucleation or crystal growth. Anti-agglomerators, although still surfactant based, modify the hydrate crystals, such that a degree of nucleation and growth is allowed, but prevent hydrate agglomeration and deposition. Thus, a transportable hydrate slurry is formed.

Kinetic inhibitors have been trialed in the UK sector of the North Sea[8-10]. Here two mature fields, one offshore and one onshore, have replaced their ageing mono ethylene glycol systems with the use of a low dose kinetic inhibitor. The offshore field has been using the kinetic inhibitor for over one year and has saved $45m in capital (CAPEX) and operating (OPEX) costs. In addition the ETAP development, (200,000 BOPD, 450 mmscf/d) has used kinetic inhibition as a base design case to avoid the capital cost of a methanol supply system and will be using a kinetic inhibitor from day one of production. It is anticipated that a $40m CAPEX saving will be realised p.a.

This paper is primarily concerned with the application of surfactant based chemistries that are employed in the inhibition of gas hydrates. Attention will also be given to the mechanisms by which the materials work.

2 GAS HYDRATES

Natural gas hydrates are crystalline structures, known as clathrates, of polyhedral cavities in which 20 or more water molecules form a hydrogen bonded network (the host) filled with gas molecules (the guest) occupying each cavity. Typical hydrate forming guests include CH_4, C_2H_6, C_3H_8, i-C_4H_{10}, CO_2, H_2S, $CHCl_3$, and the noble gases[3,11,12]. These structures form the building blocks for growth of the hydrate which normally propagate by repeating one or more structural units.

There are three principle hydrate structures that exist, the most well known in hydrocarbon production and processing are called SI and SII. SI is a body-centered cubic structure formed with natural gases comprising molecules smaller than propane, and SII a diamond lattice within a cubic framework formed from natural gases, with or without oils, containing larger molecules than ethane but smaller than pentane. The only other structure that is proven to exist is type H of which there is significant research associated with its influence on hydrate formation.

The basic structural unit for both SI and SII is a cavity bounded by 12 pentagonal faces, abbreviated 5^{12}. Linking the vertices of 5^{12} cavities results in SI, while linking the faces of 5^{12} cavities gives SII. The interconnecting cavities are comprised of two or more hexagonal units giving the abbreviated form for S1 as $5^{12}6^2$ and SII as $5^{12}6^4$, Figure 1. Nucleation and growth of the hydrates is depicted in Figure 2. Here water and gas combine to form labile clusters which agglomerate to form stable nuclei from which nucleation and growth propagate.

From the stand point of inhibiting hydrates formation it is currently the SII form that is of major interest to the oil and gas industry and it is this structure that shall be referred to in discussions concerning surfactant interactions.

3 HYDRATE INHIBITION

Hydrate inhibitors can either be termed thermodynamic, kinetic, and anti-agglomerants. These are summarised as:

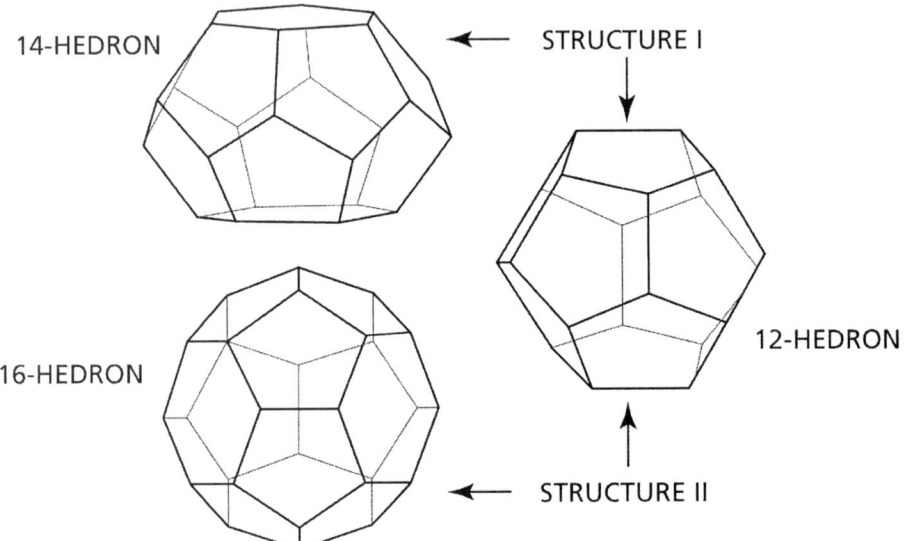

Figure 1 *Structural units for hydrate types SI and SII*

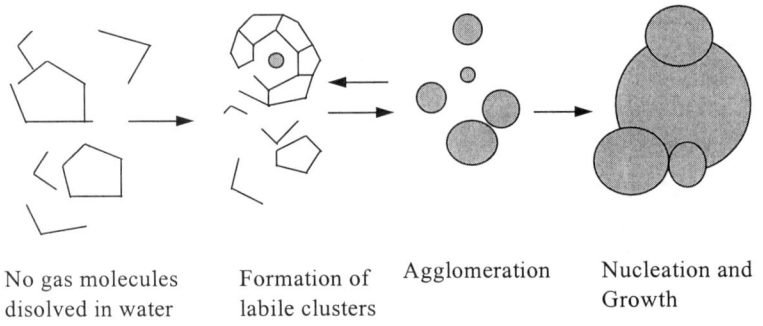

| No gas molecules disolved in water | Formation of labile clusters | Agglomeration | Nucleation and Growth |

Figure 2 *Nucleation and growth of hydrates*

Thermodynamic: These are added such that they alter the chemical potential of water in either the liquid or hydrate phase, thereby shifting the point at which hydrates form (similar principle to antifreeze). Examples are methanol or ethylene glycol which are added at concentrations between 10-60 wt %.

Kinetic: Here the emphasis is on the delay of initial hydrate nucleation. The mechanism of inhibiton does not alter the thermodynamics of hydrate formation but is a surface absorption phenomena of the inhibitor such that growth is retarded, thus delaying hydrate formation for a given time, known as induction time. Unlike thermodynamic inhibitors the general quantity sufficient for inhibition is < 1wt%.

Anti-Agglomerators: These differ from kinetic inhibitors in that they allow a certain amount of growth of the gas hydrate but then act to suppress the continued propagation and agglomeration by dispersing the hydrate in the oil phase. Thus, a transportable slurry is formed.

4 SURFACE EFFECTS

Before a discussion can be given with regard to the various low dose chemistry forms that have been adopted for the suppression of hydrates, and which replace methanol and glycol's, an understanding of the mechanistic features resulting in inhibition must first be given. Here examples of ionic and non-ionic polymers with surfactant properties and non-polymeric surfactants will be used to define the chemical principles associated with both kinetic and anti-agglomerate inhibition and how they interact at a hydrates surface.

Active molecules are found to adsorb strongly to the surface of a propagating hydrate crystal or pre-nuclear hydrate-like clusters. In this process, they change the energy of the surface of the hydrate crystal or cluster and so change its growth characteristics. This can be likened to a lock-and-key mechanism, with part of the inhibitor showing a very specific and strong interaction for the microstructures that characterise the hydrate surface or interface, while another part of the inhibitor then interferes with the continued growth process. In order to exploit this mechanism it is necessary to understand the key surface features of the hydrate involved in the process.

The topography of the hydrate surface for SII has been identified[13,14] as having potholes and channels. The potholes arise from partially formed cavities or cavities that have been cleaved, Figure 3. This figure shows an open structure where the potential adsorption sites are located at the terminal hydrate water molecules, forming hydrogen bonds. The channels in the crystalline structure have been associated with adsorption of inhibitor side groups subsequent to the main body of the inhibitor being hydrogen bonded at the pothole sites.

To optimise the use of the sites on the hydrate surface various key functions have been identified which maximise the ability of a chemical species to adsorb. Some of the optimum structures are polymers with cyclic (non-aromatic) pendant chemical groups[15] which vary in conformation from isotactic, syndotactic through atactic symmetry. Commonly, and most preferably, these cyclic groups should be five or seven membered rings. Although these rings may contain only carbon atoms it has been proven that a heterocyclic ring containing either oxygen or nitrogen with a polar moiety, such as oxygen, attached to the adjacent carbon atom provides efficient inhibition. The greatest efficiency has been observed for ester or amide linkages to a polymer 'backbone'. The polymer backbone has also been attributed to forming a barrier to further availability of hydrate former molecules on the growing surface[14,16]. A typical example of a material

Figure 3 *SII hydrate lattice showing the partially filled cavities "potholes" and channels*

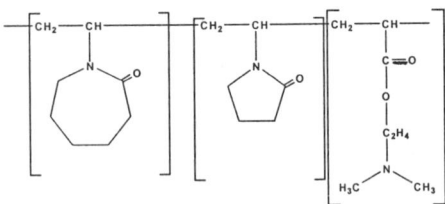

Figure 4 *A terpolymer showing the monomer structural units, left to right N-vinylcaprolactam, N-vinylpyrrolidone, and dimethylaminoethylmethacrylate*

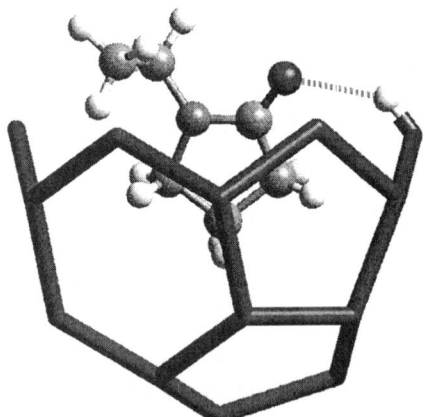

Figure 5 *Adsorption sites showing carbonyl oxygen bonding and entry of the pryolidone ring into the pothole*

that possess most of the above requirements is a terpolymer containing the groups N-vinylcaprolactam, N-vinylpyrrolidone and dimethylaminoethylmethacrylate, Figure 4. The interaction for N- vinylpyrrolidone is depicted in Figure 5. Here the carbonyl oxygen can form one or two hydrogen bonds to the hydrate water molecules. The size of the "pothole" allows the pyrrolidone ring to sit within the hole.

5. KINETIC HYDRATE INHIBITION

5.1 Theory

The concept of using polymeric surfactants derived from the discovery of fish anti-freeze proteins[17] - AFP's, Figure 6, which are able to kinetically prevent the formation of ice in fish preventing them from freezing. This finding prompted researchers to investigate the possibility of developing effective chemical structures that could kinetically inhibit hydrates produced in oil and gas recovery. The main difference between ice formation and that of hydrates is that ice is formed in the bulk phase, whereas, hydrate nucleation is dependent on the presence of hydrate formers of which the highest concentration is at the interface between water and oil or water and gas. Therefore when defining potential chemistry for kinetic inhibition there are three basic design principles;

1. *Find functional groups that bind strongly onto hydrate crystal surfaces*: Here hydrogen bonds will be the strongest, therefore carbonyl groups with lone pair oxygen atoms, and hydroxyl groups will be preferred. Structures possessing these features include: amides, esters, lactams, lactones, lactams etc. The use of polymers with lactam groups, such as pyrrolidone and capolactam, Figure 7, were identified by Lederos[7] as effective kinetic hydrate inhibitors. Here the binding may take place by first a Van der Waals interaction between the hydrophobic part of the pendant group and a vacant open cage on the hydrate surface. This interaction is further stabilised by an adjacent group that is able to hydrogen bond to the surface. Thus the hydrogen bonding from the carbonyl group of amides and lactams are able to "lock" the molecule onto the surface.

2. *Maximise the number of hydrate interactions with the inhibitor*: A backbone is selected such that the most efficient orientation and spacing of the functional groups is achieved. An example of this is for N-vinyl amides and their polymers which results in *cis* and *trans* conformations. These isomeric structures may be critical in determining the polymers interaction with the hydrate nuclei and crystals.

3. *The inhibitor must reside or sit near to the oil-water and or gas-water interface*: This may be achieved by adding hydrophobic co-monomers to the polymer either mixed in the polymer or as blocks on the ends of the polymer chains. The blocks may be preferable as they interfere less with the functional groups. The hydrophobic blocks will sit in the hydrocarbon phase and position the functional group towards the interface.

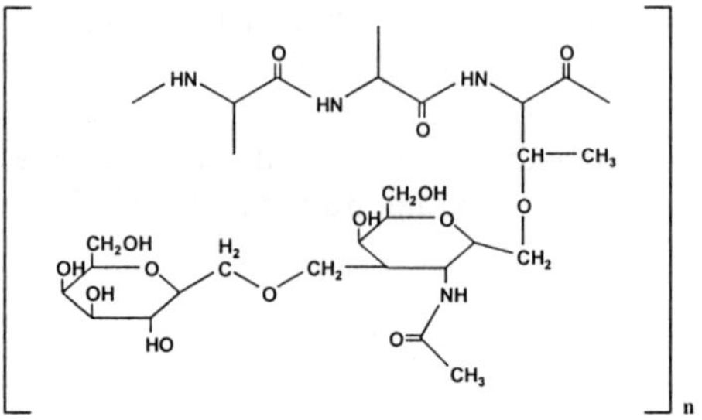

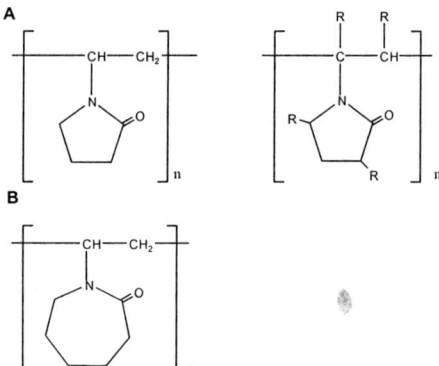

Figure 6 *The structural unit of the fish anti freeze protein - a glycopeptide.*

A

B

Figure 7 A *Structural units of poly N-vinylpyrolidone and its butyl derivative where R = C₄H₉* **B** *Poly N-vinylcaprolactam*

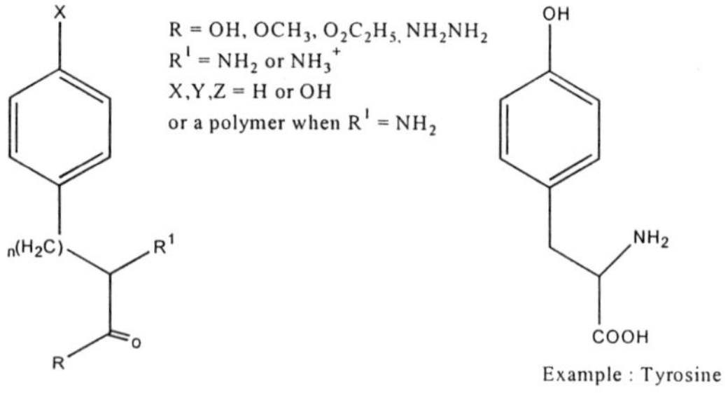

R = OH, OCH₃, O₂C₂H₅, NH₂NH₂
R¹ = NH₂ or NH₃⁺
X,Y,Z = H or OH
or a polymer when R¹ = NH₂

Example : Tyrosine

Figure 8 *Tyrosine derivatives*

5.2 Types of Inhibitor

One of the first reports detailing chemistry associated with hydrate inhibition by chemicals other than methanol or glycol was issued by Conoco in 1989. This identified the use of aminophosphonic acids, however there was no clear indication as to the kinetic effect on hydrate formation[18]. The first time that the chemistry of kinetic inhibitors was truly described was in a patent from BP, UK, laid open in 1993. Here tyrosine, particularly L-tyosine and L-tyrosine methyl ester, and examples of its derivatives, including some polymeric forms were described[19], Figure 8. The use of polymers with surface activity in preventing hydrates was followed up by Shell with a patent defining tests on many classes of polymers[20]. This patent describes polymers of N-vinylpyrrolidone, principally the homo polymer, poly(N-vinylpyrrolidone) (PVP) and its polybutyl derivative

The work on polymers with surface active properties towards kinetically inhibiting hydrate growth was extended in a patent laid open in 1994 by Sloan and co-workers at the Colorado School of Mines[15]. Here polymers with pendant groups containing five, six, and seven-membered heterocyclic ring systems were described. This included polymers of N-vinylpyrrolidone (5-ring), saccharides (6-ring) and N-vinylcaprolactam (7-ring) (Figures 4,5) Also a terpolymer combining both 5 and 7 membered rings together with an aminomethacrylate was quoted, (Figures 4,7).

Subsequent to this N-methyl-N-vinylacetamide (VIMA) copolymers were listed in a patent[21]. The VIMA copolymers were acryloylpyrrolidine (APYD) and N-substituted methylacrylamides such as N-isopropylmethacrylamide (iPMAM), Figure 9. These polymers, it is suggested, can be combined with various surfactants such as sodium n-butyl sulphate and its sulphonate analogue, or sodium n-pentanoate in a synergistic relationship with regard to kinetic inhibition. Small alkyl zwitterions, Figure 10, are also mentioned as having a beneficial effect.

The benefits of blending polymeric surfactants with other surfactants such as corrosion inhibitor bases was highlighted by BP[22]. Primarily this consisted of a blend of a polymer with N-heterocyclic carbonyl groups and a salt of a corrosion inhibitor, generally the bromide salt of a quaternised amine.

6. ANTI-AGGLOMERATORS (AA's)

6.1 Theory

When hydrates form the initial stage is formation (nucleation) of small embryo particles. These particles grow beyond the critical nuclear size forming crystals which subsequently agglomerate to larger sizes. A solid hydrate mass is then formed as a result of this process. The forces binding the agglomerates are weak and therefore they are easily disrupted by shearing and or impaction. An anti-agglomerate inhibitor (AA) is different from a kinetic inhibitor in that hydrate particles are allowed to form, but the AA binds to the hydrate particle surface, preventing agglomeration, forming a dispersion of particles in a hydrocarbon phase. Unlike kinetic inhibition the presence of a hydrocarbon

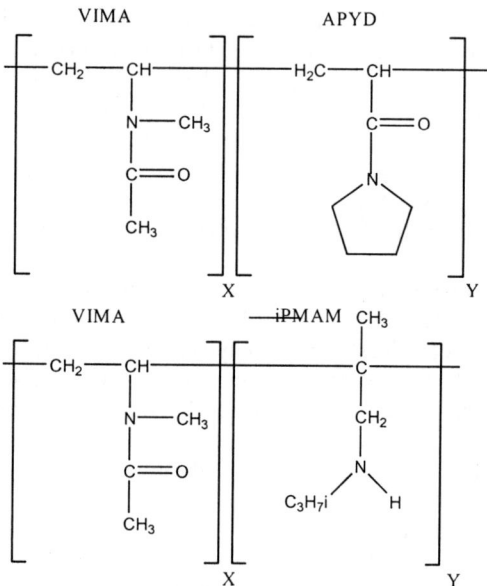

Figure 9 *N-methyl - N-vinylacetamide
(VIMA) & acryloylpyrrolidine (AYPD)
N-isopropylmethacrylamide (iPMAM)*

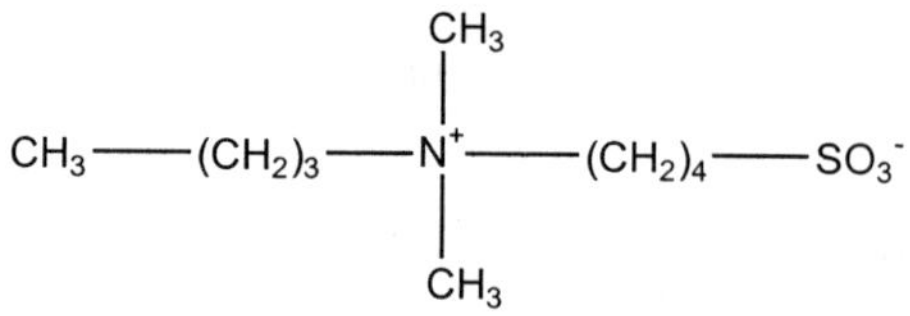

$$CH_3-(CH_2)_3-\overset{\overset{\textstyle CH_3}{|}}{\underset{\underset{\textstyle CH_3}{|}}{N^+}}-(CH_2)_4-SO_3^-$$

Figure 10 *Sulphonated alkyl zwitterion*

phase is essential for the application of AA chemistry. This allows the dispersed hydrate particles to be transported as a slurry.

The key requirements for anti-agglomeration and dispersion are;

1. Adsorption by the molecule on hydrate nuclei or crystals, controlled growth
2. Prevent hydrate crystals from agglomerating
3. Non-adsorbed part of the molecule must be able to interact with a hydrocarbon phase

Surfactants would seem the obvious choice as AA's as they posses both hydrophilic and hydrophobic properties, modify electric charges on surfaces and can introduce steric barriers which increase dispersion stability. Additionally surfactants modify the adsorption characteristics of a surface enabling enhanced adsorption of other materials.

The building block for hydrates is water. This has a high dielectric constant $(6.95 \ e^{-10} \ J^{-1} \ C^2 \ m^{-1})$, which influences the hydrate particles as they begin to form producing a small δ-ve charge due to proton loss. Any substances therefore that possess a δ+ve charge, or greater, would preferentially adsorb. Cationic surfactants are ideal for this with non-ionic surfactants containing hydrophilic groups also being preferable.

6.2 Cationic Surfactants as AAs

Common cationic surfactants are based on quaternary nitrogen carrying a positive charge (less common those containing phosphorous and sulphur) which is attracted to a wide variety of negatively charged surfaces, e.g. the surface of a hydrate. The cationic charge alone is insufficient to prevent the agglomeration of the hydrate particles. This can be achieved by incorporating a hydrophobic group with the cationic charge. This hydrophobic group will act to separate the hydrate particles by "oil-wetting" them and dispersing them into the hydrocarbon phase.

As it is the cationic group that adsorbs to the hydrate surface this group (and not necessarily the whole molecule) needs to reside in the water phase, where hydrate nucleation occurs. To achieve this the cationic group may contain small alkyl groups, which can attach to hydrate nucleation or crystal growth sites. This will be via a similar mechanism to that for KI inhibition where this alkyl group(s) is partially contained in an open, or incomplete, hydrate cage / cavity. The optimum alkyl groups in quaternary headgroups of general formula N^+R_3X are n-butyl, pentyl and iso-pentyl. Shell has exploited this idea in a recent patent[23]. Examples given in the patent are quaternary ammonium (or phosphonium salts) with simple alkyl chains ($R' = C_{12-18}$) and with $R =$ butyl, iso-butyl and iso-pentyl.

Assuming $-N^+R_3X^-$ are useful end groups where $R = C_{4-5}$ (iso or normal), then the question is how to design the rest of the molecule to prevent agglomeration. The use polymeric cationic surfactants may not be necessary if monomers adsorb strongly to the surface forming a stable membrane around the hydrate crystal. Clearly a long hydrophobic oil-wetting group is needed in the molecule. The use of a spacer group between the hydrophobe and the cationic group ensures that the cationic portion of the molecule protrudes well into the water phase, whilst maintaining its presence near the oil/water interface. Therefore, one class of AA may ideally be designed with the following features:

1. A long hydrophobe
2. A cationic group
3. A spacer group between the hydrophobe and cationic group.
4. Preferably one or more C_{4-5} (iso or normal) groups on or near the cation
5. Ability to disperse the water in the hydrocarbon fluid but not form a stable emulsion that needs extra process treatment.

These 5 factors are discussed in more detail below.

6.2.1 The Hydrophobic Group. To achieve the greatest stability of cationic surfactants in the hydrocarbon phase, long alkyl groups with a high lipophilic nature such as C_{14-18} should be used. Additionally the use of long branched alkyl groups, i.e. tri- or tetra-isobutylene are alternative options.

6.2.2 Cationic Groups. The centre cationic group is most commonly nitrogen. This group is attached to a hydrophobic group such that it points out into the water phase whereby it adsorbs on a hydrate. The main drawback for cationic surfactants in today's environmental demands is the high toxicity.

6.2.3 Spacer Groups. Spacer groups are used to extend the distance between the center of cationic charge and the hydrophilic alkyl chain group. Potentials for this could be via an extra amide or ester linkage or a polyglycol or both. The Figure below shows examples for monomeric cationic surfactants with varying spacer groups included.

$$R{-}COOH \quad \xrightarrow[R''X]{NH_2-(CH_2)_n-NR'_2} \quad R{-}CO{-}NH{-}(CH_2)_n{-}N^+R'_2R''\ X^-$$

$$R{-}COOH \quad \xrightarrow[R''X]{OH-(CH_2)_n-NR'_2} \quad R{-}CO{-}O{-}(CH_2)_n{-}N^+R'_2R''\ X^-$$

$$R{-}NH_2 \quad \xrightarrow[R''X]{HOOC-(CH_2)_n-NR'_2} \quad R{-}NH{-}CO{-}(CH_2)_n{-}N^+R'_2R''\ X^-$$

$$R{-}(EO)_m{-}COOH \quad \xrightarrow[R''X]{OH-(CH_2)_n-NR'} \quad R{-}CO{-}O{-}(CH_2)_n{-}N^+R'_2R'\ X^-$$

6.2.4 Small Alkyl Groups Near The Cation Centre. Kinetic and Anti-agglomerate inhibitors are similar in that they both have small alkyl groups that interact with incomplete hydrate cavities to prevent growth of the hydrate crystal. Under normal field conditions the largest hydrocarbon that fits a SII cavity is iso-butane. Therefore as an end group to fit into the partial cavities iso-butyl would appear to be ideal. Another possibility is iso-pentyl, however this group is too large to fully fit into a closed hydrate cage. Thus it simulates iso-butane with a CH_2 spacer group joining the iso-butyl group to the rest of the surfactant.

6.3 Polymeric Cationic Surfactants (PCS)

Dispersant molecules which are simple monomeric surfactants prevent particles from agglomerating via Van der Waals and electrostatic attraction forces. Polymeric surfactants have two additional mechanisms by which they can stabilise particle dispersions. These are depletion flocculation and steric stabilisation. Therefore, we might expect polymeric surfactants to be superior anti-agglomerators for hydrates than simple monomeric surfactants. This factor may be especially important during a pipeline shut-in where there is the potential for hydrates to form and settle. Here the polymeric surfactant must stabilise the dispersion and prevent agglomeration such that when the line is re-started the hydrate mass flows freely and does not adhere to the pipe walls.

Polymeric cationic surfactant chemistry based on poly(diallylamine) is well-known for application as flocculating agents. PCS should, therefore, adsorb well onto hydrate surfaces, however they do not contain the hydrophobic groups that can interact with the oil phase for preventing hydrate agglomeration.

6.4 Non-Ionic Surfactants as AAs

Another group of chemicals that potentially adsorb onto hydrate surfaces are non-ionic surfactants with neutral hydrophilic groups that bind via hydrogen bonding. The most common nonionic groups are the hydroxyl groups (R-OH) and the ether groups (R-O-R'). These have oxygen atoms with lone pairs for hydrogen bonding to hydrogen atoms on the surface of the hydrate particle. Possible advantages of a non-ionic over a cationic surfactant are potential lower toxicity and that the non-ionic AA activity will not be so dependent on the ionic strength of the brine (i.e. salinity),

Examples of non-ionic surfactants are alkyl polyglycosides, Figure 11, patented by Shell[25]. Shell[24] also patented the anionic alkyl aromatic sulphonates, Figure 12. Alkylphenylethoxylates[26], Figure 13 have also shown potential as inhibitors, however, the concentration levels required are considerable (up to 7 wt%). Work carried out by RF-Rogaland Research[27] indicated that alkyl polyglucosides are capable of preventing agglomeration of hydrates allowing the formation of a slurry. Urdahl *et al*[26] suggest that water-in-oil emulsions may be a suitable property of AAs. Forming stable emulsions is probably not an advantage as it may need extra processing on arrival at a process plant, such as the addition of demulsification agents which may compromise any advantage environmentally in the use of low dose additives.

The conclusion so far is that simple non-ionic surfactants have potential especially oil-soluble surfactants that form water-in-oil dispersions but not stable emulsions.

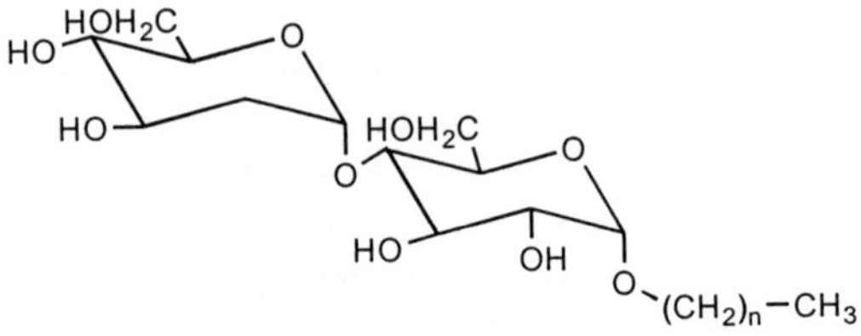

Figure 11 *Structure of an alkyl polyglycoside*

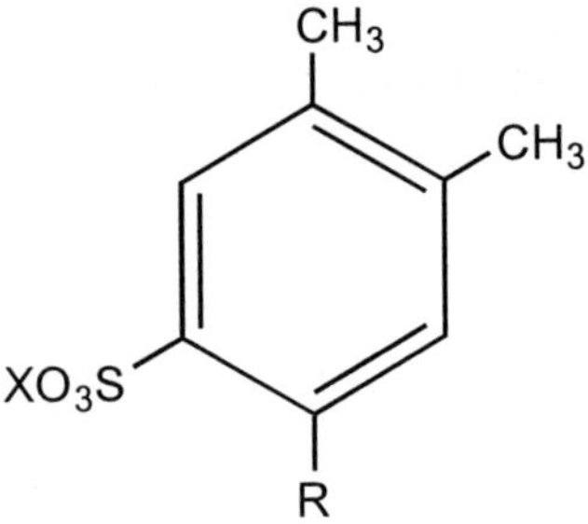

Figure 12 *Alkyl aryl sulphonic acids:*
X = H, K, Na; R = C$_8$-C$_{22}$

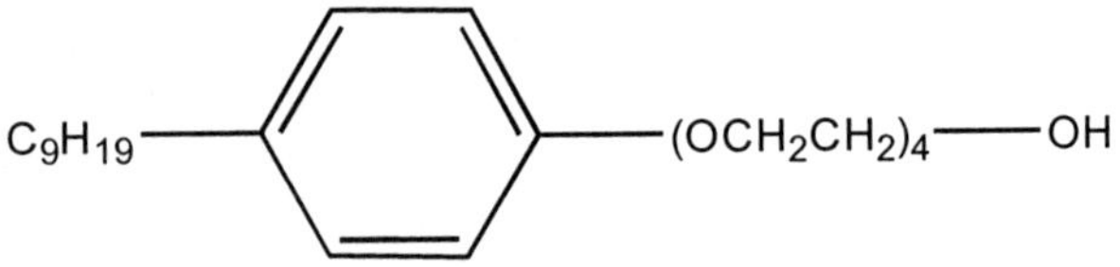

Figure 13 *Tetraloxyethyleneonylphenylether*

6.5 Polymeric Non-Ionic Surfactants

Dispersant molecules which are simple monomeric surfactants prevent particles from agglomerating via Van der Waals and electrostatic attraction forces. In hydrate control, the additional dispersion stabilisation mechanisms using polymers can give added benefit for shutdown situations where you must prevent the hydrate particles from agglomerating after they have settled. However, monomeric surfactants are initially faster at wetting surfaces than polymers which need longer time to orient themselves. This may make monomeric surfactants superior for initial adsorption onto hydrate surfaces, but a polymeric surfactant might give better long-term dispersion stability. Therefore, a mixture of a monomeric and polymeric surfactant would be a useful combination.

Polymeric non-ionic surfactants can be classified into 5 classes:

- *Random copolymer.*
  This has been shown not to be as effective a dispersant as block or graft copolymers.

- *Single chain polymer with a terminal function group.*
  This is really a monomeric surfactant as only one group from the molecule interacts with dispersed particles.

- *AB block copolymer.*

- *ABA block copolymers.*

- *"Comb type" (also graft) copolymers.*
  These molecules have the advantage over AB block copolymers in that they can have multiple interactions with both the solvent hydrocarbon and hydrate particles.

There are several 'comb' type polymers the backbones of which could be the hydrophilic part interacting with water and hydrate, with hydrophobic side chains that interact with the oil. An example of this is polysaccharide. Here hydrocarbon chains are attached at hydroxy sites by ester linkages. A further example could include sidechains made up of hydrophobic chains tipped with hydrophilic groups i.e. polyamine with polypropoxylate sidechains tipped with polyethoxylates groups. Another polymer in this class could be a polyalkylenecarboxylic acid and its polyethylene glycol (PEG) esters. This could be built up in a variety of synthetic ways. Examples are PEG esters of styrene-maleic anhydride copolymer and poly-α-olefin or polyisobutylene succinic esters (used in oil lubricant additives). Yet another comb polymer can be made from vinyl polymerisation of 4-hydroxymethylstyrene, followed by tipping the hydroxy groups with polyethoxylate groups.

7 CONCLUSION

It has now been well established that conventional hydrate treatment chemicals, methanol and glycol's, can be replaced with low dose inhibitor surfactant based additives. The use of these additives has enabled the expansion of both oil and gas recovery with significant cost savings and logistical improvements.

Investigation to date has examined how these materials interact at a growing hydrate surface and thereafter how they prevent further growth. There is certainly great scope for advancing this technology of which surfactant chemistry will continue to play a vital role. However, there are still many fundamental challenges facing the research community as more and more fields are designed with increasingly severe conditions that the inhibitors must overcome. Certainly as these challenges are met the economic rewards for all concerned are considerable.

8 ACKNOWLEDGMENTS

We would like to acknowledge Clariant GmbH for their support and development of this exciting new application of surfactant chemistry. Additionally, we would like to thank Dr. Mark Rodger of Reading University for his help and assistance.

REFERENCES

1.   Hammerschmidt, E.G., Ind Eng. Chem. **26**, 851 (1934).
2.   'A critical review of hydrate formation phenomena' Health and Safety executive, offshore technology report No. OTH 93, 413, (1994).
3.   Sloan, E.D., Clathrate hydrates of natural gases. Marcel Dekker Inc. New York (1990.)
4.   Long, J. *et al.*: 'Kinetic inhibitor of natural gas hydrates', 73rd Annual GPA Convention, New Orleans, LA, March 7-9 (1994).
5.   Lund, A,. Urdahl, O., Giertsen, L.,Kirkhorn, S., and Fadness, F.: , *Proc.* Of the 2nd Int. Conference on Natural gas hydrates, Toulouse, Jun. 2-6 407-414, (1996).
6.   Panchalingham, V., and Sloan, E.D.: 'Hydrate kinetic inhibition chemicals', p. 171, *Proc.* Of the 2nd Int. Conference on Natural gas hydrates, Toulouse, Jun. 2-6 171-180, (1996).
7.   Lederos, J.P., Long, J.P, Sum, A., Cristiansen, R.L., and Sloan, E.D.: 'Kinetic inhibitors of natural gas hydrates', *Proc.* 73rd Gas Processors Association Annual Convention, p. 85, New Orleans, LA, Mar. 7-9, (1994).
8.   Bloys, B and Lacey,C.: 'Laboratory Testing and Field Trial of a New Kinetic Hydrate Inhibitor', 27th Annual Offshore Technology Conference Houston, Texas, May 1-4, 691-700, (1995).

9.   Corrigan, A., Duncum, S.N., Edwards, A.R., and Osbourne, C.G., paper SPE 30696 presented at the 1995 SPE Annual Technical Conference, Dallas, Oct. 22-25.

10.   Argo, C.B, Blain, R.A Osbourne, C.G., and Priestly, I.D., Paper SPE 37255 presented at the 1997 SPE International Symposium on Field Chemistry, Houston, Feb. 18-21.

11.   Mandelcorn, L., Chem Rev., **59** 827 (1959).

12.   Rodger, P.M, Chem. Br., **26** 1090 (1990).

13.   Carver, T.J., Drew, M.G.B., and Rodey, P.M., J.Chem.   Soc., Faraday Trans., **93** (24) 5029-5033 (1996).

14.   Carver, T.J., Drew, M.G.B., and Rodger, P.M., J. Chem. Soc. Faraday Trans. **91** (19) 3449-3460 (1995).

15.   Sloan, E.D, World Pat, WO 94/12761, 1994.

16.   Mokoyam, T.Y., Colorado School of Mines. Thesis (1996).

17.   Freeney, R.E., Food Technology, **82**, (1993) and references therein.

18.   Matthews, R.R and Clark C.R. Eur, Pat 309. 210, 1989.

19.   Duncum, S.N, Edwards, A.R and Osborne, C.G, Eur. Pat 536, 950, 1993.

20.   Anselme, M.J., Reijnhout, M.J., Muijs, H.M and Klomp, U.C, World Pat. WO 93/25798, 1993.

21.   Colle, K.S *et al*, UK Pat. GB 2301825 1996.

22.   Duncum, S.N., Edwards, A.R and James, K. World Pat. WO 96/29501 1996.

23.   Klomp, U.C, World Pat. WO 96/3417 1996.

24.   H.M. Muijs, N.C Beers, N.M Van Om, C.E. Kind and M.J. Anselme, Can. Pat 2,036,084, 1992.

25.   M.J Reinhout, C.E. Kind and U.C. Klomp, Eur. Pat. 526,929, 1993.

26.   O. Urdahl, A. Lund, P. Moerk and T-N. Neilson, Chem. Eng. Sci., **50** (5), 863, 1995.

27.   M.A Kelland, T.M. Svartaas and L. Dybvik. Paper SPE. 30695 Presented at the Annual Technical Conference , Dallas 22-25 Oct. 1995.

Influence of a Silicone Surfactant Structure on the Burning Behaviour of Polyurethane Foam

A. Weier, Dr. G. Burkhart and M. Klincke

TH. GOLDSCHMIDT AG, GOLDSCHMIDTSTRASSE 100, D-45127 ESSEN, GERMANY

1 SUMMARY

For many applications polyurethane foam has to match certain flammability requirements. Since the material itself is flammable, also minor ingredients like the used silicone surfactants should be optimized. It is well known that silicone surfactants have an effect on the burning behaviour of polyurethane foam. The size of this, normally moderately effect, depends on the use level but even more so on the chemical structure of the surfactants. In this paper we present a model for the negative influence of the silicone surfactant on the FR-performance and experiments supporting this model. Also some unexpected results about the influence of solid inorganic materials on the burning behaviour of polyurethane foams are presented.

2 INTRODUCTION

A polyurethane is the reaction product of a polyol and a di- or polyisocyanate.

$$R-N=C=O \ + \ R'-OH \ \longrightarrow \ R-NH-\underset{\overset{\|}{O}}{C}-O-R'$$

Urethane

$$2\,R-N=C=O \ + \ H_2O \ \longrightarrow \ CO_2 \ + \ R-NH-\underset{\overset{\|}{O}}{C}-NH-R \ + \ Heat$$

Urea

Figure 1 *Basic reactions during foam processing*

This reaction, catalysed by amine or tin catalysts, results in a plastic material, but not necessarily in a polyurethane foam. The foam is characterized by the formation of gas bubbles in the liquid reaction mixture before the material is totally crosslinked and loses its flowability. Therefore the second important reaction to consider during the generation of polyurethane foams is the reaction of water and isocyanates resulting in the generation of carbon dioxide, urea, and heat. The gas formed by this process, or the evaporation of low boiling liquids by the heat of the exothermic water isocyanate reaction leads to the generation of gas bubbles. These result in a reduction of the density of the polymeric material. To prevent the material from defoaming is one of the tasks of the silicone surfactants.

Rigid Foams:
➡ **appliance, construction, packaging etc.**

Flexible Foams:
➡ **carpet backing, cushioning, matresses etc.**

Figure 2 *Use examples for polyurethane foams*

The resulting polyurethane foams can vary from flexible to rigid materials, from high to low density materials, from hydrophilic to hydrophobic materials. The main types of foams are flexible polyurethane foams and rigid polyurethane foams. As both of these types can be found in every day use, for example flexible foams as mattresses or as cushioning materials in furniture and rigid foams in the construction industry there is a considerable amount of legislation on the flame retardent (FR-) properties of such foams.

➡ **use of aromatic polyols**

➡ **production of polyisocyanurate foam**

➡ **incorporation of liquid and/or solid flame retardants**

Figure 3 *Technologies for the Production of Flame retarded polyurethane foams*

To decrease the flammability of polyurethane foams several technologies are common practice:

1. using polyol systems which contain higher amounts of aromatic units
2. introducing isocyanurate structures by trimerization of isocyanates in the case of rigid foams
3. incorporation of liquid reactive or non-reactive flame retardants based on halogenated and/or phosphorous compounds
4. using solid flame retardants.

3 EXPERIMENTAL

Of the many different flame tests the less time consuming ASTM.-D.1692 was used for the following evaluation. This horizontal burn test can be done with reasonable effort although it is definitely not the most severe test around. Anyway it is useful for the judgement of trends **although numerical flame spread ratings given here are not intended to reflect hazards presented by the tested materials under actual fire conditions**. The ASTM-test was used for a comparison of different influences of chemical substances. As some of the materials tested were not compatible with a foam formulation a part of the work was done with external application of the materials on a flame retarded foam specimen. This procedure was chosen to ensure that no change of morphology or porosity affected the burn length so that the results represented the influence of the chemical by itself.

4 THEORY

As the topic of this talk is the burning behaviour of polyurethane foam let's have a look at a theory for flame spread development on the next slide.

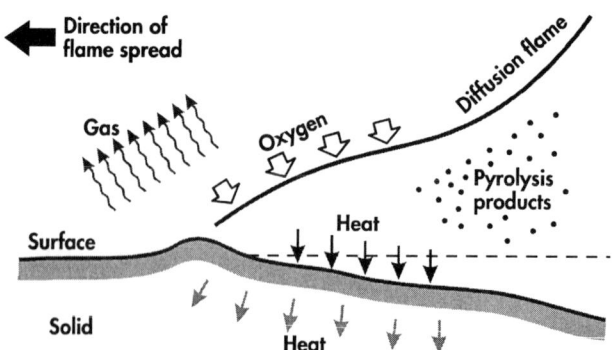

Figure 4 *General assumption about flame spread*

In this simplified illustration it is shown that heat decomposes the organic material at the surface during an endothermic procedure. Pyrolysis products are created and by an exothermic reaction with oxygen at the boundary layer of the flame the formation of even more reactive decomposition products takes place. These aggressive radicals are responsible for an accelerated degradation of the polymeric surface. As long as the result of the energy balance is positive there is a thermal feedback to the endothermic process at the surface and the combustion keeps going. It is believed that the formation of hydrogen radicals and hydroxyl radicals from the organic material are the key factors initiating and supporting the combustion phenomena.

The working mechanisms of flame retardants are manyfold. According to common assumptions halogen containing compounds act as radicals scavengers. During decomposition of the polyurethane surface the halogenated compound also breaks down .The released halogen radicals penetrate the pyrolysis gases to react with the combustion supporting radical in this phase and therefore acting like a scavenger interrupting the above-mentioned cycle. The energy balance of the combustion therefore decreases.

Phosphorous compounds in contrast can initiate a catalytic splitting of the polyurethane and lead - through dehydrogenation and dehydration reactions - to a carbonized, protective surface hindering the escape of the stream of pyrolysis gas and the heat to be transferred to the solid material below the char. Again, the above described circle is interrupted.

5 DISCUSSION

Now, which role does the silicone surfactant play in the polyurethane foam production? And what influence does it have on the burning behaviour of such foams?

5.1 Silicone surfactants

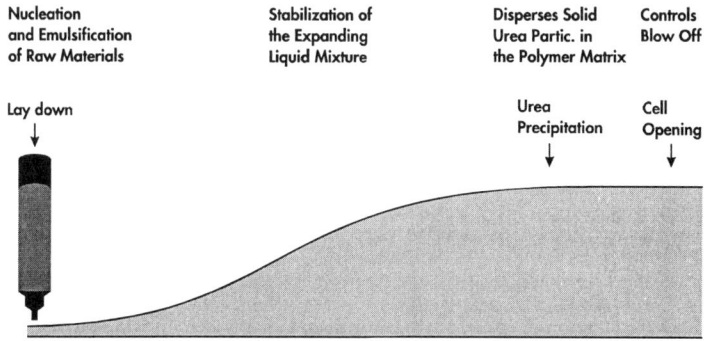

Figure 5 *Function of the surfactant in flexible slabstock production*

As you can see in this slide for the example of a flexible slabstock foam the silicone surfactant has to fulfil quite a number of different tasks throughout the foaming process. At the beginning of the foaming process besides the emulsification of the raw materials it has to promote nucleation. Nucleation is the delivery and stabilization of the nuclei for the formation of the gas bubbles within the liquid reaction mixture. Otherwise a coarse, non elastic foam of little practical use would result.

During the expansion phase of the foam the surfactant decreases the surface tension and stabilizes the resulting foam structure. At a later stage urea particles from the water - isocyanate - reaction are formed and have to be dispersed in the polymer matrix. After further viscosity increase and crosslinking of the polymer chains the formerly closed gas cells have to open by a kind of precisely timed defoaming process. Otherwise the enclosed gas could not move within the foam structure and an inflexible or even shrinking foam would result.

While this explains why a surfactant is necessarry for polyurethane foam production, it does not yet explain whether the silicone surfactant affects the burning behaviour of a polyurethane foam.

The rationalization why in this respect also the silicone surfactant structures are a point to consider is not mainly due to the fact that they are in this formulation, especially since they are normally representing less than 1 % of the plastic material. It is due to the fact that in any case the decomposition of a polyurethane foam starts at the surface. Due to the necessary surface activity of the foam stabilizers it is easy to rationalize their enrichment on the surface and it is the surface that is the most decisive part of the polymer regarding the flame spread development.

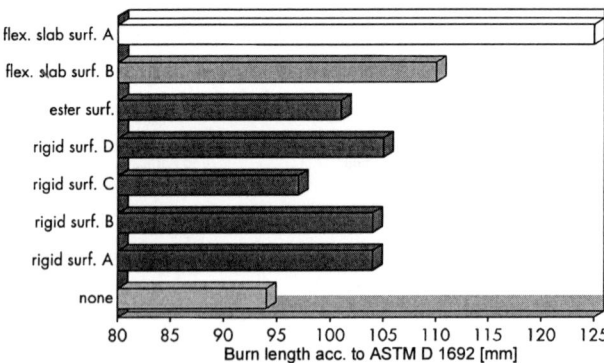

Figure 6 *Influence of Surfactants on fr-performance*
external application of 1% surfactant

This point can be further highlighted if we look at the influence of surfactants on the FR performance of a polyurethane foam with 1 wt.-% of surfactant externally applied. To characterize the burn performance in this case a horizontal burn test, the ASTM D 1692 test, was used and the burn length was listed. It can be seen that all the different types of silicone surfactants used do have a negative influence on the burn performance of this foam. To understand this generally accepted fact it is helpful to have a look at a general structure of a silicone surfactant.

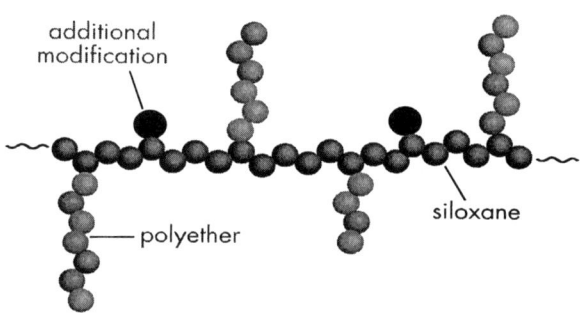

Figure 7 *Building blocks of a surfactant*

These surfactants are generally polydimethyl- polyalkylene oxide-copolymers consisting of a polydimethylsiloxane backbone (represented by the dark grey spheres) and pendant polyether side chains attached. These polyether side chains are generally formed by the polymerization of ethylene oxide and propylene oxide in different amounts and ratios in the presence of a starting alcohol. As you can see on the slide it is not only possible to have different ratios of ethylene oxide and propylene oxide in these polyethers but also the succession in which these monomers were added onto each other does result in different structures.

5.2 Polyether Structures

To identify whether these possible different structures have an effect on the burn behaviour of a polyurethane foam, polyethers with different propylene oxide content were externally applied on test specimens and subjected to an ASTM D 1692 burn test.

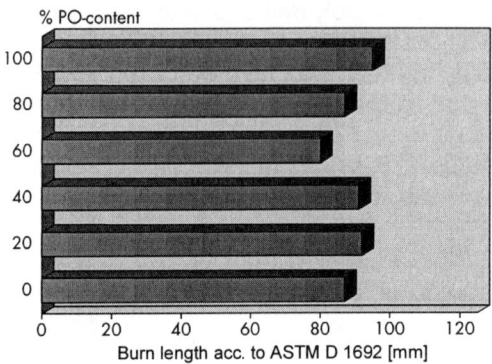

Figure 8 *Polyethers with different oxide content*
1% externally applied

As you can see there are only minor variations in the burn length seen with polyethers of different propylene oxide content. Therefore it has to be assumed that these variations are mainly due to statistical errors within the burn test.

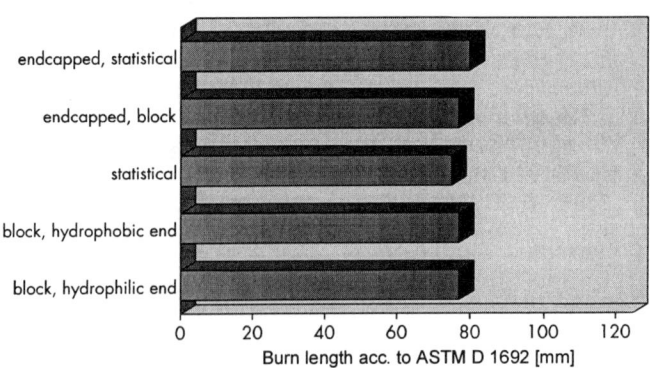

Figure 9 *Polyethers with different structures*
1% externally applied

The same is true if polyethers with the same propylene oxide content and the same molecular weight but different succession of the monomers and different polarity distributions along the chains are tested.

5.3 Siloxanes

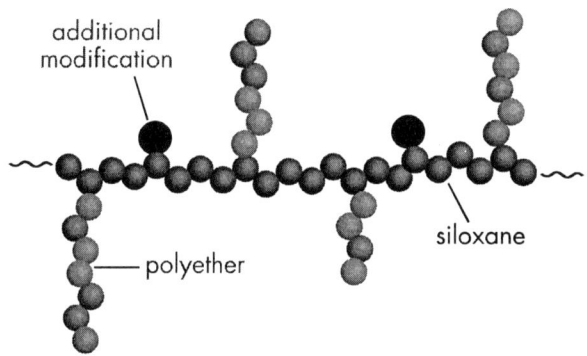

Figure 10 *Building blocks of a surfactant*

Obviously a second necessary part in the synthesis of a silicone polyether copolymer is the polydimethyl siloxane backbone of such a molecule. To characterize the effect of the polydimethyl siloxane chain used as the backbone of such a structure here again some tests were made.

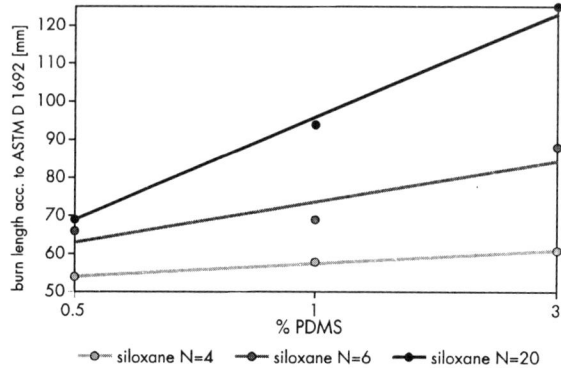

Figure 11 *Effect of polydimethylsiloxanes (PDMS)*
1% externally applied

It is not surprising that increasing amounts of externally applied polydimethyl siloxanes lead to increasing burn length in an ASTM D 1692 test. More astonishing might be that this effect is obviously correlated to the average chain length of the applied siloxane. It can be seen that with longer chain length of the polydimethyl siloxane the detrimental effect on the FR performance on the foam grows. This effect is by far outbalancing the pure weight effect that would rationalize why a higher amount of polydimethyl siloxane leads to higher burn length values, due to the larger amount of material supporting the decomposition. To address the question whether the different volatilities of those different polydimethyl siloxane chains is the main reason for this effect an additional test with silica was carried out.

5.4 Solids

➡ **acting as heat sink**

➡ **free radical scavenger**

➡ **increasing char formation**

➡ **liberating non-burning gaseous decomposition products**

Figure 12   *Possible effects of solid materials*
            as ingredients in polyurethane foams

It is easy to expect that such solid materials, especially if inorganic, will have a positive effect on the FR performance of a foam and that they will lead to decreasing burn length in a burn test. Depending on their chemical nature they could be either expected to be a heat sink, a material to catch free radicals, to char the burn front of the foam or to liberate non-burning gaseous products in order to dilute the oxygen at the burn front.

In this test solid materials and their influence on the FR performance of a foam were compared. Melamine is used as a flame retardant additive in polyurethane foam and behaved according to our expectations. But it was surprising to us that calcium carbonate and even silica as a chemical inert material under these conditions increased the burn length in these tests.

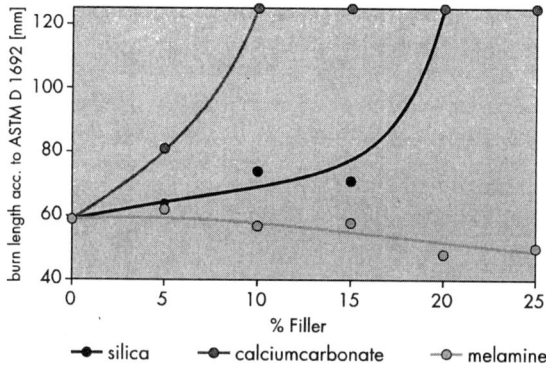

Figure 13 *Addition of Solids*

Assumption:

SiO₂ - the final oxidation product of silicone surfactants generated in the surface area (flame front) catalyses the pyrolysis process.

Figure 14 *Theoretical model*

Our assumption is that on the surface of the SiO_2 particles either catalytic effects on the heat initiated oxydation processes take place or the silica particles themselves act like a wick would act in a candle by forming a means to transport melted organic material from the surface and increase the diffusion flame front. If the formation of SiO_2 particles on the burn front of a foam is indeed the major detrimental effect of silicone containing polymers it should be helpful to decrease the average length of any unmodified polydimethyl siloxane chain part between any two modifying groups.

Standard Silicone

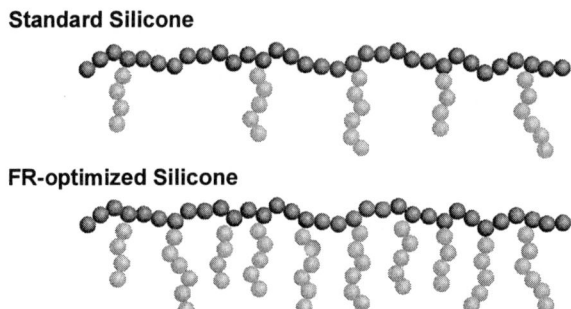

FR-optimized Silicone

Figure 15 *General structure of silicone surfactants*

This can be done either by attaching a higher number of polyethers to the siloxane backbone or by the incorporation of additional modifications that would result in a higher probability for the molecular fragments to migrate into the polymer after a thermal breakdown.

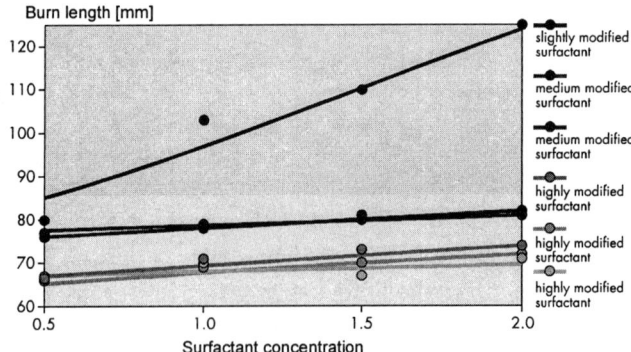

Figure 16 *Surfactant Comparison according to ASTM D 1692*

To test this relationship three classes of surfactants were synthesized. The main difference between them was the average length of unmodified polydimethyl siloxane chains between any two pendant groups as outlined previously. It can clearly be seen that there is indeed a significant correlation between the detrimental effect of a silicone surfactant and its molecular structure. That proves the structural optimizations of ingredients with relative amounts even as low as 1 % in a polyurethane foam might result in measurable improvements in respect to their FR performance. Further work in this area and on possible synergistic effects between different building blocks of such molecules is still under way.

➡ **Correlations between molecular structures of silicone surfactants and the burning behaviour of polyurethane foams were established.**

➡ **A model on the FR-effect of solid particles at the burn front was outlined.**

Figure 17 *Conclusion*

6 CONCLUSION

It has been shown that there is a correlation between the molecular structures of silicone surfactants and the burning behaviour of the resulting polyurethane foam. Structural parts of the used silicone polyether copolymers were classified according to their influence on the burning behaviour of polyurethane foams. The relationship between the length of the siloxane backbone of the molecules and the number of attached groups has been identified as the main characteristic for a surfactant to change the FR performance of the resulting polyurethane foam. Additionally a model on the effect of solid particles at the burn front was proposed.

Application of Surfactants in the Textile Industry

P. Mallinson

BROOKSTONE CHEMICALS LIMITED, BARNFIELD CLOSE, LEEK, STAFFS. ST13 5SG, UK

1.   Understanding the Textile Industry

Although the title to this paper indicates "The Textile Industry", a better description would be the textile industries. From rearing of merino sheep in the Australian outback, to the production of printed carpets in Belgium, to the washing of denim jeans in Blackburn - all are different sectors in the world wide production of textile fibres, yarns and fabrics.

A simplified look at these sectors reveals the following:

The Textile Industries

Natural fibre production	sheep rearing cotton growing silk production
Man-made fibre production	polyester, nylon, acrylics viscose, acetate, polypropylene
Yarn manufacture	cotton spinning woollen and worsted yarn synthetic continuous filament synthetic staple yarns
Fabric production	warp and weft knitting weaving non-woven fabric carpets
Scouring, dyeing and printing	cotton bleaching wool scouring dyeing of fibre, yarn and fabric printing of fabrics

Finishing	softening
	resin finishing
	water proofing
Garment manufacture	"stone washing" of denims
	garment dyeing
	garment dipping

Each of these sectors have their own individual requirements with respect to processing, geographic location, sourcing and environmental issues when it comes to the chemical usage and as such, many different types of surfactants are used world wide in these textile industries. It is this diversity of application which means in general the involvement of many surfactants being used in small to medium tonnages, rather than 5 - 10 products being consumed in vast quantities.

2.   Types of Surfactants Used

Undoubtable, every surfactant known to man will, at some time or another, have been evaluated for use in one or more sectors of the textile industry. The list given below is therefore not intended to be a complete listing, but rather an indication of the more commonly used products, on a day to day basis.

Surfactants commonly used in textile processing

Nonionics	ethoxylates)	of alcohols, alkyl phenols, amines,
	propoxylates)	amides, glycerides, acids, esters etc.
	block co-polymers	
	alkylolamides	
	amine oxides	
	esters	
	amides	
Anionics	sulphates	
	sulphonates	
	phosphate esters	
	phosphonates	
	carboxylates	
	soaps	
	sulphosuccinates	
	lignin derivatives	
	sarcosinates	
	anionic polymers	

Cationics   amino-amides
                             amines
                             quaternaries
                             cationic polymers

Amphoteric   acetates
                             betaines
                             modified amine ethoxylates

Others   silicones
                             fluorocarbons
                             protein derivatives
                             polyurethanes
                             lanolin derivates

Whilst some surfactants in this list have been specifically researched, developed, improved, marketed and patented for the textile industry, eg cationic softening agents, many have been "borrowed" from other industries and their use adapted to fulfil a role in textile application. Examples would come from:

-       water treatment industry   (cationic polymers)
-       household detergents / rinse aids   (nonionic non-foaming detergents)
-       lubricants (silicone lubricants / softeners)

3.   General Formulation of Textile Auxiliary

A textile auxiliary is a chemical, or blend of chemicals, which when used in a textile process improves the efficiency or economics of the process, or improves the quality of the textile materials involved.

Rarely will one single surfactant fulfil all the requirements needed of a textile auxiliary. Hence the manufacturers of textile auxiliaries are blenders and formulators - we shall use as many different components as necessary to achieve the required physical and chemical properties for our products. It also makes it more difficult for competitors to copy individual products.

Not only are different surfactants blended together, eg to make a nonionic / anionic detergent, but many other chemicals could be used in the formulia. Typically, the following chemicals could be used in formulating a typical textile auxiliary:

Blending of Surfactants

Surfactants blended with   -   solvents (white spirit, perchloroethylene)
                                       carriers (eg butyl benzoate)
                                       sequestrants (EDTA, NTA)
                                       anti-microbial agents
                                       enzymes (α-amylase, cellulase)
                                       dyes / pigments
                                       paraffin waxes, polyethylenes, mineral oils, silicones

> inorganics (alkalis, electrolytes)
> inert chemicals (silicas)
> other surfactants (eg nonionic / anionic detergent)
>
> fatty acids, alcohols
> foam stabilisers
> anti-foams

Further, having decided which chemicals are needed to give the auxiliary product its functional properties, we may then need to modify it even more to allow for:

Final formulation to contain: stabilisers
                                              sequestrants (hard water)
                                              pH regulator
                                              viscosity regular (higher or lower)
                                              odour masker
                                              identification tint
                                              anti-setting agents (cold weather)
                                              preservatives (hot weather)

Finally, we have to consider its ease of use, packing, transportation, disposal, environmental impact, price as shown below

Environmental / Safety Considerations

Biodegradability	local / regional legislation
Toxicity	fish
Safety	COSHH regulations flash points corrosiveness
Transportation	dangerous good classification packing group air / sea freight restrictions
Disposal	waste disposal sewerage treatment
Industry standards (eg. Oekotex)	metal content pesticide content
Irritancy	patch tests allergenic tests

4.   Properties of Textile Auxiliaries

So far, we have considered general matters relating to the use of surfactants in the textile industries.   Now, we wish to study in more detail three typical textile auxiliaries to understand their properties relating to specific uses in a particular textile application.

a.   Scouring Agent for Cotton

When formulation a textile auxiliary, for the scouring of cotton fabrics, generally under alkali conditions, the following properties would be studied:

Cotton Scouring Agent

Properties

wetting out
emulsification
dispersing
anti-redeposition
solubilisation
hard water dispersing effect
foaming (30 - 100°C) / defoaming
stability to   oxidation / reduction
                acids
                alkalis
                electrolyte
                other ionic species
                temperature

What is more important is the effect that the formulated auxiliary would have on the scoured fabric. As a textile technologist, we would look at the fabric and assess it for the following:

Effect

cotton seed removal
absorbency (eg. towels)
residual surfactant (medical use)
residual inorganics
whiteness
size / oil removal
fabric handle
yarn friction
dye uptake / penetration
print penetration
evenness of dyeing

b.   **Dye Levelling Agent**

The role of a dye levelling agent is to ensure that a piece of fabric is evenly dyed, face to back, side to side, end to end, evenly penetrated and has not effected the dye fastness of the fabric after drylng.

The points to observe are as follows:

Dye Levelling Agent

Properties	dispersing	)	
	migration	)	
	retarding	)	
	penetration	)	dye levelness
	build up	)	
	reserving	)	
	rinsability		
	anti-crease		
	anti-precipitation		
	foaming		

Effect	even dyeing	-	hue depth and brightness
		-	edge to edge
			front to back
			side to side
		-	fibre to fibre
		-	free of spots, marks, patches, bars, lines
	good fastness	-	water, washing, perspiration, light rubbing
	dye shade	-	no change of shade to dyestuff
		-	no leaching during rinsing

c.   **Softening Agent**

A softening agent is an auxiliary that when applied to a textile material brings about an alteration in handle resulting in the goods being more pleasing to the touch.

However, it must have many other properties as applied to yarns or fabrics:

Textile Softening Agent

Properties   stability to   heat (up to 210°)
                                      acids
                                      shear and pumping
                                      pH
                                      chemicals on the fabric
                                      hard water
                                      residual dyestuff

                  ease of dissolving

During application to the textile, the following must be considered during its formulation:

Effect   primary   -   handle   (surface friction
                                                     bulkiness
                                                     bending / stiffnes
                                                     stretch recovery)

                             -        soft, dry, wet, tacky, smooth, flowing, cool etc.

              secondary   -   effect on whiteness / shade
                               -        tensile properties, abrasion, garment life
                               -        absorbency
                               -        stretch recovery
                               -        stitchability
                               -        pilling
                               -        anti-static
                               -        seam slippage
                               -        soiling

5.   The Future

As textiles are one of the three essentials of living, food and a home being the other two, there is an obvious future for the textile industry (where it will be is another matter). The factors which are brought to bear on a textile auxiliary manufacturer when planning his products into the 21st century are as follows:

Consider   fashion (easy care finishes)
                                machines (foaming)
                                environment (biodegradability)
                                health (non allergenic, Oekotex)
                                performance (outdoor use, non-smelling)
                                geographic location

Subject Index